Plumb, R.T. & Thresh, J.M.,
(eds), 1983.

D1187888

~ology

Acc no. J. 51
UDC no. CH632.35:632.9

Donated by Dr L.J. Rosenburg

WITHDRAWN
FROM
UNIVERSITIES
AT
MEDWAY
LIBRARY

THE LIBRARY,
TROPICAL DEVELOPMENT AND RESEARCH INSTITUTE
COLLEGE HOUSE, WRIGHTS LANE
LONDON W8 5SJ

Plant Virus Epidemiology

THE SPREAD AND CONTROL OF INSECT-BORNE VIRUSES

UNIVERSITY OF GREENWICH LIBRARY

632.
35
PLU

40116077

Plant Virus Epidemiology
THE SPREAD AND CONTROL OF
INSECT-BORNE VIRUSES

Edited for the
Federation of British Plant Pathologists
by
R. T. PLUMB
Rothamsted Experimental Station
Harpenden, Herts, UK

and
J. M. THRESH
East Malling Research Station,
East Malling, Maidstone, Kent, UK

BLACKWELL SCIENTIFIC PUBLICATIONS
OXFORD LONDON EDINBURGH
BOSTON MELBOURNE

©1983 by Blackwell Scientific Publications
Editorial offices:
Osney Mead, Oxford, OX2 0EL
8 John Street, London, WC1N 2ES
9 Forrest Road, Edinburgh, EH1 2QH
52 Beacon Street, Boston, Massachusetts 02108, USA
99 Barry Street, Carlton, Victoria 3053, Australia

All rights reserved. No part of this publication may be
reproduced, stored in a retrieval system, or transmitted in any
form or by any means, electronic, mechanical, photocopying, recording
or otherwise without the prior permission of the copyright owner.

First published 1983

British Library
Cataloguing in Publication Data

Plant virus epidemiology
  1. Virus diseases of plants
  I. Plumb, R T II. Thresh, J M
  III. Federation of British Plant Pathologists
  581.2'34   SB736

  ISBN 0-632-01028-2 √

DISTRIBUTORS

USA
  Blackwell Mosby Book Distributors
  11830 Westline Industrial Drive, St. Louis, Missouri 63141

Canada
  Blackwell Mosby Book Distributors
  120 Melford Drive, Scarborough, Ontario, M1B 2X4

Australia
  Blackwell Scientific Book Distributors
  31 Advantage Road, Highett, Victoria 3190

Printed in Great Britain

Contents

Preface

The stimulus to produce this book arose from the current interest in
plant virus epidemiology that is evident from the recent decision of the
International Society for Plant Pathology to form a special committee on
this topic. The first activity of the committee was to foster an international conference in July 1981 at the University of Oxford arranged by
the Federation of British Plant Pathologists (F.B.P.P.) and the Association of Applied Biologists. Immediately after the conference a workshop
was held, under the same sponsorship but with the additional support of
the International Soybean Program (INTSOY), on pathogens transmitted by
whiteflies. Both meetings attracted a large international attendance of
virologists and those concerned with vectors.

Many chapters of the book are detailed accounts of work outlined at
the Oxford meetings, whereas others were specially invited from some of
those unable to be present. The book is concerned almost exclusively
with the insect-borne viruses and their vectors. This selectivity was
deliberate because insects are the class of invertebrates that transmit
most viruses and because nematode, mite and fungus vectors have been
covered adequately in recent texts. However, much of the material in
the general chapters and in those that consider specific viruses or
insect vectors has wider relevance in the search for general epidemiological principles.

On behalf of the Federation we thank the contributors for their co-
operation in the preparation of this volume. We also thank Susan Jellis
as production editor, Eileen M. Ennever for preparing the camera-ready
typescript, Margaret J. Howe who compiled the index and Ian Cooper who
made all local arrangements for the Oxford meetings.

This book is the fifth and last of the F.B.P.P./Blackwell series as
the Federation was discontinued at the end of 1981 with the formation of
the British Society for Plant Pathology which is to publish a new series
with Blackwell Scientific Publications.

R.T. PLUMB
J.M. THRESH
Federation of British Plant Pathologists

Epidemiology of plant virus diseases: a prologue

B.D. HARRISON
Scottish Crop Research Institute,
Invergowrie, Dundee, UK

One reason I am pleased to be contributing this introduction is that it gives me the opportunity to express, as President of the Association of Applied Biologists, my pleasure that the AAB has been actively involved, through the Virology Group of the FBPP, in the ISPP Virus Epidemiology Committee's first major venture - the Oxford meeting which led to the production of this book. A second reason is that although the ecology of plant viruses with soil-inhabiting vectors has received much attention in recent years (Lamberti et al., 1975; Harrison, 1977; Scott & Bainbridge, 1978) I feel that comparable work on viruses with aerial vectors has been less intensively reviewed and discussed, and I hope that this book will help to fill the gap. In this introduction I outline briefly some of the main achievements of earlier work on the epidemiology of plant virus diseases, together with a few views of my own, as a background to the newer findings described by other contributors.

SOME IMPORTANT ADVANCES

Knowledge of the epidemiology of plant virus diseases has advanced more by the steady acquisition of information and careful analysis of phenomena than by sudden dramatic leaps. However, it is not difficult to pick out a few developments that mark important stages in the evolution of the subject. First came the evidence, beginning at the end of the 19th century, that many viruses are transmitted from plant to plant by vectors. Leafhoppers were the first to be implicated (Fukushi, 1969), then other arthropods including aphids, thrips, whiteflies, beetles and eriophyid mites. Nematodes and fungi were added to the list much later.

This was followed by the recognition of vector specificity: the realization that the vectors of individual viruses typically are confined to one major taxon, and in some instances to one species. Also, as mechanisms of virus transmission by vectors were studied, their remarkable variety became apparent, and more recently the key role of virus coat proteins (Harrison, 1981a) or helper proteins (Pirone, 1977)

Plumb R.T. & Thresh J.M. (1983) *Plant Virus Epidemiology.* Blackwell Scientific Publications, Oxford.

in several of these mechanisms has emerged. The epidemiology of many
plant virus diseases is thus determined by the several interactions of
virus, host and vector, and the effects of environmental conditions on
these interactions.

A few plant viruses, however, have long been known to spread from
plant to plant by contact. Tobamoviruses and potexviruses usually
spread in this way and do not have vectors in the normal sense. In
addition, other viruses, notably ilarviruses, can pass from infected
pollen to the plant pollinated (e.g. George & Davidson, 1963). Epidemiology is therefore simpler for diseases caused by these viruses than for
diseases caused by viruses with specific vectors.

An important advance of a different kind was the application of mathematical methods to the analysis of patterns of virus spread in space
and time. The concept of the gradient of disease incidence resulting
from dispersal from a source was applied to plant viruses (Gregory &
Read, 1949), and spread of viruses into and within crops was analysed
in terms analogous to "simple" and "compound" interest (Vanderplank,
1960). Thus it became possible to quantify the effects of environmental factors on virus spread and to compare the behaviour of different
viruses. For example, the influence of factors such as direction of the
prevailing wind and vector type on the steepness of gradients of disease
incidence can now be assessed more accurately. Also, the importance of
vector numbers, and in some instances of vector activity, was shown by
statistical analyses of large amounts of data such as those relating to
the incidence of yellowing viruses in sugarbeet crops (Watson & Healy,
1953).

Establishing the effects of environmental conditions on vector
numbers and activity was a further advance, with aerial vectors being
most affected by air temperature and movement, and soil-inhabiting
vectors being most responsive to soil moisture and temperature. In a
few instances, effective ways were found of using data on the weather
experienced before crops are planted to predict vector abundance and/or
behaviour, and hence virus spread. For instance, the incidence of
yellowing viruses in English sugarbeet crops can be forecast from the
number of freezing days in the previous January to March, and the
temperature in April (Watson et al., 1975), and the August incidence of
viruses transmitted non-persistently by aphids in pepper crops in the
Niagara Peninsula of Canada was accurately predicted from the temperature and sunshine records for the previous April (Kemp & Troup, 1978).
Such forecasts have great practical value.

MORE RECENT TRENDS

More recently, attention has turned to virus survival systems as seen
from an ecological viewpoint (Thresh, 1980; Harrison, 1981a; Gibbs,
1982). With some viruses, notably those with soil-inhabiting vectors,
a well-developed ability to survive at a site seems to compensate for
limited ability to spread to new sites. Conversely other viruses,

especially some of those transmitted in the persistent manner by aerial
vectors, spread readily to new sites but perennate inefficiently at
existing ones. Possession of effective means both of spreading and of
perennating is a feature of some of the most consistently prevalent
viruses.

Because viruses in the same taxonomic group tend to have similar
survival systems, and different types of plant community probably favour
different kinds of survival system, it is to be expected that specific
groups of viruses will be best adapted to specific types of plant
community. Crops are favourable environments for the contact-transmit-
ted potexviruses and tobamoviruses, which produce high concentrations
of relatively stable particles in their hosts. Other cultivated-plant-
adapted viruses (CULPAD viruses; Harrison, 1981a) include the ilarviruses, which are mainly found in woody species and spread in pollen
to the plant pollinated, and the fungus-transmitted viruses grouped with
barley yellow mosaic virus, which persist in the resting spores of their
vectors and are favoured by monoculture. At the other extreme are the
wild-plant-adapted viruses, or WILPAD viruses, which include many nepoviruses, tobraviruses, geminiviruses and luteoviruses. Typically these
viruses have wide host ranges and long persistence in their vectors,
properties that fit them for survival in communities that contain many
plant species.

The ecology of plant viruses and hence the epidemiology of plant
virus diseases must not, however, be thought of as having an unchanging
pattern. This may alter in response to many factors such as long-term
weather trends and changes in methods of crop production. For instance,
the growth in use of translucent plastic sheeting in several countries
has affected the incidence of viruses in crops protected by it, and in
England the increasing replacement of spring-sown by autumn-sown barley
has been accompanied by increases in the number and prevalence of
viruses causing problems in this crop (Plumb, this volume).

In addition to these side-effects of changes in agricultural methods,
there are the more direct effects of attempts to control viruses. For
example, the planting of tomato cultivars that have the gene *Tm-1* for
resistance to tobacco mosaic virus led to the selection and spread of
resistance-breaking strains (Pelham *et al.*, 1970). However, these
strains apparently had a limited biological fitness and, when the
selection pressure imposed by the resistant cultivar was removed, they
decreased in prevalence.

More typically, plant viruses counter resistance that is controlled
by one or two host genes more slowly or not at all. This is perhaps
because their genomes have a limited coding capacity and their gene
products have multiple functions, or because their modes of transmission
restrict the extent to which plants with resistant genotypes are exposed
to inoculum (Harrison, 1981a). In this 'struggle of the genes' (Gibbs,
1982), resistance-breaking variants may fail to appear or, as already
mentioned, may be relatively unsuccessful. The converse of this interaction, the selective effect of viruses on their hosts, is usually only

evident after many years. Thus genes for tolerance of or resistance to infection are usually found in cultivars grown in areas where plants have been continuously or regularly exposed to virus infection for long periods - maize resistant to maize streak virus in Réunion Island (Bock, 1980) and potatoes resistant to several viruses in the Andean region of South America (Jones, 1981).

POINTERS TO THE FUTURE

In epidemiology, the geographical range over which factors operate is very important, because as the scale of events increases, so also does their complexity. There is much to be gained from the detailed study of virus spread at specific sites. Quiot et al. (1979) have made just such a multidisciplinary study of the spread of strains of cucumber mosaic virus in successive years in a field in France. Other promising areas for study are isolated irrigated farms in arid regions and small isolated islands. It is hardest to understand the epidemiology of a virus in a large continental land mass where factors operating hundreds of kilometres away can have important effects, and where the annual incidence of infection may fluctuate greatly. In such areas, recent advances in methods of following the long-distance movement of air masses should be particularly helpful for tracking the migrations of aerial vectors. Indeed, such a way of studying the influence of one part of a continent on another is exemplified in this volume by Rosenberg & Magor's contribution on the long-distance movement of the rice brown planthopper.

Other recently introduced techniques too should be helpful in analysing the causes and courses of epidemics. Large numbers of plants can now be rapidly indexed for virus infection by sensitive serological techniques such as ELISA (Clark, 1981). Progress is also being made in detecting viruses serologically in individual vector organisms collected in the field (Harrison, 1981b), and different genotypes of a virus can be distinguished by hybridization with complementary DNA, thus facilitating work on the comparative ecology of different virus strains (Randles et al., 1981). Computer analysis can be expected to help utilize the massive volumes of data generated by multidisciplinary studies and by continuous records of environmental conditions, and to assist in the vital advance from explaining events to predicting them.

In addition to advances arising from the use of these aids to more intensive epidemiological investigations, the application of less sophisticated methods in less well-studied areas of the world and to less well-studied vectors should produce many other interesting and useful results. Much work therefore remains to be done, and there is good reason for doing it, because a sound understanding of the epidemiology of plant virus diseases should be the key to rational control measures.

REFERENCES

Bock K.R. (1980) Crop virology research project: final report. United
Kingdom Overseas Development Administration and Kenya Ministry of
Agriculture. XVIII + 24 pp.

Clark M.F. (1981) Immunosorbent assays in plant pathology. *Annual
Review of Phytopathology* 19, 83-106.

Fukushi T. (1969) Relationships between propagative rice viruses and
their vectors. *Viruses, Vectors and Vegetation* (Ed. by K. Maramorosch),
pp. 279-301. Interscience Publishers, New York.

George J.A. & Davidson T.R. (1963) Pollen transmission of necrotic ring
spot and sour cherry yellows viruses from tree to tree. *Canadian
Journal of Plant Science* 43, 276-88.

Gibbs A.J. (1982) Virus ecology - struggle of the genes. *Physiological Plant Ecology. III. Responses to the Chemical and Biological
Environment* (Ed. by O.L. Lange, P.S. Nobel, C.B. Osmond & H. Ziegler),
(in press). Springer-Verlag, Berlin.

Gregory P.H. & Read D.R. (1949) The spatial distribution of insect-
borne plant virus diseases. *Annals of Applied Biology* 36, 475-82.

Harrison B.D. (1977) Ecology and control of viruses with soil-inhabiting
vectors. *Annual Review of Phytopathology* 15, 331-60.

Harrison B.D. (1981a) Plant virus ecology: ingredients, interactions
and environmental influences. *Annals of Applied Biology* 99, 195-209.

Harrison B.D. (1981b) Two sensitive serological methods for detecting
plant viruses in vectors and their suitability for epidemiological
studies. *Proceedings of the 11th British Insecticide and Fungicide
Conference, 1981* 3, 751-757. BCPC Publications, Croydon.

Jones R.A.C. (1981) The ecology of viruses infecting wild and culti-
vated potatoes in the Andean region of South America. *Pests,
Pathogens and Vegetation* (Ed. by J.M. Thresh), pp. 89-107. Pitman,
London.

Kemp W.G. & Troup P.A. (1978) A weather index to forecast potential
incidence of aphid-transmitted virus diseases of peppers in the
Niagara peninsula. *Canadian Journal of Plant Science* 58, 1025-8.

Lamberti F., Taylor C.E. & Seinhorst J.W. (Eds.) (1975) *Nematode
Vectors of Plant Viruses*. Plenum Press, London.

Pelham J., Fletcher J.T. & Hawkins J.H. (1970) The establishment of a
new strain of tobacco mosaic virus resulting from the use of
resistant varieties of tomato. *Annals of Applied Biology* 65, 293-7.

Pirone T.P. (1977) Accessory factors in non persistent virus transmission. *Aphids as Virus Vectors* (Ed. by K.F. Harris & K. Maramorosch),
pp. 221-35. Academic Press, New York.

Quiot J.B., Marrou J., Labonne G. & Verbrugghe M. (1979) Ecologie et
épidémiologie du virus de la mosaique du concombre dans la Sud-Est
de la France. Description du dispositif expérimental. *Annales de
Phytopathologie* 11, 265-82.

Randles J.W., Palukaitis P. & Davies C. (1981) Natural distribution,
spread, and variation, in the tobacco mosaic virus infecting
Nicotiana glauca in Australia. *Annals of Applied Biology* 98,
109-19.

Scott P.R. & Bainbridge A. (Eds.) (1978) *Plant Disease Epidemiology*.
Blackwell Scientific Publications, Oxford.

Thresh J.M. (1980) An ecological approach to the epidemiology of plant virus diseases. *Comparative Epidemiology* (Ed. by J. Palti & J. Kranz), pp. 57-70. Centre for Agricultural Publishing and Documentation, Wageningen.

Vanderplank J.E. (1960) Analysis of epidemics. *Plant Pathology*, Vol. 3 (Ed. by J.G. Horsfall & A.E. Dimond), pp. 229-89. Academic Press, London.

Watson M.A. & Healy M.J.R. (1953) The spread of beet yellows and beet mosaic viruses in the sugar-beet root crop. II. The effects of aphid numbers on disease incidence. *Annals of Applied Biology* 40, 38-59.

Watson M.A., Heathcote G.D., Lauckner F.B. & Sowray P.A. (1975) The use of weather data and counts of aphids in the field to predict the incidence of yellowing viruses of sugar-beet crops in England in relation to the use of insecticides. *Annals of Applied Biology* 81, 181-98.

Plant virus ecology; the role of Man, and the involvement of governments and international organizations

L. BOS
Research Institute for Plant Protection,
Wageningen, The Netherlands

INTRODUCTION

Awareness of the risks of relying on chemicals to control plant pests and diseases has increased rapidly during the past few decades. Thus, interest in the ecology of pests and pathogens to facilitate control through some form of ecosystem management is thriving. Because chemo-therapy of virus diseases is not yet possible, the main approach to their control has always been preventive. This approach is now increasingly being considered in ecological terms.

Ecological studies have improved understanding of the factors under-lying the development of epidemics in crops. They have also shown the inevitable part played by Man, who remains part of nature, yet often unintentionally contributes to the occurrence of virus diseases, despite attempting to prevent them. This chapter deals with Man's ambivalent role in the ecology of viruses. It discusses the involvement of governments and international organizations in efforts to overcome the detrimental effects of human activity and to improve crop health.

A DIAGRAMMATIC REPRESENTATION

Virus diseases do not result merely from a simple "disease triangle" involving pathogen, host and environment. Fig. 1 presents the main groups of factors and their complex interactions (Bos, 1981a, 1982). After including the *viruses*, that increasingly tend to be disseminated everywhere, the hosts are differentiated into "target" *crops* to be protected and *sources of infection*. Moreover, the *biological environment*, including sources of infection and the *vectors* or other means of spread, is distinguished from the *physical environment* of soil, water and climate. This diagram shows clearly the involvement of Man as a factor in the ecology of plant viruses, and it also summarizes the opportunities for disrupting the disease cycle.

Plumb R.T. & Thresh J.M. (1983) *Plant Virus Epidemiology.*
Blackwell Scientific Publications, Oxford.

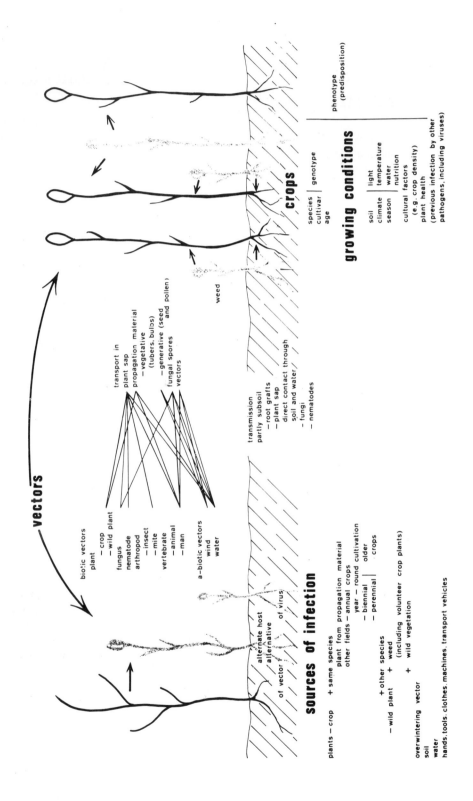

Figure 1. Diagram of the ecology of plant viruses, of the groups of factors that are involved apart from the viruses themselves, and of their interrelationships (After Bos, 1981a, 1982).

MAN-MADE VIRUS DISEASES

Whether a crop will suffer from virus disease or not depends on its predisposition (i.e. vulnerability), on the presence of sources of infection and, frequently, on the availability and behaviour of vectors.
All these are determined or greatly influenced by husbandry. Moreover,
Man himself often disseminates viruses and many, if not most, virus
epidemics of crops are in some way man-made.

Crop vulnerability

The losses resulting from infection by virus depend on plant susceptibility and sensitivity. These, in turn, depend on the host's genotype as
conditioned phenotypically by environment and any previous infection(s)
by other pathogens, including viruses. The opportunity for spread
within crops and for epidemic development depends on the type of
planting (mixed crop, land race, heterogeneous cultivar, homozygous
line, vegetatively-propagated clone, etc.).

    Modernization of agriculture and the demand for uniform produce has
led to increased crop homogeneity. Modern crops, if susceptible, are
commonly infected rapidly. Consequently, the effects of virus infection
are often drastic.

    In the developed countries, there is often a rapid turn-over of
cultivars. The introduction of new genotypes and their encounters with
endemic viruses (Bennett, 1952; Bos, 1981a) may lead to sudden outbreaks
of "new" diseases if the introductions are genetically vulnerable.
Examples are the severe outbreaks of turnip mosaic virus in new lettuce
cultivars resistant to *Bremia lactucae* after they were widely planted
in California (Zink & Duffus, 1975), and of turnip and cauliflower
mosaic viruses in crops of a new, highly productive F_1 hybrid Brussels
sprout in Britain (Tomlinson & Ward, 1981). The same happens as
high-yielding, short-stemmed rice cultivars are introduced into tropical
Africa and encounter the endemic rice yellow mottle virus (Bakker,
1974; Raymundo & Buddenhagen, 1976). Most of the 500 imported lines
tested in Nigeria were sensitive, whereas tolerance was common in old,
African upland cultivars (Anon., 1980).

    New crops planted over vast areas may also affect vector population
density and the spread and incidence of virus infection. Thus, the
recent increase in whitefly-borne golden-yellow mosaics of legumes in
Brazil has been associated with the tremendous increase in area of
soyabean, which is a favourable host for the vector *Bemisia tabaci*
(Costa, 1975).

Type of cultivation

Shifts in date of sowing or planting may have a drastic effect on
virus incidence and yield loss. Early-autumn sowing of winter wheat
in Alberta, Canada, led to severe damage by mite-borne wheat streak
mosaic virus (Slykhuis *et al.*, 1957). Successional sowings of vege-

table crops, to extend the period of harvesting for industrial processing, generally leads to high incidence of aphid-borne viruses.

The trend to year-round cultivation of leek (*Allium porrum*) is considered the main cause of epidemics of leek yellow stripe in western
Europe (Bos *et al.*, 1978). Overwintering large areas of sugarbeet
became established practice in California in 1960-70 to extend the
harvest period and make better use of processing facilities. It also
permitted the epidemic development of virus diseases in the crop until
a "close season" was introduced (Duffus, this volume).

Inadequate weed control or mechanized harvesting, with the risks of
increasing the number of crop volunteers, as happened with potato
(Doncaster & Gregory, 1948), is another change in farming practice of
considerable epidemiological consequence.

A final example of the effects of changing husbandry on disease
incidence is the recent problem caused by lettuce big vein in glasshouse-grown lettuce in Great Britain. The introduction of the
nutrient film technique apparently aided the spread of zoospores of
the fungus vector *Olpidium brassicae* (Tomlinson & Faithfull, 1979).

Man as vector

The ease of spread of very stable viruses such as tobacco mosaic virus,
cucumber green mottle mosaic virus and potato virus X on clothes, hands
and tools is well known. Spread of cocksfoot mottle virus in leys
by forage harvesters was inferred by Upstone (1969) and, more
recently, red clover mottle virus has been shown to be spread on
cutting machines (Rydén & Gerhardson, 1978).

Much more important and far-reaching is the role of Man as a vector
through traffic in plant propagules. The risks are obvious for
vegetatively multiplied material and many viruses are seed-borne
(Bos, 1977). With the rapid internationalization of production and
trade in plant material, often for breeding purposes, there is an
ever-increasing risk of virus spread (Bos, 1981a).

Of 1277 vegetative plant introductions to the United States of
America during 1957-67, 62% were infected by one or more viruses
(Kahn *et al.*, 1967). Later, 66% of 551 introductions of *Solanum* spp.
were reported to contain viruses, but only one third showed symptoms;
on average 40% of the imported wild specimens were infected (Kahn &
Monroe, 1970). The risk of importing potato viruses with tubers or
seed from the Americas causes much concern in western Europe, and
Andean latent and Andean potato mottle viruses have already been
intercepted in Germany (Koenig & Bode, 1977, 1978). Citrus tristeza
illustrates the devastating effect a virus may have once it becomes
established in new regions with large areas of a sensitive host and
a numerous, efficient vector (Bar-Joseph *et al.*, this volume).

Viruses that are endemic but harmless, and thus may occur unnoticed

in certain areas, may cause symptoms and damage in crops when the area
is first cultivated. This occurred in Scotland, where sensitive crops
revealed nematode-borne viruses that existed unobserved in the
indigenous vegetation (Murant, 1970). As plant viruses are increasingly spread throughout the world, an "inevitability-of-establishment"
theory has emerged that suggests that every virus may eventually get
everywhere (Kahn, 1977).

CONTROL

With his better knowledge of virus ecology, Man is increasingly able
to interfere with nature and recognize his own harmful activities.
There are various ways of avoiding the harmful effects of viruses by:-

     1.   removing or avoiding sources of infection,
     2.   decreasing spread through vector control,
     3.   increasing crop resistance (Bos, 1982; Buddenhagen,
          this volume).

     Fig. 1 demonstrates that many control measures require action out-
side the crops to be protected. Moreover, the incidence of viruses
in these crops, whether or not they develop symptoms, influences the
well-being of neighbouring crops or those of other regions or even
other parts of the world. This is especially true for crops grown to
produce propagation material. Hence, public interests are at stake and
that is why public organizations, mainly government agencies, and even
international organizations become involved. This is also true for
other pathogens and pests but is most obvious for viruses because they
are more often hidden, are particularly hard to detect, and are often
highly contagious.

GOVERNMENT INVOLVEMENT

The administrative control of plant pathogens and pests has recently
been assessed in detail (Ebbels & King, 1979). Government involvement
with virus diseases has evolved gradually as information has increased.
Several countries already have sophisticated infrastructures to cope
with plant virus problems and intervene to prevent attacks by viruses
in crops from becoming calamities. Their very effectiveness may lead
to misconceptions by strengthening the impression of some growers that
virus problems are negligible. Important aspects of government involvement are legislative and executive, but these cannot be effective without extension, education and research.

Legislation

When control is necessary, the parties involved may be in conflict
and some measures may have to be enforced. Therefore most governments adopt phytosanitary, import/export and plant propagation laws.

Execution

Most countries have special government bodies to carry out certain tasks directly. Plant protection services usually play a central role.

  Inspection at import and export, especially of propagation material, is an important means of preventing the international dissemination of viruses and other harmful organisms. But many pathogens, notably viruses, are difficult to detect in propagation material which is often transported when dormant. By international agreement (see below) emphasis is on *certification* by the plant protection service of the exporting country. This mainly involves field inspection during the growing season, but is always supplemented by bulk inspection at export. Detailed examination at import is time-consuming and of little value if the material is dormant. Such systems are certainly not foolproof for viruses and they are not meant to substitute fully for detailed tests on arrival. Imports of some plant material from certain countries are therefore prohibited or subject to quarantine.

  Quarantine means the detention in pathogen and pest-proof isolation of imported plant material for observation for freedom from disease, or in some instances, for testing for freedom from viruses and other harmful organisms. Detention must be for sufficiently long to allow suspected pathogens to complete their incubation period. Quarantine aims to exclude hazardous viruses and organisms while admitting healthy plant material. Healthy material can be true seed or other propagation material obtained from plants shown to be virus-free or freed from viruses by heat treatment and/or meristem culture, as with citrus (Button, 1977). For further details see Mathys & Baker (1980).

  Viruses create special quarantine problems because many infections are symptomless, they often have long incubation periods, e.g. symptoms may only be expressed in adult plants or on fruits, and many viruses are difficult to detect. Hence, quarantine is costly, requires specialist supervision and is not infallible. Expensive vector-proof glasshouses and screenhouses are usually required but, for some low-risk pathogens, cultivation in the open during a vector-free period or when the crop is not grown, or in an isolated area or even on the importer's premises may be acceptable. Detention and examination is mostly in the importing country (*post-entry quarantine*), but risks can be further reduced with assistance from a third country to check the material while in transit (*intermediate quarantine*).

  So far, quarantine has been used mainly to protect regions such as Australia that are naturally isolated and have a limited number of ports of entry. It is increasingly practised in developing regions of the world with vulnerable economies heavily dependent on a single crop. It is impractical in countries with a long history of intensive international traffic, practically open borders and many ports of entry, as in western Europe. However, even advanced countries are now increasingly aware of the importance of some sort of quarantine, e.g. for valuable germplasm.

Crops have in the past also been protected from alien pathogens by the prohibition of imports from countries where hazardous pathogens were first detected. This has often discouraged active research. Moreover, certain importing countries have occasionally sought to exclude pathogens that may already occur.

Decisions concerning which viruses warrant quarantine should be based on risk analysis (Kahn, 1979) and on exact knowledge of which viruses and virus strains already occur in the importing country. The risks depend on the availability of means of spread and of susceptible plants. But vectors are also increasingly spread internationally, most plant viruses infect several hosts and wild vegetation may facilitate virus establishment (Bos, 1981a).

Once potentially harmful viruses have entered a country, timely *eradication* may prevent their establishment. Eradication may be essential to protect national production and avoid import restrictions or total embargoes by other countries. One example is the removal of citrus trees to control tristeza disease in California and Israel (Bar-Joseph *et al.*, this volume).

Sharka (plum pox) is a destructive virus disease of fruit crops and, since it was first observed in Bulgaria *c*. 1918, it has gradually spread across Europe (Thresh, 1980). Efforts to contain the disease and to safeguard exports of planting material to North America have been costly. In two areas of West Germany alone 14 000 infected trees were removed during 1963/4 (Niemöller, 1965).

The most striking attempt at disease containment by eradication is with cocoa swollen shoot in West Africa. In Ghana alone *c*. 183 million diseased cocoa trees or their contacts have been cut out since 1946, at one stage at a rate of 15 million a year. Eradication efforts are now mainly restricted to areas where outbreaks are few and scattered (Owusu, this volume). For further examples of eradication campaigns, see Allen (this volume) and Egan & Hall (this volume).

Surveys of crops for viruses and assessment of their economic or potential importance are needed to decide whether, when, which, and to what extent phytosanitary measures, including quarantine, are justified. A prerequisite is reliable *diagnosis* including the development of routine tests. For virus diseases this still largely belongs to the domain of research (see below and Bos, 1976a). Data on economic importance are also indispensable in setting research priorities, but crop loss assessment for virus diseases is still in its infancy (Bos, 1981b).

The *production* and *distribution* of *virus-free* or *virus-tested propagation material* and its *inspection* and *certification* to guarantee its quality has already proved of great importance in controlling virus diseases, by reducing or avoiding initial within-crop sources of infection and spread between regions.

Schemes start with the production of *nuclear, basic* or *foundation stock* derived from a few virus-free mother plants. Such plants may have been selected from partially infected crops by testing individual plants. If vegetatively propagated cultivars are completely infected, then virus-free source plants may be obtained by heat-treatment (Nyland & Goheen, 1969) or by aseptic (*in vitro*) culture of apical meristems or stem tips, as practised either alone or combined with heat or cold treatment (Quak, 1977; Mellor & Stace-Smith, 1977; Lizzaraga *et al.*, 1980). Virus-free stock must be maintained in isolation to prevent reinfection. This may be difficult for vegetatively propagated mother plants. Insect-free glasshouses or screenhouses may then be used as a repository, but maintenance in tissue (aseptic plantlet) culture is increasingly proving useful and there are prospects for low-temperature storage (long-term freeze-preservation) of such tissue cultures (Kartha & Gambor, 1978; Henshaw, 1979).

The subsequent large-scale production of *commercial propagation material* from the basic stock cultures obtained is usually by normal propagation methods. At this stage there is always the risk of re-infection, but this can be reduced by suitable isolation and strict hygiene. Again, modern aseptic techniques of tissue culture are being used increasingly to assist in large-scale propagation (Boxus *et al.*, 1977).

Inspection of the final commercial material to assess its health and guarantee its trueness to type and freedom from pathogens (*certification*) is based on (*a*) the absence of specific symptoms in the crop that produced the propagation material (field inspection) and in the eventual product (bulk inspection), or in plants raised from random samples (growing-on tests), or (*b*) the absence of virus in such samples or the resulting plants when tested in the laboratory (by infectivity or serology). Since guaranteeing freedom from viruses would be pro-hibitively expensive, some infection is tolerated. This represents a compromise between the costs of further improvement and the risks of damage and may vary with growing conditions, season or region, and depend on expected vector density. Thus, lettuce seed with 0.1% infec-tion by lettuce mosaic virus is acceptable in many countries but in California, where aphids are particularly numerous, seed stocks have to contain no infected seed in a sample of 30 000 (Grogan, 1980). However, for diseases subject to quarantine no infection of plant material is permitted.

Depending upon the country, some or all of the above tasks are done by government agencies, or start there because of the expertise and facilities required. Production of commercial stock may be by private growers or companies, but inspection and certification should be by government or government-controlled inspection or certification services (Hiddema, 1972; Cumming, 1979).

There are many examples of the losses caused by viruses only becoming apparent when the introduction of virus-free material allowed comparison with the infected clones grown previously, as with Baccara roso (Pool *et al.*, 1970), apple (Peerbooms, 1971) and strawberry (Aerts, 1977). Even freeing from latent or semi-latent infections may lead to yield increases, as with potato virus M (Kassanis, 1965).

For some virus diseases the use of resistant cultivars is the only practical means of control, and this is particularly important in countries where farmers lack knowledge of hygiene and where there is no means of producing virus-free propagation material. An early example of successful *breeding for resistance* was the development in the early 1930s of sugarbeet cultivars resistant to curly top (Duffus, this volume). Breeding resistant cultivars is now increasingly used as a method for virus control (Russell, 1978; Buddenhagen, this volume). Resistance breeding is ideally for immunity, but is often for resistance to infection or to virus multiplication and sometimes for tolerance; more recently, resistance to vectors has also been sought.

Resistance to viruses may be monogenic or governed by several genes that are effective only against certain strains, as with *Phaseolus vulgaris* and bean common mosaic virus (Drijfhout, 1978). Thus, introduction of resistant cultivars may affect virus evolution and the use of resistant cultivars may not provide a permanent solution.

Special epidemiological risks arise from the introduction of tolerant cultivars that, though they are not damaged themselves, may, in at least some instances, greatly increase the amount of inoculum available and thus endanger the health of nearby, sensitive cultivars or other crops.

In many countries breeding is done by government institutes, whereas in others it is almost exclusively the responsibility of private enterprises, with support from government institutes in locating resistant germplasm and in developing reliable screening techniques and inocula. Government bodies must finally evaluate claimed resistance to viruses.

Extension and education

Growers and others involved in disseminating plant material should be made aware of the importance of viruses and phytosanitary control measures. Hence most countries have federal or state agricultural advisory or extension services and growers as well as extension officers are given plant pathological training.

Research

The information currently available on plant viruses and their effects on crop production is inadequate, yet studies on the ecology and epidemiology of viruses are still being neglected. Research must fill this gap if it is to keep up with the continuing changes being made in agriculture. Part of such research may be basic and concerned solely with the nature of viruses in the hope of eventually finding ways of direct control. There is considerable need, however, for applied research to solve immediate practical problems (Bos, 1976a). The emphasis should be on:-

1. surveying crops for viruses and the identification of viruses and strains,
2. detailed study of the ecology of viruses of actual or potential economic importance,
3. the development of control measures.

Such applied research cannot avoid some service work, such as routine diagnosis, antisera production and assistance in breeding for resistance. Work done on a "customer-contractor" basis has some advantages in that it should keep research workers aware of the virus problems being encountered in agriculture. It should also be realized that agricultural practice and research mutually influence each other. Plant virology helps to define practical problems and also reveals problems previously unrecognized.

ROLE OF INTERNATIONAL ORGANIZATIONS

Many problems are international because viruses are spread by man or in a persistent manner by insect vectors that may fly or be blown far as in North America (Gill, 1975) and Asia (Rosenberg & Magor, this volume). Since some non-persistently transmitted viruses are retained for up to 12 h during post-acquisition fasting (Conti & Caciagli, 1979), these viruses may also be carried relatively far. Wind also facilitates the long-distance transport of some seed-borne viruses (Bos, 1982).

Virus dissemination by Man has increased rapidly during recent years, and sometimes in unexpected ways (Mink, this volume). This justifies the various measures intended to delay international spread of viruses and other pathogens and pests.

International plant protection

There is an increasing tendency towards international and even global approaches to plant protection, although international organizations usually lack legislative or executive power. Hence, governments have to be convinced of the value of certain measures to be taken for their own and mutual benefit. The 1951 International Plant Protection Convention has now been signed by 79 countries (Chock, 1979). It stimulated the establishment of national plant protection services and of eight co-ordinating organizations, including the European and Mediterranean Plant Protection Organization (Mathys & Baker, 1980). Much attention is given to certifying the health of internationally transferred plant propagation material and to the necessity for certain phytosanitary regulations to be included in national legislation. Directives to harmonize phytosanitary export and import regulations within the European Community are binding and execution by national governments can be legally enforced by the European Court.

Checking seed lots for viruses in international commercial traffic presents difficulties and is seldom attempted. However, a working group of the International Seed Testing Association was set up for seed-borne viruses in 1975 and modern techniques like enzyme-linked immunosorbent assay (ELISA) and serologically specific electron microscopy have created new opportunities for improving the current unsatisfactory position.

International co-operation in plant virus research

Various working groups have been set up on virus diseases of specific crops or groups of crops. These groups are partly independent and partly under the auspices of other organizations. Such international bodies cannot enforce co-operative programmes but, by improving direct contacts between scientists, they do stimulate collaboration and a natural division of labour, and greatly assist newcomers.

Plant viruses and development aid

Concern for the virus problems of third-world countries is increasing rapidly since viruses have such drastic effects on the production of food and raw materials. It is also in the interests of developed countries to help with these problems because of the risks of introducing viruses with plants or plant products. The present efforts to improve crop production in developing areas of the world are creating new virus problems. This emphasizes the urgent need for improvements in plant virus research in the developing countries to counteract, or avoid, man-made diseases (Bos, 1976b) and much effort is being made by the *Food and Agriculture Organization* (FAO) of the United Nations Development Programme. Most developing countries cannot afford specialized virus research units or have insufficient specialists. Thus virus pathologists have to work alone, mostly have to deal with a wide range of crops and cannot readily keep abreast of technical developments.

Temporary solutions are often provided through *bilateral aid* by developed countries that provide visiting experts or consultants, either directly or through FAO, or other assistance in virus laboratories of the home country. However, the latter necessitates transfer of virus-infected material with the attendant risks of escape and there may be undue emphasis on the molecular biology of viruses at the expense of work on their ecology.

Some countries are pursuing historical ties with former colonies, as with the British-aided Plant Virus Laboratory of the Kenya Agriculture Research Institute, or the French virus laboratory at Adiopodoumé, Ivory Coast.

Progress in initiating national research on plant viruses has remained slow in most developing countries. Appreciation by administrators of the role of viruses in crop production is poor and there is a lack of information on the losses caused by viruses. Research infra-

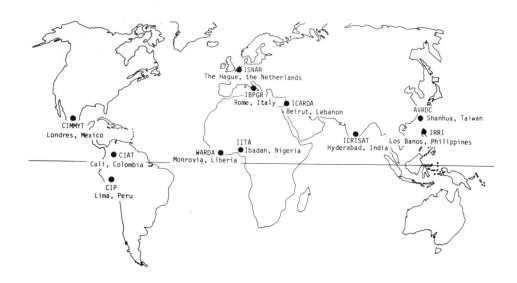

IRRI   : International Rice Research Institute, 1960.
CIMMYT   : Centro Internacional de Mejoramiento de Maíz (maize) y
            Trigo (wheat), 1966.
CIAT   : Centro Internacional de Agricultura Tropical, 1967.
IITA   : International Institute for Tropical Agriculture, 1968.
CIP   : Centro Internacional de la Papa (potato), 1971.
ICRISAT : International Crops Research Institute for the Semi-Arid
            Tropics, 1972.
AVRDC   : Asian Vegetable Research and Development Center, 1972.
WARDA*   : West Africa Rice Development Association, 1972.
IBPGR*   : International Bureau for Plant Genetic Resources, 1974.
ICARDA   : International Center for Agricultural Research in the
            Dry Areas, 1976.
ISNAR*   : International Service for National Agricultural Research,
            1979.

* concentrate on coordination

Figure 2. CGIAR-supported international agricultural research
centres dealing with crops and AVRDC (not funded by CGIAR), listed
in order of establishment.

structures are still poor and research programmes often lack continuity. Nevertheless, changes in the agricultural ecosystems of developing nations to improve crop production are rapid. Such changes are further enhanced by the activities of the international agricultural research institutes funded by the *Consultative Group on International Agricultural Research* (CGIAR) (Fig. 2). Emphasis is on crop improvement through breeding and improved farming systems (Bunting, 1981). Breeding programmes are supported by the International Bureau for Plant Genetic Resources established at FAO, Rome in 1974, to collect, conserve, evaluate, document, and make available for use the heritable diversity of economic plants and their wild relatives. Immense germplasm collections often of thousands of entries per crop species exist at CGIAR institutes. Though most of these institutes have only recently been established, promising results have been obtained already in selecting or producing material resistant to viruses and other pathogens.

Through these activities and their involvement in changing crop ecosystems, such as the increasing crop uniformity over large areas, CGIAR institutes, and other development programmes, are increasingly being confronted with problems caused by viruses and other pathogens and pests. Risks resulting from the accumulation and transfer of germplasm for such programmes are now increasingly being recognized (Hewitt & Chiarappa, 1977).

Several CGIAR institutes have already embarked upon plant virus research. Emphasis is on work to support breeding programmes and to safeguard the international exchange of germplasm. But this cannot be done without gathering reliable information on the detection, identity and ecology of the viruses involved. Manpower for virus research on important food crops, for which such institutes have global responsibility, is still very small as compared with that for single crops, such as maize or potato in single countries of the northern hemisphere. The CGIAR institutes could act as "centres of excellence" for stimulating plant virus research in surrounding countries.

CONCLUSIONS

Evidence is increasing that the best approach to understanding and solving virus problems in crops is ecological. Man himself continues to play an important role in inducing virus epidemics by changing crop ecosystems and by disseminating viruses. Through an increased understanding of the ecology of viruses, Man has also developed various phytosanitary methods to control virus diseases and they are often part of crop management.

However, human involvement with nature is ambivalent and interference always entails risks. This is true for any change in agro-ecosystems with the aim of improving crop productivity and for virus control itself. Breeding for resistance introduces the risks of:-

1. introducting genetic susceptibility and/or sensitivity to
    endemic viruses previously hidden in the natural vegetation,
2. stimulating the emergence of new virus strains through a
    genetic change in the crop,
3. introducing new viruses and other pathogens and pests with
    alien germplasm.

This partly explains why breeders and plant pathologists can be in
conflict.

Finally, it should be realized that the acceptance of risks is
unavoidable and without taking risks there will be no progress.
Civilization, including modernization of agriculture, is penetrating the
remotest corners of the world and plant viruses tend to follow in the
wake of these human activities. This has led to the pessimistic
"inevitability-of-establishment" hypothesis, but there are ways of
keeping ahead of nature and of reducing the risks. However, the ecology
of viruses will remain dynamic and Man will never be entirely successful in controlling virus diseases. Even breeding for resistance cannot
provide the final solution. Man must deal carefully with nature and
act as a steward rather than a monopolizer. There is a continuing need
for research to support the changes necessary for crop improvement.

REFERENCES

Aerts J. (1977) De betekenis van virusvrij plantmateriaal voor de
    glasaardbeienteelt. *Mededelingen van de Faculteit Landbouwweten-
    schappen Rijksuniversiteit Gent* 42, 1135-9.
Anon. (1980) Rice. *Annual Report for 1979, International Institute
    for Tropical Agriculture, Ibadan, Nigeria*, pp. 116-28.
Bakker W. (1974) Characterization and ecological aspects of rice
    yellow mottle virus in Kenya. *Agricultural Research Reports* 829,
    152 pp. Centre for Agricultural Publishing and Documentation,
    Wageningen.
Bennett C.W. (1952) Origin and distribution of new or little known
    virus diseases. *Plant Disease Reporter*, Supplement 211, 43-6.
Bos L. (1976a) Applied plant virus research: an analysis, and a scheme
    of organization. *Netherlands Journal of Plant Pathology* 82, 24-38.
Bos L. (1976b) Research on plant virus diseases in developing
    countries; possible ways for improvement. *FAO Plant Protection
    Bulletin* 24, 109-18.
Bos L. (1977) Seed-borne viruses. *Plant Health and Quarantine in
    International Transfer of Genetic Resources* (Ed. by W.B. Hewitt &
    L. Chiarappa), pp. 36-9. CRC Press, Cleveland, Ohio.
Bos L. (1981a) Wild plants in the ecology of virus diseases. *Plant
    Diseases and Vectors: Ecology and Epidemiology* (Ed. by K.
    Maramorosch & K.F. Harris), pp. 1-33. Academic Press, New York.
Bos L. (1981b) Experimental aspects and assessment of losses caused
    by viruses. *Crop Loss Assessment Methods* (Ed. by L. Chiarappa),
    Supplement 3, pp. 79-84. Commonwealth Agricultural Bureaux,
    Farnham Royal.

Bos L. (1982) Ecology and control of virus-induced diseases of plants;
a critical synopsis. *Applied Biology* 7 (in press).
Bos L., Huijberts N., Huttinga H. & Maat D.Z. (1978) Leek yellow
stripe virus and its relationship to onion yellow dwarf virus;
characterization, ecology and possible control. *Netherlands Journal
of Plant Pathology* 84, 185-204.
Boxus Ph., Quoirin M. & Laine J.M. (1977) Large scale propagation of
strawberry plants from tissue culture. *Applied and Fundamental
Aspects of Plant Cell, Tissue and Organ Culture* (Ed. by J. Reinert &
Y.P.S. Bajaj), pp. 130-43. Springer-Verlag, Berlin.
Bunting A.H. (1981) CGIAR: The first ten years. *Span* 24, 3-5, 15.
Button J. (1977) International exchange of disease-free citrus clones
by means of tissue culture. *Outlook on Agriculture* 9, 155-9.
Chock A.K. (1979) The international plant protection convention.
*Plant Health. The Scientific Basis for Administrative Control of
Plant Diseases and Pests* (Ed. by D.L. Ebbels & J.E. King), pp. 1-11.
Blackwell Scientific Publications, Oxford.
Conti M. & Caciagli P. (1979) Retention of infectivity of some cucumo
and potyviruses by the aphid vector *Myzus persicae* Sulz. *Abstracts
3rd Conference ISHS Working Group on Vegetable Viruses 29-31 August
1979, Bari, Italy*, pp. 33-4. Edizioni Quadrifoglio, Bari.
Costa A.S. (1975) Increase in the population density of *Bemisia tabaci*,
a threat of wide-spread virus infection of legume crops in Brazil.
Tropical Diseases of Legumes (Ed. by J. Bird & K. Maramorosch),
pp. 27-49. Academic Press, New York.
Cumming R.W.R. (1979) The role of nuclear stock associations in the
dissemination of healthy planting material in the United Kingdom.
*Plant Health. The Scientific Basis for Administrative Control of
Plant Diseases and Pests* (Ed. by D.L. Ebbels & J.E. King), pp. 149-
54. Blackwell Scientific Publications, Oxford.
Diener T.O. & Raymer W.B. (1971) Potato spindle tuber 'virus'. *CMI/
AAB Descriptions of Plant Viruses* 66, 4 pp.
Doncaster J.P. & Gregory P.H. (1948) The spread of virus diseases in
the potato crop. *Report Series of the Agricultural Research Council*
7, 189 pp.
Drijfhout E. (1978) Genetic interaction between *Phaseolus vulgaris*
and bean common mosaic virus, with implications for strain identifi-
cation and breeding for resistance. *Agricultural Research Reports*,
872, 98 pp. Centre for Agricultural Publishing and Documentation,
Wageningen.
Ebbels D.L. & King J.E. (Eds.) (1979) *Plant Health. The Scientific
Basis for Administrative Control of Plant Diseases and Pests.*
Blackwell Scientific Publications, Oxford.
Gill C.C. (1975) An epidemic of barley yellow dwarf in Manitoba and
Saskatchewan in 1974. *Plant Disease Reporter* 59, 814-8.
Grogan R.G. (1980) Control of lettuce mosaic with virus-free seed.
Plant Disease 64, 446-9.
Henshaw G.G. (1979) Plant tissue culture: its potential for disse-
mination of pathogen-free germplasm and multiplication of planting
material. *Plant Health. The Scientific Basis for Administrative
Control of Plant Diseases and Pests* (Ed. by D.L. Ebbels & J.E. King),
pp. 139-47. Blackwell Scientific Publications, Oxford.

Hewitt W.B. & Chiarappa L. (Eds.) (1977)   *Plant Health and Quarantine in International Transfer of Genetic Resources*. CRC Press, Cleveland, Ohio (second printing 1978).

Hiddema J. (1972)   Inspection and quality grading of seed potatoes. *Viruses of Potatoes and Seed-Potato Production* (Ed. by J.A. de Bokx) pp. 206-15.   Centre for Agricultural Publishing and Documentation, Wageningen.

Kahn R.P. (1977)   Plant quarantine: principles, methodology, and suggested approaches.   *Plant Health and Quarantine in International Transfer of Genetic Resources* (Ed. by W.B. Hewitt and L. Chiarappa), pp. 289-307.   CRC Press, Cleveland, Ohio.

Kahn R.P. (1979)   A concept of pest risk analysis.   *EPPO Bulletin* 9, 119-30.

Kahn R.P. & Monroe R.L. (1970)   Virus infection in plant introductions collected as vegetative propagations.   I. Wild vs. cultivated *Solanum* species.   *FAO Plant Protection Bulletin* 18, 97-102.

Kahn R.P., Monroe R.L., Hewitt W.B., Goheen A.C., Wallace J.M., Roistacher C.N., Nauer E.M., Ackermann W.R., Winters H.F., Seaton G.A. & Pifer W.A. (1967)   Incidence of virus detection in vegetatively propagated plant introductions under quarantine in the United States, 1957-1967.   *Plant Disease Reporter* 51, 715-9.

Kartha K.K. & Gamborg D.L. (1978)   Meristem culture techniques in the production of disease-free plants and freeze-preservation of germ plasm of tropical tuber crops and grain legumes.   *Diseases of Tropical Food Crops. Proceedings International Symposium held at Université Catholique de Louvain* (Ed. by H. Maraîte & J.A. Meyer), pp. 267-73.   Université Catholique de Louvain, Louvain-la-Neuve.

Kassanis B. (1965)   Therapy of virus-infected plants.   *Journal of the Royal Agricultural Society of England* 216, 105-14.

Koenig R. & Bode O. (1977)   In Westeuropa bisher nicht vorkommende Viren aus südamerikanischen Kartoffeln und ihr hochempfindlicher Nachweis mit dem serologischen Latextest.   *Mitteilungen aus der Biologischen Bundesanstalt für Land- und Forstwirtschaft* 178, 102 pp.

Koenig R. & Bode O. (1978)   Sensitive detection of Andean potato latent and Andean potato mottle viruses in potato tubers with the serological latex test.   *Phytopathologische Zeitschrift* 92, 275-80.

Lizzaraga R.E., Salazar L.R., Roca W.M. & Schilde-Rentschler L. (1980)   Elimination of potato spindle tuber viroid by low temperature and meristem culture.   *Phytopathology* 70, 754-5.

Mathys G. & Baker E.A. (1980)   An appraisal of the effectiveness of quarantines.   *Annual Review of Phytopathology* 18, 85-101.

Mellor F.C. & Stace-Smith R. (1977)   Virus-free potatoes by tissue culture.   *Applied and Fundamental Aspects of Plant Cell, Tissue and Organ Culture* (Ed. by J. Reinert & Y.P.S. Bajaj), pp. 616-35. Springer-Verlag, Berlin.

Murant A.F. (1970)   The importance of wild plants in the ecology of nematode-transmitted plant viruses.   *Outlook on Agriculture* 6, 114-21.

Niemöller A. (1965)   Massnahmen zur Bekämpfung der Pockenkrankheit (Scharka) in Rheinhessen und Pfalz 1963 und 1964.   *Gesunde Pflanzen* 17, 69-74.

Nyland G. & Goheen A.C. (1969)   Heat therapy of virus diseases of perennial plants. *Annual Review of Phytopathology* 7, 331-54.

Peerbooms H. (1971) Met virusvrije klonen betere resultaten bij Colden Delicious. *Fruitteelt* 61, 94-6.

Pool R.A.F., Wagnon H.K. & Williams H.E. (1970) Yield increase of heat-treated 'Baccara' roses in a commercial greenhouse. *Plant Disease Reporter* 54, 825-7.

Quak F. (1977) Meristem culture and virus-free plants. *Applied and Fundamental Aspects of Plant Cell, Tissue and Organ Culture* (Ed. by J. Reinert & Y.P.S. Bajaj), pp. 598-615. Springer-Verlag, Berlin.

Raymundo S.A. & Buddenhagen I.W. (1976) A rice virus disease in West Africa. *International Rice Commission Newsletter* 25, 58 pp.

Reinert J. & Bajaj Y.P.S. (Eds.) (1977) *Applied and Fundamental Aspects of Plant Cell, Tissue and Organ Culture.* Springer-Verlag, Berlin.

Russell G.E. (1978) *Plant Breeding for Pest and Disease Resistance.* Butterworths, London.

Rydén K. & Gerhardson B. (1978) Rödklöver mosaikvirus sprids med slättermaskiner. *Växtskyddsnotiser* 42, 112-5.

Slykhuis J.T., Andrews J.E. & Pittmann U.J. (1957) Relation of date of seeding winter wheat in southern Alberta to losses from wheat streak mosaic, root rot and rust. *Canadian Journal of Plant Science* 37, 113-27.

Smith I.M. (1979) The work of a regional plant protection organization, with particular reference to phytosanitary regulations. *Plant Health. The Scientific Basis for Administrative Control of Plant Diseases and Pests* (Ed. by D.L. Ebbels & J.E. King), pp. 13-22. Blackwell Scientific Publications, Oxford.

Thresh J.M. (1980) The origins and epidemiology of some important plant virus diseases. *Applied Biology* 5, (Ed. by T.H. Coaker), pp. 1-65. Academic Press, London.

Tomlinson J.A. & Faithfull E.M. (1979) The use of surfactants for the control of lettuce big-vein disease. *Proceedings of the 1979 British Crop Protection Conference. Pests and Diseases* 1, 341-6.

Tomlinson J.A. & Ward C.M. (1981) The reactions of some Brussels sprout F1 hybrids and inbreds to cauliflower mosaic and turnip mosaic viruses. *Annals of Applied Biology* 97, 205-12.

Upstone M.E. (1969) Epidemiology of cocksfoot mottle virus. *Annals of Applied Biology* 64, 49-55.

Zink F.W. & Duffus J.E. (1975) Reaction of downy mildew-resistant lettuce cultivars to infection by turnip mosaic virus. *Phytopathology* 65, 243-5.

Crop improvement in relation to virus diseases and their epidemiology

IVAN W. BUDDENHAGEN

Department of Agronomy and Plant Pathology,
University of California, Davis, CA 95616, USA

INTRODUCTION

The virus epidemiologist naturally considers fields or plantations as
the basis for study but the data obtained and the conclusions drawn
depend on the particular crop genotype grown. It follows that epidemiological data collected for one cultivar differ from those for another.
The cultivar characteristics influence both vector and virus population
dynamics and the incidence, severity and importance of disease in the
crop. Of considerable importance to the goal of reducing losses is the
potential magnitude of the difference in virus/vector epidemiological
development between existing cultivars and those that could be developed
through breeding.

In a sense, those who develop new cultivars determine the findings of
epidemiologists and there are enormous opportunities for influencing the
future prevalence of virus diseases. Diseases may be changed from insig-
nificance to importance by introducing new cultivars, or vice versa.
The introduction of new cultivars may reveal "new" virus diseases that
occurred on old cultivars but so infrequently or with so little effect
that they were not known or at least not brought to the attention of
agricultural scientists. This happened with tungro disease of rice in
India and the Philippines in the 1960s, and with maize rough dwarf
disease on new American maize hybrids in Italy in the 1940s. The key to
future progress is to develop new crop cultivars that lower the incidence and/or severity of known viruses and do not give rise to "new"
viruses in that crop or in others nearby.

A good research team could probably alter the host genotype sufficiently within a few breeding cycles to render any virus disease insignificant. There have been many successes (Russell, 1978), yet many
viruses remain important. I believe this is due not only to the complexities of working with vector-borne pathogens, but also to the acceptance
of their inevitability. As a result insufficient funds, effort, creative
research and teamwork are directed towards breeding them into insignifi-
cance. It should be useful to compare the successes of virus resistance breeding with the cases where losses continue, and thereby develop

Plumb R.T. & Thresh J.M. (1983) *Plant Virus Epidemiology*.

approaches to making the latter also insignificant.

An interesting example is the case of tungro virus in relation to the high-yielding rice cultivars introduced in India. Recently, 89 high-yielding cultivars already released by State and Central Government organizations in India or by the International Rice Research Institute were evaluated for field resistance to tungro for three years. Ratings were based on severity of symptoms and yield depression. Only seven cultivars were judged "resistant", 27 were "intermediate" and 55 were "susceptible" (A. Anjaneyulu, personal communication). Thus, in India, tungro is being maintained as a potential threat by breeders and other decision-makers who develop and release new, vulnerable cultivars. It would be interesting to analyse how this paradox has arisen. Amazingly, it is still not known if tungro is caused by only one virus (Saito, 1977; Ling, this volume), yet it is potentially a serious threat to millions of hectares of rice in tropical Asia, an area which is chronically short of food.

ORIGIN, COEVOLUTION AND NON-COEVOLUTION

An important first step is to determine if the crop/virus system being studied has coevolved or if it is the result of a recent encounter (Buddenhagen, 1977). This requires consideration of the origins and spread of crops and of their viruses and vectors. Many crop virus diseases are the result of new encounters arising from man's exploration and colonization, especially since 1500 AD. For instance, the viruses causing cassava mosaic and maize streak in Africa could not have originated in cassava or maize since they do not occur in the Americas where these crops evolved and from where they were transported to Africa.

Viruses reported from rice provide another example. Many viruses infect rice (Ling, 1972) but only one or two appear to be viruses *of* rice, in the sense of the crop being the original host. Hoja blanca of rice in the Americas probably originated from wild grasses and giallume of rice (a strain of barley yellow dwarf virus) in Italy probably originated similarly. The well-known viruses of rice in temperate Japan, including stripe, dwarf and black-streaked dwarf, are unlikely to have originated in rice since they and their leafhopper or plant-hopper vectors have a wide host range in the Gramineae, and rice is a fairly recent introduction to Japan.

Coevolution or recent encounter in host-pathogen relationships has obvious implications in disease ecology and epidemiology. It is also crucial in considering strategies for resistance breeding. For co-evolved systems the centre of origin can be expected to be a source of coevolved genes for vertical resistance to vector and/or virus, which is not durable. This is because such genes interact with the pathogen in a gene-for-gene system. Different strains carry different genes, and mutations to overcome a host resistance gene will be retained in a specific virus strain. In spatially separated and genetically variable natural host populations the pathogen population will be genetically

variable and have vertical gene plasticity. Also genes for tolerance
and tolremicity* (Buddenhagen, 1981a) are likely to occur, which should
slow epidemics and reduce the severity of infection. It should be
useful to search for individuals with these various characteristics in
the regions where coevolution occurred. For example, resistance to rice
tungro involving tolerance as well as tolremicity comes from tungro-
endemic areas in and near Bengal where the causal agent(s) coevolved
with rice.

For non-coevolved systems it is likely that resistance will be found
in material genetically quite different from that in the affected area.
Such "accidental" resistance may act vertically or horizontally, but be
due to normal constitutive genes present for unknown reasons, which also
confer resistance. It is unnecessary to search the centre of origin of
the crop host for such genes. Moreover, one would not expect resistance
due to such genes to break down, even if the hosts are immune. This is
because no strains and no gene-for-gene relationships have evolved for
the crop/virus system. Any apparent host/pathogen relationship is
either false or distant. For example, resistance to the temperate rice
virus diseases of Japan comes from an introduced Indica (tropical)
subspecies that is genetically remote from Japonica (temperate) cultivars. Similarly, resistance to rice hoja blanca of the American tropics
is found in introduced Japonica germplasm that is genetically and geographically distant from the region of the virus problem. If, by
chance, Japonica rather than Indica rice had been planted in tropical
America the disease would probably be unknown and there would be no need
to increase resistance.

Other sources of resistance to non-coevolved pathogens can occur in
material "resistant" to other pathogens present elsewhere. Thus,
"resistance" to A may be effective also against B, in what is regarded
as a fortuitous way but which probably is not. More likely is that
these are constitutive genes, useful in maintaining health when challenged by various agents. It is a mistake to consider them genes for
specific resistance to A or B, even if genetic tests suggest this.

A more remote, but interesting, question, and one that is almost
totally neglected, concerns the origin of viruses of land plants. The
apparent paucity of virus diseases of Gymnosperms and lower plants
suggests that plant viruses originated with the appearance and radiation
of the Angiosperms (Doyle, 1978) and their associated insects in

---

* The term "tolremic" has been introduced to represent that part of
resistance that operates to slow disease spread in a field or area, thus
reducing final incidence. For systemic diseases, including virus
diseases, it is useful to separate this characteristic from tolerance,
which is the individual plant reaction to being infected. In this
usage, individual plant "tolerance" combined with cultivar "tolremicity"
together constitute "resistance". (For polycyclic fungal or bacterial
diseases, "tolremicity" can be applied to build-up of disease with time
on a single plant as well as to population incidence).

Cretaceous times or earlier (100-140 million years ago). This means that continental drift (Schuster, 1976) should have influenced the evolution and original distribution of plant viruses and their vectors. An even earlier origin has been proposed for the Tobamoviruses (Gibbs, 1980). This subject cannot be elaborated here but it is relevant to the development of a coherent understanding of virus evolution and of genetic strategies for reducing virus damage to crops.

VIRUS STRAIN VARIATION

Despite the interest of virologists in strains and minor differences in plant reaction to different virus isolates in laboratory and glasshouse studies, breeding crops for resistance to viruses has often resulted in durable resistance *sensu* Johnson (1981).

For some viruses there are strains in different locations that affect cultivars differently and necessitate different resistances. In the indigenous crop area one would expect to find diverse strains or entirely different viruses, not all of which have been exported else-where. This is so with potatoes in the Andes (Moreira *et al.*, 1980; Jones, 1981).

Resistance developed to strains in one location cannot be expected to operate against strains occurring elsewhere, and growing a cultivar in a new area may lead to serious epidemics. This is especially likely for virus resistance based on hypersensitivity in clonal crops. The evolution and selection *in situ* of resistance-breaking strains that also survive well is less common. In raspberry ringspot virus there are "Scottish" and "English" strains; a new strain in Scotland virulent against newly developed resistant cultivars apparently has characteristics which militate against its survival (Murant *et al.*, 1968; Harrison, 1978). For tomato mosaic virus, which is readily transmitted by contact, selection of more virulent types occurred soon after the introduction of resistance genes (Pelham *et al.*, 1970; Pelham, 1972), but such cases seem rare.

In terms of pathogen/host/vector coevolution, it is difficult to see why the introduction of major resistance genes into the host has been so effective in controlling many virus diseases. The numbers of virus particles in a crop are so enormous that even a low mutation rate should provide ample opportunity for virulent mutants of increased epidemiological potential to occur. They could be extracted from the sites of origin by vectors and could be expected to replace previous strains. For reasons that are not yet clear this seldom seems to occur. It is possible that a virus has so few genes and these genes are so finely tuned for survival that any change reduces fitness. Cauliflower mosaic virus has only six genes (Gardner *et al.*, 1981) and probably not all determine epidemiological parameters. Even the large virion of tobacco mosaic virus has only four or five genes.

It is interesting to compare viruses with the obligate rust fungi,

those fungi most closely resembling viruses in their biotrophy, which
readily overcome host resistance. Rusts must have hundreds or even
thousands of genes, although less than 100 may govern host specificity.
The genetic and biological complexity of vector transmission of viruses
combined with difficulties in perennation are the areas that especially
justify investigation of the stability of virus resistance. Addition-
ally, because they invade the host systemically, the increase in infections during an epidemic is less than with typical leaf spot fungi
(Vanderplank 1959). With rates of increase of only 10 to 10 000 fold
per season as compared with a billion fold for some leaf spot fungi,
there is that much less opportunity for the selection of resistance-
breaking strains of viruses. I have discussed elsewhere (Buddenhagen,
1965, 1977) the paucity of pathogens that are epidemiologically competent and the difficulty pathogens face in becoming more virulent and more
competent. The view that there is an inexorable evolutionary trend
towards greater aggressiveness and thus greater disease is not valid.

ECOLOGY AND BREEDING FOR RESISTANCE

Diseases can be reduced in severity by the development of less susceptible cultivars and some diseases can thus be eradicated, for example,
smallpox was eradicated by immunologically breaking the epidemiological
cycle. For viruses of crops, the same objective can be approached by
making the appropriate genetic changes in the host.

With plant pathogens the cycle can be broken before the epidemic or
early or late in the epidemic, and the pre-epidemic stage is most
vulnerable to attack. With viruses of annual crops this may mean consideration of alternative weed hosts, and development of resistance in
the crop host should then take account of the strains and vectors
surviving in weeds. Weeds may act as a stabilizing "screen", that would
block survival of any new strains selected by new resistance genes
incorporated in the crop. It is probable that not only immunity but
also all aspects of resistance to vectors, including non-preference, as
well as resistance to virus infection, decreased susceptibility to
virus invasion and increased tolerance, can be useful. If the virus
is transmitted by the vector to its progeny and no weed hosts occur,
then resistance to insects, non-preference and resistance to infection
should become primary targets of resistance breeding. If viruses
survive through seed, then breeding for non-seed-transmissibility
becomes a logical and simple target. This has been achieved naturally
in most plants during the evolution of the seed habit but with some
exceptions that enable the perpetuation and dissemination of certain
viruses in the absence of, or as a supplement to, spread by vectors.
The potential for blocking the annual reappearance of virus diseases
through breeding for non-seed-transmissibility in annual crops has been
neglected, but some limited work has been carried out with soybean
mosaic virus (Goodman *et al.*, 1979) and barley stripe mosaic virus
(Carroll *et al.*, 1981). A full knowledge of disease biology is
required to know if such an approach is realistic (Hanada & Harrison,
1977).

In ancient, clonally-propagated horticultural crops (and even in seed-propagated perennials), selection to increase the tolerance of cultivars has proceeded unconsciously for centuries. Thus, we have symptomless cherries, potatoes, etc. which carry so-called latent viruses. Off-season survival is not then a consideration in breeding, but as host tolerance is increased, virus invasiveness is often reduced. Selection during plant propagation can sometimes result in escape from an incompletely systemic virus, as with African cassava mosaic, and with certain potato viruses (Jones, 1981). In the absence of common sources of infection or highly active vectors, new plantings can be established that may remain undamaged for years. This approach to managing cassava mosaic has been proposed (Bock *et al.*, 1976; Bock, this volume).

Understanding the details of plant virus survival is a great aid in developing strategies for breeding new cultivars which will be less damaged and restrict virus spread. Enormous flexibility and complexity in breeding cultivars that will not be seriously damaged by viruses are provided by the diversity of:

1) the obligate survival of viruses in crops, other plants, or vectors,
2) the role of alternative hosts,
3) the vector/crop relationship during the season,
4) the coevolved or non-coevolved relationship with the crop,
5) mode of perennation of virus and vector.

These complexities are seldom analysed sufficiently and resistance breeding is often approached too simplistically. Failure to analyse the complex origin of host/parasite systems and their ecology, biology and epidemiology can lead to inadequate levels of resistance or lack of durability (Buddenhagen, 1982).

THE MEANING OF "RESISTANCE"

Resistance can involve only one, a few, many, or even all of the steps of an epidemic. It is not known how many such steps are involved, especially at the physiological level, since such steps are detected only by complex genetic analyses of host/parasite interactions, and these have seldom been done. Our knowledge is thus very limited. However, a practical and useful subdivision of "resistance" is to distinguish the reaction to a pathogen of *individual plants* from the reaction of populations. The reaction of individual plants ranges from immunity, through various levels of tolerance to an intolerance resulting, at its maximum, in either death or hypersensitivity. Hypersensitivity usually precludes "systemicity" and thus is commonly considered a form of resistance. Such hypersensitive "resistance" (which is really super-susceptibility) is often fragile, both environmentally and genetically. The plant reaction will depend on pathogen dose, age of tissue and environment. For virus and other systemic diseases the individual plant reaction is simplified and a precise concept of tolerance* can be applied in breeding.

If the virus dose is sufficient for systemic infection to occur then the interaction of virus/ontogeny/host genotype/environment can be quantified as some level of tolerance. This is because these inter-actions can be subtracted from uninfected plants involving only ontogeny/host genotype/environment. The complex steps involving virus replication, systemic invasion and cellular incompatibility need not be understood for this analysis, although knowing them would be desirable. This latter area of research does, of course, receive much support, far more than efforts to reduce virus diseases through breeding.

For virus diseases it is especially important to separate severity of *individual* plant/virus interaction (tolerance) from probability of infection within a *population* (tolremicity). If this is not done in breeding, one does not know what epidemiological vulnerability to expect. For example, a system was developed in breeding for tungro virus resistance in rice whereby genetically uniform lines were assessed as resistant or susceptible on the basis of the proportion that showed symptoms after a 48 h infection feeding period by a few infective vectors (Ou, 1972). If 30% or fewer (of 29 seedlings) showed symptoms after 12 days, the line was judged "resistant". It seems obvious that if 30% of a pure breeding line can be infected in 48 h, 100% could be infected in a few more days by the same number of vectors and more quickly if vector numbers were increased. If infected plants are severely affected then such a cultivar would be very susceptible in the field although rated "resistant" based on incidence. (However, additional readings were taken on degree of stunting after six weeks which served as "a secondary criterion of resistance". Exactly which readings the breeders accepted is not clear.) This example is one of many and it is the kind of problem that is fundamental to concepts of resistance and susceptibility in resistance breeding and to effective communication between pathologists and breeders.

What does "resistance" mean epidemiologically? This is especially important for virus diseases since *incidence* in a field will involve all aspects of the host genotype's ability to restrict transmission by resistance to vectors and the ease with which virus is acquired from, and transmitted to, plants. Moreover, the level of tolerance can

---

* I confine "tolerance" to an expression of performance by a host when infected by pathogens causing systemic disease. A plant is tolerant to some degree if it is less damaged than its neighbours when infected by a systemic pathogen. Tolerance and intolerance are relative terms and as defined here tolerance can be selected for easily in segregating generations in a breeding programme (Buddenhagen, 1981). Tolerance as defined by Schafer (1971) is based on leaf spot disease and is a different concept, essentially equal damage but less yield reduction. It is confounded by varietal ontogeny, harvest index differences and cryptic error in disease measurement over time and is almost impossible to select for in a breeding programme.

greatly influence disease progress. In a recent study in Nigeria with a maize cultivar developed in a screening system which detected only plant *tolerance*, field *incidence* during an epidemic was reduced from 63% in the intolerant control to only 10% in the tolerant cultivar (Soto *et al.*, 1982). Thus, changing the host's tolerance of virus alone was a principal cause of a reduction in virus spread.

UNPREDICTABILITY

Pathogen/host/environment systems which result in patchy and unpredict-able outbreaks are characteristic of many virus diseases. Such systems may be considered as having high survival value for the pathogen and, in a sense, being a successful strategy for its survival. It is such systems that escape the selection routines of breeders. Thus, the breeder may produce new cultivars that appear to develop "new" virus diseases (i.e. enhance existing, undetected ones), or convert existing virus diseases of minor importance to serious problems by changing levels of tolerance or tolremicity.

If virus incidence in an area is consistently high, breeders' plots will be challenged naturally and the breeder will automatically select tolerant or immune plants. This must have happened naturally in primi-tive agriculture where virus diseases are endemic, such as tungro virus on rice in Bengal and many potato viruses in the Andes. When scientists come to such areas and apply recombinant techniques, the development of resistance is enhanced. Thus productive rice cultivars were developed in Indonesia in the 1930s that withstood the presence of endemic tungro, even though it was not then known to be a virus problem (Buddenhagen, 1982).

Conversely, if local viruses were patchy or sporadic in incidence or low in severity on old cultivars, breeders could at first ignore them and develop new, higher-yielding cultivars by concentrating on plant type and resistance to important fungal diseases. Such cultivars could turn out to be much more vulnerable to minor or unknown viruses. This is probably why barley yellow dwarf virus with its erratic transmissi-bility and dependency on unusually favourable environmental conditions for epidemic development (Thresh, 1980; Carter *et al.*, 1980) became important in the 1950s, and remains important, even though the disease is probably ancient and certainly widespread.

The continued appearance of "new" viruses (Duffus, 1977; Ling *et al.*, 1978; Shikata, 1979) is, I believe, not just a matter of increased scientific awareness but also a response to the development of new cultivars which facilitate the survival and spread of previously "hidden" viruses not present or considered at breeding stations. In the quest for higher-yielding cultivars with resistance to the principal known endemic diseases, breeding against occasional diseases is neglec-ted and their control is left to chemicals. Since breeding for resist-ance to virus diseases is so easy, there is considerable scope for preventive breeding, especially in annuals, for precluding the appearance of "new" viruses and for decreasing the importance of sporadic, unpredictable, but sometimes severe, known viruses, such as barley yellow dwarf virus. More field surveys of disease and feedback to the breeder are needed, as are efforts to make "occasional" diseases prevalent in breeders' plots, so as to facilitate selection.

I have advocated "preventive breeding" to avoid the increase of "hidden" or "low level" diseases, especially where old cultivars are in balance with their pathogens but are low-yielding. For example, in West Africa, rice yellow mottle virus occurs widely but sparsely and is not a problem on old rice cultivars. Introduced high-yielding Asian cultivars have little tolerance of this virus and the disease has become more obvious. But *incidence* is still too low to make this disease of economic importance. Thus, to the administrator or routine breeder, breeding for resistance has low priority. However, past experience suggests that if cropping intensity is increased disease *incidence* will also increase on vulnerable cultivars; it is only low *incidence* which keeps such a disease unimportant on cultivars with little tolerance. It is easy to prevent this disease from ever becoming important, through "preventive breeding". It is even easier to do nothing and allow such an obscure disease to become important as has happened already with rice yellow mottle virus in Kenya (Bakker, 1974).

It is important that breeding cultivars with resistance to diseases should begin before damaging epidemics have occurred. There should be a strategy of preventing pests and diseases from becoming important in a specific agroecosystem. To do this the breeder needs much help from and contact with plant pathologists and epidemiologists; teamwork is the real solution.

PERENNIAL CROPS

Different problems and opportunities are presented by perennial crops grown at permanent sites (cherry, etc.) or as annuals (tuberous perennials). Clonal propagation confers immortality on selected seedlings and also on any systemic pathogens that are present, including viruses. Thus, ancient clones have to be tolerant (or immune) to systemic pathogens or they would have been discarded. Breeding new cultivars of ancient, clonally propagated crops (such as yams) that have been selected for virus tolerance for aeons can unlock much susceptibility, often hidden in the polyploid genetic "reservoir" of the clone.

Progress may be slow, especially if recombination is difficult, host genetic reserves are low, or the disease is the result of a new encounter, as with the viruses of taro in the Solomon Islands (Jackson & Pelomo, 1979) and of American yams in the Caribbean (Martin, 1976). However, in potatoes, with their coevolved viruses from the Andes, much variability exists and both tolerance and immunity are available for genetic manipulation. With the "degeneration" of potato cultivars in Britain and other parts of Europe due to viruses in the 19th century,

great improvement was obtained simply from field selection of plants
obtained from an initial collection of self-set seed. Patterson's
Victoria was a cultivar selected in this way before viruses were known
to exist and its resistance genes are still utilized (Davidson, 1980).

I see no reason why breeding should not make virus diseases of
potatoes insignificant within a few years. Their importance has been
maintained largely because of a reluctance to breed for tolerance,
the main breeding concern being for quality and yield. Tolerance has
been rejected largely because certification schemes involve inspection
of plants and rejection of those with symptoms, thus susceptibility is
desirable. However, now techniques such as enzyme-linked immunosorbent
assay (ELISA) are available it should be possible to utilize highly
tolerant cultivars and still maintain low virus incidence. Tolerance to
virus is probably already an important factor in maintaining potato
production in the Andes (Jones, 1981) and in other areas such as the
Philippines and Turkey (C.R. Brown, personal communication). A problem
that sometimes may occur with tolerance is the threat posed by a large
reservoir of virus to other nearby crops that have little tolerance
(Simons, 1959). This should not arise if tolerance is associated with
low virus content, reduced systemicity, and slow spread in the field,
as with maize streak in Africa. Such highly tolerant cultivars would
then pose less threat to neighbouring crops than the less tolerant ones
desired for ease of certification.

For clonal orchard crops where quality is most important it has been
common breeding practice to give minimal consideration to diseases and
this explains their continuing importance. Since great genetic diversity exists in many orchard species, especially at their centres of
origin (Simmonds, 1976), there is considerable scope for increasing the
tolerance and tolremicity of new cultivars to virus diseases.

A virus such as plum pox virus, which has spread extensively in
Europe in the last 70 years but is not yet known in Asia, Australasia
or America (Thresh, 1980), is a prime candidate for resistance breeding.
If the main plum, peach and apricot cultivars of California and the
United States of America in general have little tolerance of this virus,
serious losses could occur if the pathogen were introduced. Tolerant
(or immune) cultivars would then be invaluable.

The rootstock/scion relationship becomes critical in virus diseases
of crops such as grapevine or citrus that are grafted or budded. This
is well illustrated by tristeza, which is mainly a problem of sweet
orange on sour orange rootstocks (Bar-Joseph et al., this volume). This
complicates the breeding of clonally propagated perennial crops. Nevertheless, if early horticulturalists with limited germplasm and no
knowledge of viruses could make so much progress in developing virus
tolerance, even to the extent of obtaining symptomless clones, what
could dedicated breeders accomplish with present knowledge? There is
also a need for greater effort and less acceptance of the status quo of
virus diseases in these crops.

CONCLUSION

The importance of any crop disease depends on the genotype being grown. The seriousness of disease and the course of epidemics are a reflection of the interaction between the particular crop genotype, the environment, and the virus and vector genotypes. The potential for influencing the system by altering the crop genotype is very great. For virus diseases this potential is under-utilized due to inadequate conceptual approaches and inadequate development of reliable methods in breeding and selection of resistance. A reluctance to select for tolerance combined with inadequate efforts to analyse the evolution, ecology and epidemiology of pathosystems inhibit progress. Most crop species have so much genetic variability that, if it is manipulated wisely, new cultivars should become such poor hosts of viruses and their vectors that losses due to virus diseases are greatly decreased.

REFERENCES

Bakker W. (1974) Characterization and ecological aspects of rice yellow mottle virus in Kenya. *Agricultural Research Report* 829,152 pp. Centre for Agricultural Publishing and Documentation, Wageningen.

Bock K.R., Guthrie E.J. & Seif A.A. (1976) Field control of cassava mosaic in coast province, Kenya. *Proceedings of the Fourth Symposium of the International Society for Tropical Root Crops, CIAT, Cali, Colombia.* (Ed. by J. Cock, R. MacIntyre & M. Graham), pp. 160-3. International Development Research Centre, Ottawa.

Buddenhagen I.W. (1965) The relation of plant-pathogenic bacteria to the soil. *Ecology of Soil-borne Plant Pathogens* (Ed. by K.E. Baker & W.C. Snyder), pp. 269-84. University of California Press, Berkeley.

Buddenhagen I.W. (1977) Resistance and vulnerability of tropical crops in relation to their evolution and breeding. *Annals of the New York Academy of Science* 287, 309-26.

Buddenhagen I.W. (1981) Conceptual and practical considerations when breeding for tolerance or resistance. *Plant Disease Control* (Ed. by R.C. Staples & G.H. Toenniessen), pp. 221-34. John Wiley, New York.

Buddenhagen I.W. (1982) Durable resistance in rice. *Proceedings FAO/NATO Conference, Martina Franca, Italy, October 1981.* Plenum Publishing Corporation, New York (in press).

Carroll T.W., Zaske S.K. & Hockett E.A. (1981) Development of a barley germ plasm resistant to the seed transmission of barley stripe virus. *Phytopathology* 71, 865.

Carter N., McLean I.G., Watt A.D. & Dixon A.G. (1980) Cereal aphids: A case study and review. *Applied Biology* 5 (Ed. by T.H. Coaker) pp. 271-348. Academic Press, London.

Davidson T.W. (1981) Breeding for resistance to virus disease of the potato (*Solanum tuberosum*) at the Scottish Plant Breeding Station. *Annual Report of the Scottish Society for Research in Plant Breeding 1980*, pp. 100-8.

Doyle J.A. (1978) Origin of angiosperms. *Annual Review of Ecology and Systematics* 9, 365-92.

Duffus J.E. (1977) Aphids, viruses, and the yellow plague. *Aphids as Virus Vectors* (Ed. by K.F. Harris & K. Maramorosch), pp. 361-83. Academic Press, New York.

Gardner R.C., Howarth A.J., Hahn P., Leudi-Brown M., Shepherd R.J. & Messing J. (1981) The complete nucleotide sequence of an infectious clone of cauliflower mosaic virus by M13mp7 shotgun sequencing. *Nucleic Acids Research* 9, 2871-88.

Gibbs A. (1980) How ancient are the Tobamoviruses? *Intervirology* 14, 101-8.

Goodman R.M., Bowers G.R. & Paschall E.H. (1979) Identification of soybean germplasm lines and cultivars with low incidence of soybean mosaic virus transmission through seed. *Crop Science* 19, 264-7.

Hanada K. & Harrison B.D. (1977) Effects of virus genotype and temperature on seed transmission of nepoviruses. *Annals of Applied Biology* 85, 79-92.

Harrison B.D. (1978) The groups of nematode-transmitted plant viruses, and molecular aspects of their variation and ecology. *Plant Disease Epidemiology* (Ed. by P.R. Scott & A. Bainbridge), pp. 255-64. Blackwell Scientific Publications, Oxford.

Jackson G.V.H. & Pelomo P.M. (1979) Breeding for resistance to diseases of taro, *Colocasia esculenta*, in Solomon Islands. *Proceedings International Symposium on Taro and Cocoyam, Viscaya, Baybay, Leyte, Philippines, September 1979*, pp. 287-98. International Development Research Centre, Ottawa.

Johnson R. (1981) Durable resistance: definition of, genetic control and attainment in plant breeding. *Phytopathology* 71, 567-8.

Jones R.A.C. (1981) The ecology of viruses infecting wild and cultivated potatoes in the Andean region of South America. *Pests, Pathogens and Vegetation* (Ed. by J.M. Thresh), pp. 89-107. Pitman, London.

Ling K.C. (1972) *Rice Virus Diseases*. International Rice Research Institute, Los Baños.

Ling K.C., Tiongco E.R., Aguiero V.M. & Cabauatan P.Q. (1978) Rice ragged stunt disease in the Philippines. *International Rice Research Newsletter* 16, 25 pp.

Martin F.W. (1976) Selected yam varieties for the tropics. *Proceedings of the Fourth Symposium of the International Society for Tropical Root Crops, CIAT, Cali, Colombia* (Ed. by J. Cock, R. MacIntyre & M. Graham), pp. 44-9. International Development Research Centre, Ottawa.

Moreira A., Jones R.A.C. & Fribourg C.E. (1980) Properties of a resistance-breaking strain of potato virus X. *Annals of Applied Biology* 95, 93-103.

Murant A.F., Taylor C.E. & Chambers J. (1968) Properties, relationships and transmission of a strain of raspberry ringspot virus infecting raspberry cultivars immune to the common Scottish strain. *Annals of Applied Biology* 61, 175-86.

Ou S.H. (1972) *Rice Diseases*. Commonwealth Mycological Institute, Kew.

Pelham J. (1972) Strain-genotype interaction of tobacco mosaic virus in tomato. *Annals of Applied Biology* 71, 219-28.

Pelham J., Fletcher J.T. & Hawkins J.H. (1970) The establishment of a new strain of tobacco mosaic virus resulting from the use of resistant varieties of tomato. *Annals of Applied Biology* 65, 293-7.

Russell G.E. (1978) *Plant Breeding for Pest and Disease Resistance.*
    Butterworths, London.
Saito Y. (1977) Rice viruses with special reference to particle morph-
    ology and relationship with cells and tissues. *Review of Plant
    Protection Research* 10, 83-90.
Schafer J.F. (1971) Tolerance to plant disease. *Annual Review of
    Phytopathology* 9, 235-52.
Schuster R.M. (1976) Plate tectonics and its bearing on the geographi-
    cal origin and dispersal of angiosperms. *Origin and Early Evolution
    of Angiosperms* (Ed. by C.B. Beck), pp. 48-138. Columbia University
    Press, New York.
Shikata E. (1979) Rice viruses and MLOs, and leafhopper vectors.
    Leafhopper Vectors and Plant Disease Agents (Ed. by K. Maramorosch &
    K.F. Harris), pp. 515-27. Academic Press, New York.
Simmonds N.W. (1976) *Evolution of Crop Plants*. Longmans, New York.
Simons J.N. (1959) Potato virus Y appears in additional areas of
    pepper and tomato production in south Florida. *Plant Disease
    Reporter* 43, 710-1.
Soto P.E., Buddenhagen I.W. & Asnani V.L. (1982) Development of
    streak virus resistant maize populations through improved challenge
    and selection methods. *Annals of Applied Biology* 100, 539-46.
Thresh J.M. (1980) The origins and epidemiology of some important
    plant virus diseases. *Applied Biology* 5 (Ed. by T.H. Coaker),
    pp. 1-65. Academic Press, London.
Vanderplank J.E. (1959) Some epidemiological consequences of systemic
    infection. *Plant Pathology: Problems and Progress 1908-1958*
    pp. 566-73. The University of Wisconsin Press, Madison.

A simple convolution method for describing or comparing the distributions of virus-affected plants in a plant community

ADRIAN GIBBS
Virus Ecology Research Group, Research School of
Biological Sciences, Australian National University,
Canberra, ACT 2601, Australia

The distribution of virus-infected individuals in a plant community, such as a crop, can provide useful clues to the sources and mode of spread of a virus. However, such two-dimensional patterns are time-consuming to record and I know of no simple published method for analysing them. Reports of such patterns often merely present them as unanalysed maps or photographs, or use such descriptive phrases as "the yellows-infected plants were mostly in randomly distributed clumps".

Nevertheless, two-dimensional patterns can be analysed. For example, ecologists study plant distributions and have mostly used quadrat and "nearest neighbour" methods to analyse patterns (Greig-Smith, 1964; Pielou, 1969), though there now is increased interest in more direct measures such as "radial distribution functions" (Ripley, 1977; Emmerick, 1979; Diggle, 1981). By contrast, those studying repetitive features in electron micrographs of macromolecules have had greatest success with optical or computed Fourier methods (Klug & Berger, 1964; De Rosier & Klug, 1968). Most of these methods are indirect and depend on skill in mathematics or the ability to think in reciprocal space; uncommon attributes among plant pathologists. It seemed to me that the principle of superimposing patterns, first used by Galton (1878), combined with that of the "convolution camera" used by Elliott *et al*. (1968) to examine patterns in electron micrographs, could perhaps provide a method for comparing and describing patterns of virus-infected plants in crops, as it would require no specialized apparatus or skills and its results would be simple to comprehend.

The basic data for the convolution method of analysis is a map, or maps, of the distribution of the plants in a community showing virus symptoms on one or on successive occasions. Convolution diagrams are derived from the maps, and the pattern in these will indicate whether the diseased plants in the primary maps were randomly distributed, or in what way they were not random.

Plumb R.T. & Thresh J.M. (1983) *Plant Virus Epidemiology.*
Blackwell Scientific Publications, Oxford.

COMPARISONS OF THE DISTRIBUTION PATTERNS OF TWO TYPES OF VIRUS-AFFECTED
PLANTS

It is simplest to describe the convolution method as it would be used
to compare the patterns of diseased plants in a crop grown from broad-
cast seed, such as a pasture, and mapped on two occasions, in order to
determine whether plants that developed symptoms in the interim were
clumped around those originally affected. This is done by plotting the
distribution of all of the newly affected plants ("B" plants) around
each of the affected plants found originally ("A" plants), and superimposing *all* of those patterns. Thus a summary is obtained of the
positions of the "B" plants relative to all the "A" plants; this
summary is the convolution diagram. A clumping of the points around the
centre of the convolution diagram indicates that the "B" plants are, on
average, clumped around the "A" plants in the primary map. If the "B"
plants are randomly distributed around the "A" plants then the points in
the convolution diagram will also be randomly distributed.

   The convolution diagram is most simply prepared from a single
primary map showing the positions of the affected plants found on *both*
occasions. A sheet of tracing paper is marked with a central point,
which is then positioned successively over each of the "A" plants. On
each occasion, the positions of the surrounding "B" plants are plotted.
If the map is large and sufficiently detailed, compared with the
pattern being sought, then it is best to mark on the primary map a
suitably sized margin within which the "A" plants are ignored. Thus
"edge effects" are avoided: the margin width should be half the edge
length chosen for the convolution diagram.

     Figs. 1a, 2a and 3a are maps simulating contrasting patterns of
virus-affected plants found in crops. Each map contains twenty randomly
distributed circles representing the "A" plants, and 80 points representing the "B" plants. In Fig. 1a all the plants are randomly distributed. In Fig. 2a each "A" plant is the centre for four "B" plants
placed in random directions but at distances that are normally distributed with a standard deviation that is 5% of the map edge length; this
simulates randomly distributed clusters of virus-affected plants. In
Fig. 3a the "B" plants are clustered as in Fig. 2a, but are randomly
distributed to the right or left of each "A" plant, and at a randomly
chosen angle, x (between 0 and 180) to the vertical, transformed to
$2(\sin(x/2))$. This simulates assymmetrical clusters of "B" plants, as
might be found when a virus with an aerial vector has spread mainly in
the direction of a southerly prevailing wind.

   Figs. 1b, 2b and 3b are the convolution diagrams of the maps in
Figs. 1a, 2a and 3a. Each diagram represents an area that is one
sixteenth that of the corresponding primary map. To avoid edge effects,
each map has a margin that is one eighth of its edge length wide.
These three convolution diagrams were kept in the same orientation
to the map during recording.

(a)

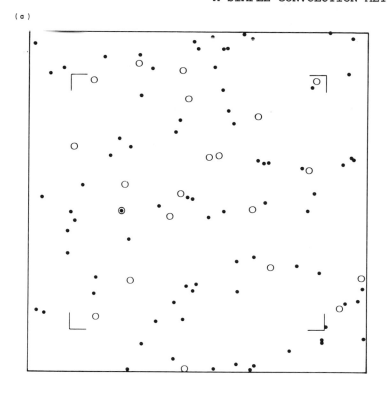

(b)

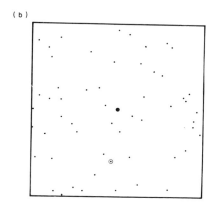

Figure 1
(a) "Map" of randomly distributed "virus-affected plants";
     20 represented by open circles (O) ("A" plants), 80 by smaller
     discs (●) ("B" plants) of which two have the same co-ordinates
     (the circled discs). The right angles indicate the inner edge
     of the margin.
(b) Convolution diagram of Fig. 1a summarizing the distributions
     of "B" plants around the "A" plants that are within the inner
     edge of the margin. The diagram is twice the magnification of
     Fig. 1a (i.e. its edge length is one quarter that of the map,
     twice that of the map's margin).

Figure 2

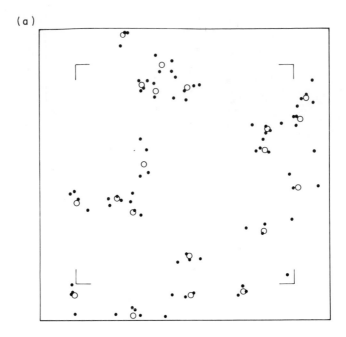

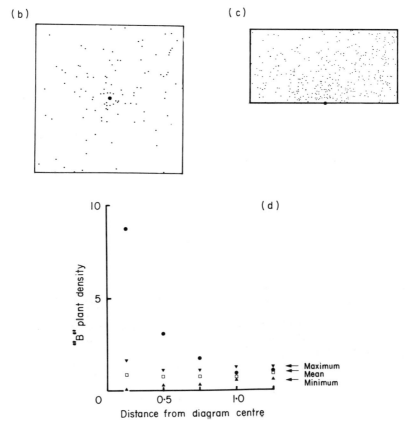

In Fig. 1b, as expected, the points show no obvious clustering. This can be shown more clearly by dividing the diagram into sixteen cells of equal sizes. The number of "B" plants in each cell is as follows:-

```
4    2    4    2
4    3    2    5
3    2    2    5
4    6    3    2
```

In the map in Fig. 1a five of the "A" plants were in the discarded edge and so the patterns around only 15 "A" plants were recorded in the convolution diagram. If the edge length of the primary map is 10 units, then the mean "B" plant density in the convolution diagram is 0.56 "B" plants/"A" plant/unit area ("B"s/"A"/unit area) with maximum and minimum values of 0.85 and 0.34, respectively. I have been advised by M. Adena, D.J. Gates, J.C. Gower and M. Westcott that direct statistical analysis of convolution diagrams is not at present feasible, and that the simplest way to test these diagrams for randomness is by "Monte Carlo" analysis. This involves the repetitive analysis of directly comparable random data to give estimates of the extent of random variation in known numbers of trials. Thus 20 random maps analogous to Fig. 1a produced maximum, mean and minimum point densities of 1.57, 0.78 and 0 "B"s/"A"/ unit area, respectively. The densities found in Fig. 1b are within these limits.

By contrast, in Fig. 2b, the "B" plants are clustered around the centre and, when the diagram is divided into 16 equal cells, these contain the following number of "B" plants around the 11 "A" plants:-

```
2    2    4    3
3   12   13    2
9   12   18    4
8    9    4    1
```

Figure 2 (opposite)
(a) "Map" of randomly distributed clusters of "virus-affected plants".
(b) The convolution diagram of Fig. 2a, summarizing the distributions of "B" plants around "A" plants. Magnification as in Fig. 1b.
(c) Half the convolution diagram of Fig. 2a obtained using every point as an "A" plant. Magnification as in Fig. 1b.
(d) Graph showing the radial distribution of "B" plants (●) in Fig. 2b, also the maximum (▼), mean (□) and minimum (▲) plant density in a "Monte Carlo" simulation of Fig. 2b. The units of the abscissa are based on a map edge length of 10 units, and a convolution diagram with an edge length of 2.5 units. The units of the ordinate are "B" plants/"A" plant/unit area.

Figure 3

(a)

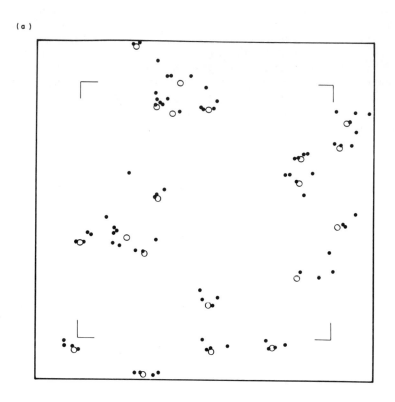

(b)

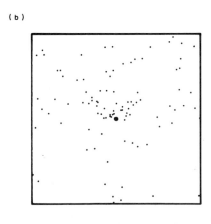

(c)

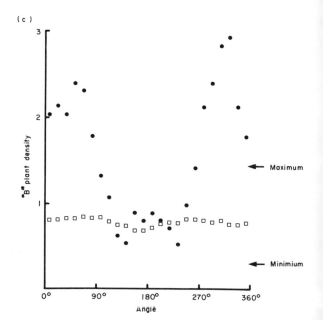

The maximum, mean and minimum "B" plant densities in the 16 cells are
4.19, 1.54 and 0.23 "B"s/"A"/unit area, respectively, much greater than
the densities found in the "Monte Carlo" simulation. Most of the
increased density is in the centre of the convolution diagram, and the
four centre cells have an average of 3.20 compared with 0.99 "B"s/"A"/
unit area in the outer 12 cells. This central clustering is better
illustrated by determining the radial distribution of points in the
convolution diagram by dividing the diagram into a series of concentric
annuli. Fig. 2d shows the radial densities of the "B" plants in Fig. 2b
and data from comparable "Monte Carlo" simulations.

The angular distribution of "B" plants in the convolution diagram may
also be used to indicate whether there is any directional bias in their
distribution. In Fig. 3b, the largest circle that would fit within the
convolution diagram was divided into 24 15° sectors and the number of
"B" plants in each sector recorded (clockwise from "noon"). To accentu-
ate trends all "running sums of 5 sectors" were calculated and plotted
against the angular direction of the central sector of each five,
together with the analogous data from the "Monte Carlo" simulations
(Fig. 3c). The angular bias is obvious.

DESCRIPTION OF THE PATTERN OF ONE TYPE OF VIRUS-AFFECTED PLANT

The convolution method may also be used to describe the pattern of
virus-affected plants in a crop mapped on only one occasion. This is
done by treating each individual as both an "A" and a "B" plant. The
diagram is centered successively over each affected individual and the
positions of all the other affected plants within the area chosen for
the convolution diagram are recorded. One consequence of this is that
the resulting convolution diagram is symmetrical about any line drawn
through its centre. This is because the positional relationship of
each pair of virus-affected plants, say X and Y, is recorded twice, i.e.
once when X is the "A" plant, and in the symmetrically opposite position,
when Y is the "A" plant. Thus all the information is contained in half
the convolution diagram. Fig. 2c is half such a diagram produced from
Fig. 2a by treating all the "plants" as equivalent. It shows clearly
the "contagious" distribution of infected plants in Fig. 2a, indicating
that the virus has probably "spread" between plants. However, the
evidence is less clear than when the patterns on two occasions are
compared (Fig. 2b).

Figure 3 (opposite)
  (a) "Map" of assymetrical clusters of "virus-affected plants".
  (b) Convolution diagram of Fig. 3a, the distribution of "B" plants
      around the "A" plants. Magnification as in Fig. 1b.
  (c) Angular distribution of "B" plants (●) around the centre of
      Fig. 3b, also the mean (□), maximum and minimum (arrowed)
      angular distributions in a "Monte Carlo" simulation of Fig. 3b.
      Units of the ordinate are "B" plants/"A" plant/unit area.

Figure 4. Aerial photograph of a New Zealand gannetry with (inset) a half convolution diagram of an enlargement of part of the above.

The convolution method is useful not only for indicating contagious distributions but also for showing regular or "overdispersed" distributions. Fig. 4a, for example, is an aerial photograph of a gannet colony, and Fig. 4b the half convolution diagram obtained by comparing the positions of each gannet with every other gannet. The diagram shows clearly that each nest is in the centre of a small territory, the radius of which is presumably slightly greater than one "gannet lunge"! Further information on the position of the gannet nests can be obtained by rotating the convolution diagram, when constructing it, so that after centering the diagram on each gannet, a near neighbour of that gannet is positioned on one particular radius of the diagram. Thus it can be shown that the gannet nest positions form a hexagonal lattice.

ROW CROPS

In the "crops" and gannetry discussed above, there were, in theory, no constraints on the positions of the individuals. Consequently the convolution diagrams of Figs. 1a, 2a and 3a reflect the pattern of "virus occurrence" randomly sampled by the plants. Most crops, however, are sown in rows or in a regular lattice, and hence convolution diagrams of virus-affected plants in crops reflect the pattern of virus distribution superimposed on the array of plants. This makes the "Monte Carlo" analysis more complex as it must be based on a map representing the positions of all the plants in the crop, and the "virus-affected" plants must be randomly distributed between all those positions.

Mapping a row crop accurately is very time-consuming and hence it is probable that only relatively small plots would be examined. It would then be best not to exclude a margin of the primary map from the analysis. Instead, all "A" plants in the primary map should be used to make a complete convolution map, four times the area of the primary one. This is illustrated with maps from a study of the spread of cauliflower mosaic virus from a single infected plant to surrounding cauliflower plants (Broadbent, 1957). Figs. 5a and 5b show the positions of the affected plants on two occasions, and Fig. 5c the convolution diagram from those two maps. Fig. 5d shows the radial distribution of plants in Fig. 5c and the results of a "Monte Carlo" simulation of the data using five random maps. The "contagious" distribution of plants in the plot is obvious, as too is the steady decrease in density of "B" plants with distance from the centre of the convolution diagram. This results from recording round all "A" plants and not excluding those in the margin (see, by contrast, Figs. 1b, 2b and 3b).

DISCUSSION

The tests described illustrate the value and simplicity of the convolution method for describing and comparing distribution patterns of virus-affected plants. Convolution diagrams may also be used as the first stage in producing other measures of pattern, such as radial and angular distribution functions.

Figure 5

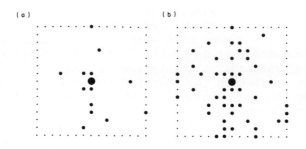

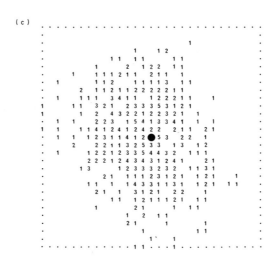

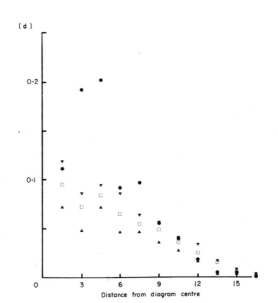

Distance from diagram centre

In practice, various factors will affect the sensitivity of the method. For example, the more "A" plants that are used in the preparation of the convolution diagram then the more representative the diagram will become. However, when more than a fifth of the plants are affected, especially if such plants are closely clumped, then "multiple infections" (Gregory, 1948) will probably bias the pattern obtained. Obviously too, the choice of the size of the convolution diagram is crucial and depends on the scale of the pattern being sought. The method is most sensitive if the diagram is slightly larger than one unit of the pattern (e.g. one clump of a clumped distribution). If the pattern is not obvious then trial diagrams of different sizes using a sample of the data may indicate the optimal size. Finally, it must be stressed that not all virus-infected plants show overt symptoms and would be recorded by someone mapping virus-affected plants.

Only two stages of a convolution analysis are time-consuming. Firstly, mapping the virus-affected individuals, and secondly, the "Monte Carlo" simulation may become tedious if an adequate number of random maps, 25 or more, are analysed. However, this part of the analysis can be easily done by a computer. Similarly, if many convolution diagrams are to be prepared, a digitizing table, computer and plotter are a great help. The inherent flexibility of computer analysis is also valuable when the scale of pattern being sought is unknown and convolution diagrams of different sizes are tried.

Computer programmes to prepare and analyse convolution diagrams and radial distribution functions of patterns are being developed collaboratively by Dr. Jelena Emmerick and Dr. David Green (Department of Biogeography and Geomorphology, Australian National University), who would be pleased to supply details.

Acknowledgements

I am much indebted to Greta Morris and Pat Miethke for help with the preparation of the data and figures, to the Wildlife Research Section

---

Figure 5 (opposite)
(a) and (b) Maps of a plot of cauliflower plants. The central plant was the first to be infected with cauliflower mosaic virus. 40 days (Fig. 5a) and 90 days (Fig. 5b) later the positions of other plants showing symptoms of the virus (●) were recorded.
(c) A complete convolution diagram of Figs. 5a and b summarizing the distribution of plants newly affected between 40 and 90 days around those affected at 40 days.
(d) Graph showing the radial distribution of plants (●) in Fig. 5c, also the maximum (▼), mean (□) and minimum (▲) radial distribution of plants in a "Monte Carlo" simulation of Fig. 5c.

of the Department of Internal Affairs, New Zealand for aerial photographs of gannetries, and to Dr. David Green and Sue House for helpful criticism of the manuscript.

REFERENCES

Broadbent L. (1957) Investigations of virus diseases of brassica crops. *Agricultural Research Council Report Series* No. 14. Cambridge University Press.
De Rosier D.J. & Klug A. (1968) Reconstruction of three dimensional structures from electron micrographs. *Nature, London* 217, 130-4.
Diggle P.J. (1981) Statistical analysis of spatial point patterns. *New Zealand Statistician* 16, 22-41.
Elliott A., Lowy J. & Squire J.M. (1968) Convolution camera to reveal periodicities in electron micrographs. *Nature, London* 219, 1224-6.
Emmerick J. (1979) Plant pattern analysis and radial distribution functions. *Ph.D. Thesis, University of New South Wales.*
Galton F. (1878) Composite portraits. *Nature, London* 18, 97-100.
Gregory P.H. (1948) The multiple infection transformation. *Annals of Applied Biology* 35, 412-7.
Greig-Smith P. (1964) *Quantitative Plant Ecology* (2nd Edn.) Butterworths, London.
Klug A. & Berger J.T. (1964) An optical method for the analysis of periodicities in electron micrographs, and some observations on the mechanism of negative staining. *Journal of Molecular Biology* 10, 565-84.
Pielou E.C. (1969) *An Introduction to Mathematical Ecology.* Wiley-Interscience, New York.
Ripley B.D. (1977) Modelling spatial patterns. *Journal of the Royal Statistical Society, Series B* 39, 172-212.

Spread of banana bunchy top and other plant virus diseases in time and space

R.N. ALLEN

Agricultural Research Centre, Wollongbar,
New South Wales, Australia 2480

INTRODUCTION

The control of many serious plant virus diseases depends upon field
inspection and prompt destruction of diseased plants. Examples include
banana bunchy top and sugarcane Fiji diseases in Australia (Eastwood
1946; Hayes, 1977), plum pox in England (Adams, 1978) and citrus
tristeza in California and Israel (Roistacher, 1976). Some of these
programmes were initially an attempt to eradicate diseases found in the
country for the first time; subsequently, not being successful, they
have increased in scale and cost as disease incidence increased.

Apart from annual reports on the incidence of diseased plants, few
attempts were made to quantify the epidemiology of these problems
(Vanderplank, 1963) until Allen (1978a, 1978b) studied the spread of
banana bunchy top in time and space. This chapter indicates how the
mathematical models developed can be applied to other plant diseases.

BANANA BUNCHY TOP DISEASE

Bunchy top disease appeared in Australian banana plantations early this
century. Although it is not known when it arrived or where it originated, one definite record was in young plants imported from Fiji in
1913 and planted near Murwillumbah, New South Wales. Between then and
1925, bunchy top devastated a large proportion of banana plantations
in northern New South Wales and southern Queensland.

Magee (1927) demonstrated that the disease is spread in planting
material and from plant to plant by the banana aphid (*Pentalonia
nigronervosa*). Legislation was introduced to prevent planting material
being taken from plantations affected by bunchy top and to require
banana growers to destroy any diseased plants. Growers co-operated to
organise an efficient system for detecting and destroying diseased
plants (Eastwood, 1946). These methods have continued for more than
50 years and have been very successful. Disease incidence decreased
from a mean of 6.6 plants/ha in 1937 to 0.5 in 1947, 0.3 in 1957 and

Plumb R.T. & Thresh J.M. (1983) *Plant Virus Epidemiology.*
Blackwell Scientific Publications, Oxford.

0.1 in 1967. Thereafter so few diseased plants occurred that for some time it became difficult to train inspection staff to recognize the early symptoms of disease and to persuade growers to finance the programme adequately. By 1977 disease incidence had increased to 0.3 plants/ha and there was recurrent infection in more than 10% of all plantations.

Spread in time

The temporal model is based on deterministic principles (Vanderplank, 1963) which state that the rate of infection (dN/dt) depends on the number of plants already infected (N), the number yet to be infected ($N_{max.} - N$) and an empirical rate constant (K), and

$$dN/dt = K N (N_{max.} - N) \tag{i}$$

The increase of disease between any two times (t_1 and t_2) is defined by

$$N_2 = \frac{N_1 N_{max.} \ exp(\ R(t_2-t_1) \ )}{N_{max.} - N_1(\ 1 - exp(\ R(t_2-t_1)) \ )} \tag{ii}$$

The constant $R = K N_{max.}$ and is the relative infection rate (Vanderplank, 1963). When N is measured as a percentage or proportion, $N_{max.}$ is 100 or 1, respectively.

The number of plants infected includes those with symptoms (S) and those that have been infected but are still symptomless. Only a proportion (δ) of plants with symptoms will be detected at each inspection and only a proportion (ρ) of these will be eliminated because some will regenerate after treatment. Thus the number of plants eradicated after an inspection at time t_1 will be:-

$$E_1 = \delta \ \rho \ S_1 = \varepsilon \ S_1 \tag{iii}$$

where ε is the eradication efficiency. The number of plants with symptoms is determined by:-

$$S_1 = \frac{N_1 N_{max.} \ exp(\ - R_q Q \ )}{N_{max.} - N_1 \ (\ 1 - exp(\ - R_q Q \ ))} \tag{iv}$$

where Q is the time interval between infection and symptom appearance, referred to hereafter as the latent period, and R_q is the relative infection rate relevant to that period. The increase of disease during control operations then becomes:-

$$N_2 = \frac{N_{max.} (N_1 - E_1) \ exp(R(t_2-t_1) \ )}{N_{max.} - (N_1 - E_1)(1 - exp(R(t_2-t_1) \ )} \tag{v}$$

with E_1 given by equations (iii) and (iv). In the special case of N_1

being much less than N_{max}., equations (ii), (iii), (iv) and (v) simplify to

$$N_2 = N_1 (1 - \varepsilon \; \underline{exp}(- R_q Q)) \; \underline{exp}(R(t_2 - t_1)) \qquad \text{(vi)}$$

This equation was used by Allen (1978a) for banana bunchy top outbreaks where the number of infected plants was kept low by roguing.

The variables in equation (vi) varied with season and were closely related to air temperature. The latent period was determined by the time taken for two new banana leaves to be produced and, since the rate of leaf emergence is correlated with the cosine curve of air temperature, the latent period at any particular season was an integral of this cosine curve. The latent period ranged from 19 days in summer to 125 days in winter.

The relative infection rate varied seasonally as a cosine curve and ranged from 0.027 plants/plant/day in summer to 0.001 in winter, which correlated well with aphid activity (Magee, 1927). Detection efficiency (δ) was determined partly by season and partly by the number of leaves with symptoms. It ranged from 0.95 in summer to 0.19 in winter. Destruction efficiency (ρ) averaged 0.86 with only slight seasonal variation.

The values determined for R and ε in one disease outbreak were used to predict the course of disease control in a separate outbreak (co-efficient of determination, 0.7). They have also been used to simulate various disease control strategies in which the ratio N_2/N_1 was used as a measure of control efficiency. A ratio >1 indicates that control is failing while a ratio <1 indicates progress towards eradication. When the ratio equals 1, critical values of Q, $(t_2 - t_1)$, R and ε can be calculated from equation (vi) to indicate what must be achieved to obtain control.

Spread in space

The spatial model assumes that aphid vectors fly more or less at random between plants. This is a useful first assumption for the banana aphid which seeks shelter in windy conditions and tends to remain with its progeny after settling. Allen (1978b) showed that the probability of a new infection occurring within a distance of x m from its source could be fitted by the distribution:-

$$P_x = 1 - \underline{exp}(-x/\bar{x}) \qquad \text{(vii)}$$

and calculated the mean distance of spread ($\bar{x}$) as 17 m in several outbreaks in several years.

Equation (vii) can be used to predict where the majority of additional infections will occur after the first diseased plants are found by inspectors. Plants that are very likely to be infected can then be destroyed before they become new sources of bunchy top (Allen, 1978b).

The probability of an aphid flying more than 100 m is < 0.001 and such an isolation distance has proved effective in practice (Allen & Barnier, 1977).

Spread in time and space

A general theory for spread in both time and space has not been developed. However, a theory based on the principles of a random walk (Daniels & Alberty, 1966) may be appropriate. The time taken for a new infection to be generated from an inoculum source is the inverse of the relative infection rate, R. The variance of the dispersion from a given infected plant is the square of the mean distance of spread (Var. $= \bar{x}^2$). The dispersion coefficient (m^2/day) can be defined as:-
$D = (R\ \bar{x}^2)/2$.

Alternatively, a numerical simulation can be done by computer using random numbers to determine the direction of aphid flight at the end of each infection period ($1/R$) and a random value of P_x to determine the length of each flight. Patterns of disease spread can then be plotted over time, starting from a single infected plant. The effects of various control strategies can then be evaluated by removing plants and altering the detection variables.

SUGARCANE FIJI DISEASE

In Australia, Fiji disease causes spectacular epidemics whenever a susceptible cultivar is grown extensively (Hayes, 1977; Egan & Fraser, 1977). Great efforts are made to control the disease by inspection and roguing until a replacement cultivar is available. However, a recent epidemic in the cultivar NCo310 at Bundaberg, Queensland, could not be controlled by inspection and roguing (Egan & Fraser, 1977; Egan & Hall, this volume).

Inspection of sugarcane is restricted by the close plant spacing adopted and by harvesting operations and it is usually done when the plants are shoulder-high in summer. The latent period is approximately 50 days at this time (Hayes, 1974), although, as in banana bunchy top, it is dependent on plant growth rate. The leafhopper vector (*Perkinsiella saccharicida*) occurs throughout the year but its main dispersal flights are in summer when "swarming" occurs. Dispersal gradients have not been studied, but the mean distance of spread is likely to be very great.

Egan & Fraser (1977) monitored disease incidence at Bundaberg in 1973-76. It increased fourfold each year in four fields where roguing was not done. A relative infection rate can therefore be estimated from their data using:-

$$R = \frac{\ln S_2 - \ln S_1}{t_2 - t_1}$$

(viii)

since disease incidence was low and inspections were made at the same
time each year (Q and ε constant, $R = R_q$). The values of R (Table 1)
were much smaller than those for bunchy top and varied considerably
between years. The data also allow the efficiency of disease eradication to be estimated using

$$\varepsilon = \underline{exp}(R\ Q)\ (1 - \frac{E_2}{E_1\ \underline{exp}(\ R(t_2-t_1)\ )}) \qquad (ix)$$

which is derived from equation (vi) using the same assumptions as in
(viii). These calculations (Table 1) show that ε is very low and
inadequate to maintain control. One inspection/year could not have
controlled the disease even if $\varepsilon = 1$ in 1975/76 when R was larger than
normal. The use of cultivars such as NCo310 that display inconspicuous
symptoms (Hayes, 1977) creates problems if control is to be achieved by
inspection and roguing alone.

Table 1. Analysis of sugarcane Fiji disease epidemics at Bundaberg,
1973-76 (from Egan & Fraser, 1977).

Year	Relative infection rate (R) (plants/plant/day)	Plants rogued (E_2/E_1)	Eradication efficiency (ε)	
			Estimated from data	Critical value ($E_2/E_1 = 1$)
1973/74	0.0029	$\frac{109\ 000}{40\ 000}$	0.06	0.77
1974/75	0.0036	$\frac{370\ 000}{109\ 000}$	0.11	0.87
1975/76	0.0051	$\frac{1\ 600\ 000}{370\ 000}$	0.45	>1.00

PLUM POX DISEASE

Plum pox was first recorded in England in 1965 and immediate but
unsuccessful attempts were made to eradicate the disease from the
country. Infection is now widespread in orchards and the main control
effort is concentrated on nurseries to ensure an adequate supply of
material for new plantings (Adams, 1978a). Plum pox is spread in
planting material and between plants by several aphid species. The
latent period in the plant under experimental conditions can be as short
as 15 days in young seedlings but can extend to 1-4 years in orchard
trees. The virus spreads slowly through the plant after inoculation
and diagnosis is difficult (Adams, 1978a). Moreover, inspections for
leaf symptoms are effective only in late spring to early summer when
leaves are abundant and symptoms are not masked by damage or other
symptoms. Orchard trees can be examined later for fruit symptoms but
these are inconspicuous in some cultivars.

Adams (1978a) found that detection efficiency was approximately 50% at any one inspection and that trees had to be removed very efficiently to prevent regeneration from root debris remaining in the ground. If it is assumed that three inspections can be made in late spring to early summer and that diseased plants are removed efficiently then $\varepsilon = 1-(1-0.5)^3 = 0.875$ for any one year. Relative infection rates may be obtained from Adams (1978a) using

$$E_2/E_1 = (1 - \varepsilon \; \underline{exp}(-R \; Q) \;) \; \underline{exp}(R(t_2-t_1)) \qquad (x)$$

which is derived from equation (vi) using the same assumptions as in (viii). Values of R can be obtained by iteration within equation (x) and have been obtained assuming latent periods of 15 days and 1 or 4 years (Table 2). Whatever the latent period, the relative infection rate is small. Roguing may be an acceptable method of control provided that infected plants are destroyed promptly and healthy nursery stock is available for re-planting. Tests on leaf samples from root suckers using enzyme-linked immunosorbent assay (ELISA) can improve detection efficiency but this method has limitations because only some trees produce suckers (Adams, 1978b).

Table 2. Analysis of plum pox epidemics in England, 1975-77 (from Adams, 1978a).

Year	Farm	(E_2/E_1)	Relative infection rate (R) when:-		
			*Q = 15 days	*Q = 1 year	*Q = 4 years
1975/76	A	104/138	0.0040	0.00130	0.00056
	B	12/18	0.0037	0.00120	0.00047
	C	5/32	0.0005	0.00009	0.00002
	D	2/6	0.0021	0.00052	0.00017
	E	6/5	0.0051	0.00200	0.00106
	F	6/11	0.0033	0.00096	0.00036
1976/77	A	40/104	0.0025	0.00130	0.00056
	B	8/12	0.0037	0.00120	0.00047
	C	3/5	0.0035	0.00107	0.00041
	F	4/6	0.0037	0.00120	0.00047

* Q = latent period

Adams (1978a) also presents a disease dispersal gradient up to 105 m from the presumed source of plum pox. The gradient is interpreted as shallow and a mean distance of $\bar{x}$ = 32 m calculated from the data confirms this conclusion.

CITRUS TRISTEZA DISEASE

Programmes to suppress tristeza disease operate in California and Israel where natural spread of the disease has occurred following the appearance of virus strains with efficient aphid vectors (Roistacher, 1976; Bar-Joseph et al., this volume). Infected trees are commonly symptomless until they decline rapidly after six or more years. Infection is detected by inspection followed by transmission tests and electron micro-scopy (Raccah et al., 1976) or by ELISA (Bar-Joseph et al., 1978). These methods, in effect, shorten the latent period because virus is detected sooner than by observing symptoms, but it is not known by how much.

Roistacher (1976) summarized data on the natural spread of citrus tristeza where disease control was not attempted. Disease incidence reached 60-90% after 9 years, giving a relative infection rate of $R = 0.0033$ plants/plant/day. Raccah et al. (1976) present results on the suppression programme at Habat Ziyyon, Israel from 1970 to 1974. In 1970/71 there was one inspection; subsequently, there were two inspec-tions/year. No estimates of R, Q or ε are given but, assuming $R = 0.0033$, ε can be estimated for different assumed values of Q using equation (ix). Table 3 shows that $\varepsilon > 1$ when $Q > 90$ days. Even if $Q = 0$, eradication efficiency is high. If $Q = 30$ days, an eradication efficiency of $\varepsilon = 0.49$ would be required to maintain control with two inspections/year.

Table 3. Analysis of citrus tristeza epidemics in Israel, 1970-74 (from Raccah et al., 1976).

Year	(E_2/E_1)	Eradication efficiency (ε) when:-		
		*$Q = 0$	*$Q = 30$ days	*$Q = 90$ days
1970/71	1.23/1.04	0.64	0.71	0.86
1971/72	0.41/1.23	0.82	0.90	>1.00
1972/73	0.21/0.41	0.72	0.79	0.97
1973/74	0.09/0.21	0.77	0.84	>1.00

* Q = latent period

A dispersal gradient for citrus tristeza disease is given by Bar-Joseph et al. (1974) and indicates $\bar{x} = 42$ m.

CONCLUDING REMARKS

Before many resources are devoted to a control programme based on inspection and roguing, consideration should be given to four important variables, the relative infection rate (R), the latent period (Q), the eradication efficiency (ε) and the mean distance of spread (x). Even if the need to eradicate a newly reported disease is urgent, these variables can be readily estimated during the early stages of the control programme.

No attempt has been made here to consider the economics of a particular control programme, although this could be done readily from the models presented. In some circumstances, but not all, there may be a direct relationship between disease incidence and yield loss. The examples discussed here concern control programmes where disease incidence is usually kept low and diseased plants are found soon after they show symptoms. Plants in the advanced stages of disease either die or lose infectivity and are therefore naturally removed from the epidemic. Predictions of loss where no control is practised will have to take this effect into consideration by using the rate equation:-

$$dN/dt = K (N - N')(N_{max.} - N) \qquad\qquad \text{(xi)}$$

where N' is the number of infected, non-infectious plants. Spatial isolation from non-infected plants may also remove some sources from consideration. Such an effect will only be appreciated by developing a general theory for spread in time and space.

REFERENCES

Adams A.N. (1978a) The incidence of plum pox virus in England and its control in orchards. *Plant Disease Epidemiology* (Ed. by P.R. Scott & A. Bainbridge), pp. 213-9. Blackwell Scientific Publications, Oxford.

Adams A.N. (1978b) The detection of plum pox virus in *Prunus* species by enzyme-linked immunosorbent assay (ELISA). *Annals of Applied Biology* 90, 215-21.

Allen R.N. (1978a) Epidemiological factors influencing the success of roguing for the control of bunchy top disease of bananas in New South Wales. *Australian Journal of Agricultural Research* 29, 535-44.

Allen R.N. (1978b) Spread of bunchy top disease in established banana plantations. *Australian Journal of Agricultural Research* 29, 1223-33.

Allen R.N. & Barnier N.C. (1977) The spread of bunchy top disease between banana plantations in the Tweed River district during 1975-76. *New South Wales Department of Agriculture, Biology Branch Plant Disease Survey (1975-76)*, pp. 27-8.

Bar-Joseph M., Loebenstein G. & Oren Y. (1974) Use of electron microscopy in eradication of tristeza sources recently found in Israel. *Proceedings of the 6th conference of the International Organisation of Citrus Virologists* (Ed. by L.G. Weathers & M. Cohen), pp. 83-5. University of California Press, Berkley.

Bar-Joseph M., Sacks J.M. & Garnsey S.M. (1978) Detection and estimation of citrus tristeza virus infection rates based on ELISA assays of packing house samples. *Phytoparasitica* 6, 145-9.

Daniels F. & Alberty R.A. (1966) *Physical Chemistry*, 3rd Edn., pp. 404-5. John Wiley & Sons, New York.

Eastwood H.J. (1946) Bunchy top disease of bananas. Controlled by co-operative effort. *Agricultural Gazette of New South Wales* 57, 643-6.

Egan B.T. & Fraser T.K. (1977) The development of the Bundaberg Fiji disease epidemic. *Proceedings of the 44th conference of the Queensland Society of Sugar Cane Technologists* (Ed. by O.W. Sturgess) pp. 43-8.

Hayes A.G. (1974) Fiji disease of sugar cane. Evidence for different strains of the causal virus at Condong, N.S.W. *Proceedings of the 41st conference of the Queensland Society of Sugar Cane Technologists* (Ed. by O.W. Sturgess), pp. 105-10.

Hayes A.G. (1977) Control of Fiji disease in sugar cane - a review of 15 years experience at Condong, N.S.W. *Proceedings of the 44th conference of the Queensland Society of Sugar Cane Technologists* (Ed. by O.W. Sturgess), pp. 49-53.

Magee C.J.P. (1927) Investigation on the bunchy top disease of the banana. *Bulletin of the Council for Scientific and Industrial Research,* No. 30, 64 pp.

Raccah B., Loebenstein G., Bar-Joseph M. & Oren Y. (1976) Transmission of tristeza by aphids prevalent on citrus, and operation of the tristeza suppression programme in Israel. *Proceedings of the 7th conference of the International Organisation of Citrus Virologists* (Ed. by E.C. Calavan), pp. 47-9. University of California Press, Riverside.

Roistacher C.N. (1976) Tristeza in the Central Valley; A Warning. *Citrograph* 62, 16-23.

Vanderplank J.E. (1963) *Plant Diseases: Epidemics and Control.* Academic Press, New York.

The epidemiology and control of citrus tristeza disease

M. BAR-JOSEPH*, C.N. ROISTACHER† &
S.M. GARNSEY‡
*Volcani Center, Agricultural Research Organization, Bet Dagan, Israel
†University of California, Riverside, Department of Plant Pathology,
Riverside, CA 92521, USA
‡US Horticultural Research Laboratory, ARS, USDA, Orlando, FL 32803, USA

INTRODUCTION

Although citrus tristeza virus (CTV) had been associated with citrus in
the Far East for some time, the first report of an outbreak of tristeza
disease, that ultimately devastated millions of trees throughout most
citrus-growing areas where sour orange was the predominant rootstock,
was in Argentina in 1930 (Zeman, 1931). This outbreak caused large
changes in horticultural practice and forced growers to adopt new, less
sensitive rootstocks. Tristeza is an unusual disease as there is
detailed information available on its history and spread in several
regions; this chapter reviews epidemiological work.

CITRUS CULTIVATION AND TRISTEZA EPIDEMICS

It seems certain that CTV originated in China where tristeza is widely
distributed. The first citrus species successfully cultivated outside
south-east Asia and the Malay archipelago was the citron (*Citrus medica*)
which was introduced to the Middle East and the Mediterranean basin,
probably before 300 BC. Subsequently the sour orange (*C. aurantium*),
lemon (*C. limon*), and sweet orange (*C. sinensis*) were also grown there.
Since the early movement of citrus was by seed and CTV is not seed-
transmitted (McLean, 1957), the virus was probably not introduced to
these new areas.

As transport improved and a worldwide interest in citrus species
developed, many cultivars were dispersed rapidly throughout the world.
Phytophthora spp. destructive to citrus were apparently moved in soil
with plants and destruction of seedlings by these pathogens was directly
responsible for the shift from unworked seedling trees to scions budded
on resistant seedling rootstocks. Sour orange (SO) was very successful
as a *Phytophthora*-resistant stock and was soon used almost exclusively
throughout the Mediterranean area and North and South America. However,
in Australia it failed, whereas trifoliate orange (*Poncirus trifoliata*)
and rough lemon (*C. limon*) were successful. It also failed in South
Africa but rough lemon was successful.

Plumb R.T. & Thresh J.M. (1983) *Plant Virus Epidemiology.*
Blackwell Scientific Publications, Oxford.

For many years the failure of sweet orange (SW) trees on SO root-stocks was attributed to incompatibility between SO and certain scion cultivars. After visiting South Africa, Webber (1925) rejected this hypothesis and suggested that a local pathogen was responsible; similar conclusions were drawn by Toxopeus (1937) in Java. In retrospect it seems possible that CTV and its aphid vector *Toxoptera citricida* were already established in these countries when the incompatibility was noticed, and infection by CTV was its cause. Between 1890 and 1935, commercial citriculture expanded and much propagation material was dist-ributed, frequently as budwood, from areas where tristeza was endemic. The majority of the early introductions were propagated on SO rootstocks and those found incompatible were propagated on SW or as cuttings. After 1930, budwood introductions to the USA declined and most subsequ-ent introductions were, as a precautionary measure, by seed.

The first disastrous outbreak of tristeza was in 1930 in Corrientas province, Argentina, where 90% of the citrus plantations were grafted on SO stock (Knorr & DuCharme, 1951). However, the disease was not named tristeza until it appeared in Brazil in 1937. The disease has subsequently been reported in many other citrus-growing areas, including Ghana, California, Florida and Spain (for a recent distribution map see Bar-Joseph *et al.*, 1979).

EPIDEMIOLOGY

Information on the main variables that affect the epidemiology of tristeza has been summarized recently (Bar-Joseph *et al.*, 1977, 1979).

Vectors

The aphids *T. citricida, Aphis gossypii* and *A. citricola* are the principal vectors of CTV. Four other vectors have also been reported, *A. craccivora, T. aurantii, Myzus persicae* and *Dactynotus jaceae*. Although no direct comparisons have been made, *T. citricida* appears to be more efficient than either *A. gossypii* or *A. citricola*. However, in recent experiments in California (Roistacher *et al.*, 1980), *A. gossypii* transmitted 40 of 63 isolates of tristeza very efficiently and this confirmed earlier reports (Bar-Joseph & Loebenstein, 1973) that *A. gossypii* is a very efficient vector for some CTV strains. Experiments in Florida (Norman & Sutton, 1969), Israel (Bar-Joseph & Loebenstein, 1973) and California (Roistacher *et al.*, 1980) all showed that CTV transmission is not affected by the geographical region or host from which *A. gossypii* is collected.

Virus spread has been correlated with aphid population densities in California and South Africa. In southern California in 1951-54 the estimated numbers of aphids flying to orange trees ranged from 185 725/ tree in the coastal area to 956 238 in the inland area of Covina and Azuza (Dickson *et al.*, 1956). In this work only *A. gossypii* was con-sidered a vector and although its populations were generally correlated with disease spread, its large numbers (3000-36 000/tree) made it

difficult to explain the low rate of new infections (2/year for each
infected tree), even though *A. gossypii* was shown to be an inefficient
vector (Dickson *et al.*, 1956). Schwarz (1965), in South Africa, related
vector populations, caught in yellow traps, to field infection of
indicator plants by CTV. The main flight of *T. citricida* was closely
correlated with the flush cycle of citrus, whereas the maximum numbers
of alate *A. gossypii*, a more polyphagus aphid, were not. In Surinam,
Klas (1980) recorded population densities and the spatial distribution
of *T. citricida*. During some weeks, there was a pronounced increase in
the numbers trapped at some sites which seemed consistent with the
patchy spread of the disease.

Transmission of CTV strains by A. gossypii

Differences in efficiency of transmission of CTV isolates by *A. gossypii*
may explain the great differences in disease spread at various times and
places in regions where *T. citricida* is absent (Bar-Joseph & Loebenstein,
1973). Recently Roistacher *et al.* (1980) clearly demonstrated
efficiently transmitted, poorly transmitted and intermediate isolates of
CTV in the University of California citrus cultivar collection. Several
of the cultivars introduced from the Far East carried virulent CTV
isolates not transmitted by *A. gossypii*, whereas very efficiently transmitted strains have recently been spreading naturally. Varying rates of
transmission from a single isolate were reported by Raccah *et al.* (1979)
but of 23 CTV isolates tested by Roistacher *et al.* (1980) only two gave
variable rates of transmission. In several countries it has been more
than 30 years after the introduction of CTV before there has been extensive natural spread by *A. gossypii* (Bar-Joseph, 1978). The dominance of
non-transmissible isolates may, through cross-protection, greatly delay
the establishment of more readily transmissible mutants.

The effect of host plants

Field observations and experimental transmission tests (Bar-Joseph *et
al.*, 1977) indicate an effect of cultivars on tristeza transmission.
Recent laboratory tests with three CTV strains, three acquisition hosts
and *A. gossypii* showed more transmission of tristeza from SW than from
grapefruit or lemon seedlings. The vector showed no host preference
but virus assay by enzyme-linked immunosorbent assay (ELISA) showed a
positive correlation between virus content and suitability as a virus
source.

Environment

In Florida, *A. gossypii* transmitted less efficiently during the summer
than in winter (Norman *et al.*, 1968) and in Israel, Bar-Joseph &
Loebenstein (1973), experimenting with SW seedlings grown at controlled
temperatures and using *c.* 100 aphids/plant, obtained transmission rates
of 61% when source plants were kept at 22 $\pm$2°C, compared with only 12%
for plants kept at 31°C. This decrease in transmission at high temperatures seems to be associated with a decrease in virus concentration
(Bar-Joseph & Loebenstein, 1974). Temperatures >30°C commonly occur

in citrus-growing areas from spring until autumn and if virus carried
in the vector is sensitive to high temperatures, this might explain why
it is efficiently transmitted in controlled (22 $\pm 2^{o}$C) tests, but spreads
only slowly in the field.

Rate of spread

Burnett (1961) reported a 410-day interval between the first positive
identification of CTV in a Valencia orange tree and the stage when all
44 samples taken from different parts of the tree gave positive
reactions. This suggests that multiple samples should be taken from
each tree, especially when indexing is done as part of an eradication
campaign.

   Results from a hot desert area in the Coachella valley of California
(Calavan & Blue, unpublished) suggest that natural thermotherapy may be
important and may restrict disease spread. It is known that there is
variability in the sensitivity of tristeza isolates to heat (Roistacher
et al., 1974).

   Inoculated nursery trees decline in 3-24 months (Bennett & Costa,
1949; Schneider, 1954), but little is known about the time taken for
mature trees to collapse on SO stock except that trees declined after
4 years in Florida (Cohen & Burnett, 1961).

   Data on disease spread into and within several individual groves in
North and South America are shown in Fig. 1 after transformation to
allow for multiple infection (Gregory, 1948). The annual infection
rates (r) were calculated according to the compound interest equation
of Vanderplank (1963). The computed values of r were 0.6-1.2 in
Argentina, with maximum annual values of 0.9-1.7 (Table 1). These
values were generally considerably larger than in Florida. However,
the maximum r for location 8 in Florida and for an "average" grove
(location 10) in Orange county, California, are only slightly smaller
than those for location 6 in Brazil. The figures for location 8 are
for the incidence of decline and may not reflect virus spread, since
all the non-declining trees indexed in these plots were infected.
Relatively large r values in certain years in areas where A. *gossypii*
occurs (locations 8 and 9) are consistent with reports of groves in
California being ruined, or almost completely infected, within 5-6
years after disease is first observed (Calavan *et al.*, 1980). The
small r value for location 7 probably reflects the unusual situation
in Florida, and possibly in other places, where trees grafted on SO do
not always decline when infected.

   Protection by avirulent isolates already established in plantations
and fluctuations in aphid populations further complicate the analysis
of field results. The rates of tree loss recorded by Garnsey & Jackson
(1975) ranged from 8 to 23%/year and extrapolation of these rates would
indicate rapid and complete destruction of these groves. However, very
little further decline has occurred since 1975 and total losses have
increased only slightly (Garnsey, unpublished).

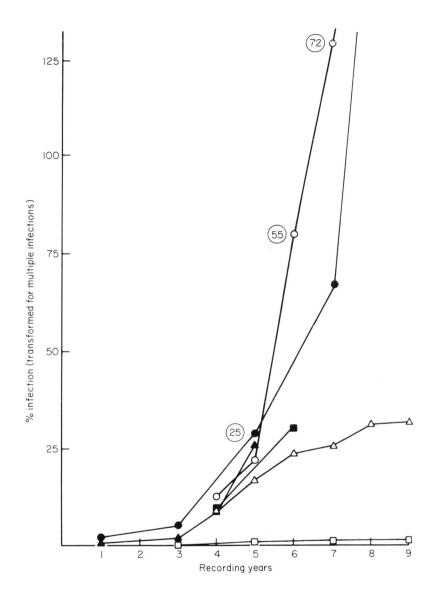

Figure 1. The spread of tristeza virus in six plantings. Data transformed to allow for multiple infection (Gregory, 1948).
●——● sweet orange and □——□ grapefruit trees in a Florida Foundation block (Bridges and Youtsey, 1972)
△——△ declining Temple orange in Florida (Cohen & Burnett, 1961
▲——▲ Parson Brown orange in Florida (Garnsey & Jackson, 1975)
■——■ Valencia orange in California (Calavan *et al.*, 1972)
○——○ Bahianinha orange in Brazil (Bennett & Costa, 1949).
The circled figures indicate % of infection before transformation.

Table 1. Data on spread of citrus tristeza virus at 12 locations

Location	Cultivar	Method*	No. of years	Total spread %	Greatest annual spread	Mean of annual r values	Max. annual r values	Refs.†
1 Argentina		D	4	5-91	25-71	0.77	0.87	1
2 Argentina		D	4	4-87	21-78	0.77	1.13	1
3 Argentina		D	4	0.2-87	0.2-10	1.18	1.74	1
4 Argentina		D	5	1-84	32-84	0.62	1.05	1
5 Argentina		D	4	1-83	6-53	0.93	1.25	1
6 Brazil	Bahianinha	D	4	12-72	19-55	0.46	0.71	2
7 Florida	Temple	D	5	8-27	8-15	0.15	0.31	3
8 Florida	Parson Brown	D	5	0.5-22	2-8	0.44	0.69	4
9 California	Valencia	D	9	0.1-34	2-8	0.31	0.58	5
10 Florida	Sweet orange	I	5	2-93	49-93	0.34	0.57	6
11 Florida	Mandarin	I	4	3-77	3-14	0.33	0.41	6
12 Florida	Grapefruit	I	4	0.5-4	0.5-2	0.15	0.30	6

* Method of detection: observing decline (D) or indexing (I)
† References: (1) DuCharme, unpublished; (2) Bennett & Costa (1949); (3) Cohen & Burnett (1961);
(4) Garnsey & Jackson (1975); (5) Nesbitt (1963); (6) Bridges & Youtsey (1972)

CTV isolates from declining and non-declining trees were very different. Inoculations from declining trees to SW/SO trees caused severe effects, while isolates from non-declining trees generally caused mild symptoms or no observable reaction.

Extremely slow spread of tristeza in grapefruit has been recorded at location 12 and elsewhere in Florida and in Spain (M. Fabregat, personal communication). This may be because certain CTV strains in SW are poorly transmitted to grapefruit seedlings whereas other strains are readily transmitted. Recently in Florida, 23% of 52 6-year old grapefruit trees on SO declined in 3 years (Youtsey, unpublished).

As a typical, mature, citrus tree presents a large catchment area to vectors and *A. gossypii* transmits CTV efficiently in experiments, it is surprising that CTV spreads so slowly compared with many viruses of annual crops. In some areas there has been no increase in disease for 2-5 years following the first record of declining trees (Nesbitt, 1963).

The infection gradients computed from 2 different years of virus indexing in a Valencia orange grove near Santa Paula, California (Calavan *et al.*, 1972), are shown in Fig. 2. The main source of inoculum was an adjoining infected Valencia/SO grove. For each year the log_{10}% infection was plotted against distance from the source and the calculated slopes reached the value of 1% (log 0.0) at distances of 304 and 383 m. Thus the disease front advanced *c.* 79 m in 2 years (Fig. 2). Tree spacing and row direction in relation to prevailing winds, differences in tree size and their effect as wind barriers all affect vector movement. Close spacing within the row might facilitate spread by apterae and this will also influence the gradient. Assessment of gradient slopes, and prediction of the "horizon of infection" (*sensu* Vanderplank, 1963) beyond which spread is so insignificant that it can be ignored, are important for effective control of disease spread, but few studies have been made.

SPREAD OF TRISTEZA IN INVADED AREAS AND THE ECONOMIC EFFECTS OF THE DISEASE

In 1930 there were *c.* 18 million trees in Argentina growing on SO; 15 years later more than 10 million trees had been killed by tristeza. In Brazil, the disease spread within 12 years to all citrus-producing areas, and killed more than 6 million trees. California lost *c.* 3 million trees on SO and hundreds of thousands of SW/SO trees have been killed or rendered unproductive in Florida. In Spain, *c.* 4 million trees were lost during 1956-75. During the last 50 years we estimate that *c.* 40 million trees were killed or became unproductive due to tristeza. In addition, new stem-pitting strains have been extremely damaging in grapefruit, sweet orange and lime in several areas.

The total losses due to tristeza are hard to assess. A calculation based on an estimated value of $10-30/tree gives a global loss during the last 50 years of $400-1200 million. However, the losses vary considerably between regions and those due to strains that cause only stem pitting are pernicious and particularly difficult to assess.

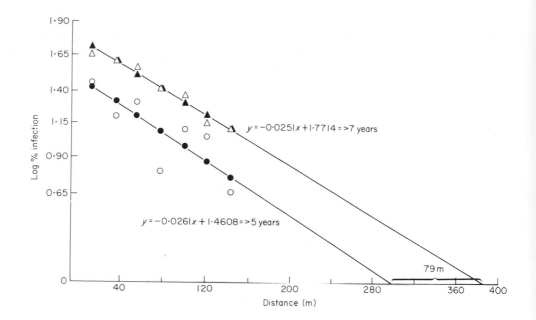

Figure 2. Tristeza infection gradients in a planting of Valencia orange on troyer rootstocks at Santa Paula, Ventura, California (Calavan *et al.*, 1972).
O and Δ are the $\log_{10}$% infection based on indexing in the 5th and 7th year after planting.
● and ▲ are computed slopes.

In most tristeza epidemics, not all damage is the result of secondary spread by aphids from primary infection foci. Infection in nurseries and movement of infected propagation material from them are also important. Thus regulations for preventing spread of infected nursery material have been introduced in California (Mather, 1968) and elsewhere, and they have been effective in restricting spread and in preventing further damage.

SUPPRESSION AND CONTROL

Limiting CTV spread into hitherto tristeza-free areas by co-ordinated eradication measures has been apparently successful in the Central Valley of California (Pratt *et al.*, 1972). Since 1945, there have been regular visual surveys by the California State Department of Plant Pathology in the San Joaquin Valley of central California (Stout, 1945). Between 1964 and 1978, some 18 000 infected SW trees were removed together with 5300 infected Meyer lemon and *c*. 1200 Satsuma trees. The

project covers 161 000 acres (Rosenberg et al., 1978) and the cost in
1981 was $285 000 (D. Cordas, personal communication); c. 2 million
trees have been indexed since 1964. The average incidence in orchards
surveyed in 1972-77 was 0.2% and there has been no evident epidemic
(Cordas, unpublished). In Coachella Valley, California, 446 infected
trees have been removed from 7500 ha of citrus, and there has been no
indication of spread by aphids (Rosenberg et al., 1978).

A similar eradication programme in Israel has been operating since
1970 (Bar-Joseph et al., 1974). In 1971-75, more than 190 000 individual and batch tests were made using Mexican lime indicator seedlings and
electron microscopy; 523 infected trees were detected and destroyed
(Raccah et al., 1976).

Following the adoption in 1977 of ELISA, 1 million symptomless SW/SO
trees were tested in 3 years (Bar-Joseph et al., 1980; Oren et al., 1981).
Several new foci of infection were detected and surveys continue. In
Spain, a similar programme (Moreno et al., 1980) has been initiated in
the southern, warmer regions, also using ELISA. So far 24 infected
trees have been found in more than 2650 tests; many infected trees
on SO stock showed no decline.

The compulsory use of virus-free propagation material throughout the
California and Israel suppression areas has minimized further distribution of CTV and an unpublished statement made by E.C. Calavan in 1965
is still valid: "Barring no significant changes in the relationship of
virus, host and vector, continuation of existing control measures should
prevent any tristeza epidemics in these suppression areas".

Cross-protection against stem-pitting forms of CTV is now being used
widely in Brazil (Müller & Costa, 1977; Costa & Müller, 1980) for SW and
acid limes, and also in India (Balaraman & Ramakrishman, 1980) for acid
limes. In these countries, CTV isolates that give adequate protection
were made by collecting budwood from vigorous trees of a specific
cultivar (oranges, grapefruits or limes) and testing these sources
extensively for protection against a number of severe challenge isolates
in the laboratory and field.

Reducing damage caused by CTV through cross protection is promising
for areas where the disease is endemic and severe isolates are prevalent,
especially if certain selected and locally desirable cultivars are to
be maintained.

CONCLUSIONS

Tristeza disease is largely a man-made problem in that it is almost
entirely a consequence of the movement of infected plant material, and
other horticultural practices. The effect of the disease on citrus
growing has been immense but there are big differences between regions
in the magnitude of the problem and when it occurred. The epidemiological studies now in progress, the efforts being made to improve the

effectiveness of control measures and the increased use of resistant or tolerant stock/scion combinations are all likely to bring substantial benefits in decreasing the losses now caused by tristeza.

Acknowledgement

The authors thank E.C. Calavan, E.P. DuCharme, C.O. Youtsey and D. Cordas for kindly providing unpublished information on tristeza epidemiology and Dr. L. Huszar, Department of Statistics, University of California, Riverside, for help in statistical analysis.

REFERENCES

Balaraman K. & Ramakrishnan K. (1980) Strain variation and cross protection in citrus tristeza virus on acid lime. *Proceedings of the 8th conference of the International Organization of Citrus Virologists* (Ed. by E.C. Calavan, S.M. Garnsey & L.W. Timmer), pp. 60-8. IOCV, Riverside.

Bar-Joseph M. (1978) Cross protection incompleteness: a possible cause for natural spread of citrus tristeza virus after a prolonged lag period in Israel. *Phytopathology* 68, 1110-1.

Bar-Joseph M., Garnsey S.M. & Gonsalves D. (1979) The closteroviruses: a distinct group of elongated plant viruses. *Advances in Virus Research* 25, 93-168.

Bar-Joseph M., Garnsey S.M., Gonsalves D. & Purcifull D.E. (1980) Detection of citrus tristeza virus. I. Enzyme-linked Immunosorbent Assay (ELISA) and SDS-immunodiffusion methods. *Proceedings of the 8th conference of the International Organisation of Citrus Virologists* (Ed. by E.C. Calavan, S.M. Garnsey & L.W. Timmer), pp. 1-8. IOCV, Riverside.

Bar-Joseph M. & Loebenstein G. (1973) Effects of strain, source plant and temperature on the transmissibility of citrus tristeza virus by the melon aphid. *Phytopathology* 63, 716-20.

Bar-Joseph M. & Loebenstein G. (1974) Effect of temperature on concentration of threadlike particles, stem pitting and infectivity of budwood from tristeza infected Palestine sweet lime. *Proceedings of the 6th conference of the International Organization of Citrus Virologists* (Ed. by L.G. Weathers & M. Cohen), pp. 86-8. University of California, Richmond.

Bar-Joseph M., Loebenstein G. & Oren Y. (1974) Use of electron microscopy in eradication of tristeza sources recently found in Israel. *Proceedings of the 6th conference of the International Organisation of Citrus Virologists* (Ed. by L.G. Weathers & M. Cohen), pp. 83-5. University of California, Richmond.

Bar-Joseph M., Raccah B. & Loebenstein G. (1977) Evaluation of the main variables that affect citrus tristeza virus transmission by aphids. *Proceedings, International Society of Citriculture* 3 (Ed. by W. Grierson), pp. 958-61. Lake-Alfred.

Bennett C.W. & Costa A.S. (1949) Tristeza disease of citrus. *Journal of Agricultural Research* 78, 207-37.

Bridges G.D. & Youtsey C.O. (1972) Natural tristeza infection of citrus species, relatives and hybrids at one Florida location from 1961-1971. *Proceedings of the Florida State Horticulture Society* 85, 44-7.

Burnett H.C. (1961) Systemic spread of tristeza in one Valencia orange tree. *Plant Disease Reporter* 45, 697.

Calavan E.C., Pratt R.M., Lee B.N., Hill J.P. & Blue R.L. (1972) Tristeza susceptibility of sweet orange on Troyer citrange rootstock *Proceedings of the 5th conference of the International Organisation of Citrus Virologists* (Ed. by W.C. Price), pp. 146-53. University of Florida Press, Gainesville.

Cohen M. & Burnett C. (1961) Tristeza in Florida. *Proceedings of the 2nd conference of the International Organisation of Citrus Virologists* (Ed. by W.C. Price), pp. 107-12. University of Florida Press, Gainesville.

Costa A.S. & Müller G.W. (1980) Tristeza control by cross protection. *Plant Disease Reporter* 64, 538-41.

Dickson R.G., Johnson M.M., Flock R.A. & Laird E.F. Jr. (1956) Flying aphid populations in southern California citrus groves and their relation to the transmission of the tristeza virus. *Phytopathology* 46, 204-10.

Garnsey S.M. & Jackson J.L. (1975) A destructive outbreak of tristeza in central Florida. *Proceedings of the Florida State Horticulture Society* 88, 65-9.

Gregory P.H. (1948) The multiple infection transformation. *Annals of Applied Biology* 35, 412-7.

Klas F.E. (1980) Population densities and spatial patterns of the aphid vector *Toxoptera citricida* Kirk. *Proceedings of the 8th conference of the International Organisation of Citrus Virologists* (Ed. by E.C. Calavan, S.M. Garnsey & L.W. Timmer), pp. 83-7. IOCV, Riverside.

Knorr L.C. & DuCharme E.P. (1951) This is tristeza - Ravager of Argentina's citrus industry. *Citrus Magazine (Florida)* 13, 17-9.

Mather S.M. (1968) Regulations affecting the control of citrus virus diseases in California. *Proceedings of the 4th conference of the International Organisation of Citrus Virologists* (Ed. by J.F.L. Childs), pp. 369-73. University of Florida Press, Gainesville.

McClean A.P.D. (1957) Tristeza virus of citrus: Evidence for absence of seed transmission. *Plant Disease Reporter* 41, 821.

Moreno P., Cambra M., Navarro L., Fernandez-Montez J., Pina J.A., Ballester J.R. & Juarez J. (1980) A survey of citrus tristeza virus in the area of Sevilla (Spain) using the ELISA method. *Proceedings of the 5th Congress of Mediterranean Phytopathology*, pp. 41-2. Hellenic Phytopathology Society, Athens.

Müller G.W. & Costa A.S. (1977) Tristeza control in Brazil by pre-immunization with mild strains. *Proceedings, International Society of Citriculture* 3 (Ed. by W. Grierson), pp. 868-72. Lake-Alfred.

Nesbitt R.B. Jr. (1963) History of the quick decline disease of oranges in Orange County. *Sunkist Pest Control Circular* No. 309, 2pp.

Norman P.A., Sutton R.A. & Burdett A.K. (1968) Factors affecting transmission of tristeza virus by melon aphids. *Journal of Economic Entomology* 61, 238-42.

Norman P.A. & Sutton R.A. (1969) Efficiency of three colonies of melon aphids as transmitters of tristeza virus. *Journal of Economic Entomology* 62, 1237-8.

Oren Y., Bar-Joseph M. & Solomon R. (1981) An epidemiologic model for the spread of citrus tristeza virus. *Phytoparasitica* 9, 244. (Abstract)

Pratt R.M., Calavan E.C. & Hill J.P. (1972) The tristeza suppression programme in California. *Proceedings of the 5th conference of the International Organisation of Citrus Virologists* (Ed. by W.C. Price), pp. 168-71. University of Florida Press, Gainesville.

Raccah B., Loebenstein G., Bar-Joseph M. & Oren Y. (1976) Transmission of tristeza by aphids prevalent on citrus, and operation of the tristeza suppression programme in Israel. *Proceedings of the 7th conference of the International Organisation of Citrus Virologists* (Ed. by E.C. Calavan), pp. 47-9. IOCV, Riverside.

Raccah B., Loebenstein G. & Singer S. (1979) Aphid-transmissibility variants of citrus tristeza virus in infected citrus trees. *Phytopathology* 70, 89-93.

Roistacher C.N., Blue R.L., Nauer E.M. & Calavan E.C. (1974) Suppression of tristeza virus symptoms in Mexican lime seedlings grown at warm temperatures. *Plant Disease Reporter* 58, 757-60.

Roistacher C.N., Nauer E.M., Kishaba A. & Calavan E.C. (1980) Transmission of citrus tristeza virus by *Aphis gossypii* reflecting changes in virus transmissibility in California. *Proceedings of the 8th conference of the International Organisation of Citrus Virologists* (Ed. by E.C. Calavan, S.M. Garnsey & L.W. Timmer), pp. 76-82. IOCV, Riverside.

Rosenberg D.Y., McEachern E.H., Blanc F.L., Robinson D.W. & Foote H.L. (1978) Regulatory measures for pest and disease control. *The Citrus Industry* Vol. 4 (Ed. by W. Reuther, E.C. Calavan & G.F. Carman), pp. 223-36. University of California Press, Berkeley.

Schneider H. (1954) Anatomy of bark of bud union, trunk and roots of quick-decline affected sweet orange trees on sour orange rootstock. *Hilgardia* 22, 567-81.

Schwarz R.E. (1965) Aphid-borne virus diseases of citrus and their vectors in South Africa. B. Flight activities of citrus aphids. *South Africa Journal of Agriculture Science* 8, 931-40.

Stout G.L. (1945) Report on a statewide survey for quick decline of orange trees in California. *California Department of Agriculture Bulletin* 34, 108-15.

Toxopeus H.J. (1937) Stock-scion incompatibility in citrus and its cause. *Journal of Pomology and Horticulture Science* 14, 360-4.

Vanderplank J.E. (1963) *Plant Diseases: Epidemics and Control.* Academic Press, New York.

Webber H.J. (1925) A comparative study of the citrus industry of South Africa. *South Africa Department of Agriculture Bulletin* 6, 106 pp.

Zeman V. (1931) Una enfermadad nueva en los naranjales de Corrientes. *Physis* 19, 410-1.

The cocoa swollen shoot disease problem in Ghana

G.K. OWUSU

Cocoa Research Institute, Tafo, Ghana

INTRODUCTION

Cocoa swollen shoot is the only virus disease of cocoa (*Theobroma cacao*) that is prevalent and damaging (Thorold, 1975). For many years it has caused serious problems in Ghana and Nigeria (Thorold, 1975) and more recently in Togo (Partiot *et al.*, 1978). The disease also occurs, but to a lesser extent, in the Ivory Coast and Sierra Leone.

Swollen shoot is most damaging in the Eastern Region of Ghana where it is endemic and where a particularly virulent strain of the causal virus is prevalent (Fig. 1). In this region, cocoa production declined from 118 110 t in 1936/37 to 38 691 t in 1955/56, largely due to the effects of swollen shoot disease (Dale, 1962). Beckett (in Legg, 1979) describes how the disease dramatically reduced yields on a representative farm in the Eastern Region.

The virus causing swollen shoot is presumed to have originated from forest trees of the order Tiliales and is transmitted by mealybugs (Pseudococcoidea, Homoptera). Many strains of the virus varying in virulence and in the symptoms produced in cocoa have been identified.

FIRST DISCOVERY OF SWOLLEN SHOOT AND INITIAL SURVEYS

Swollen shoot disease was first described in 1936 from the New Juaben district of the Eastern Region of Ghana. Eradication measures were recommended but were later discontinued until Posnette (1940) showed that the disease was graft-transmissible and inferred that a virus was the cause.

A preliminary survey of the incidence of swollen shoot in the Eastern Region was carried out in 1939 and was followed by a more systematic reconnaissance survey in 1944. The surveys revealed that the disease was widespread throughout the oldest cocoa of the Eastern Region and was also present elsewhere in Ghana. An intensive country-wide survey was started in 1946 and, by 1947, an estimated 50 million infected cocoa

Plumb R.T. & Thresh J.M. (1983) *Plant Virus Epidemiology.*
Blackwell Scientific Publications, Oxford.

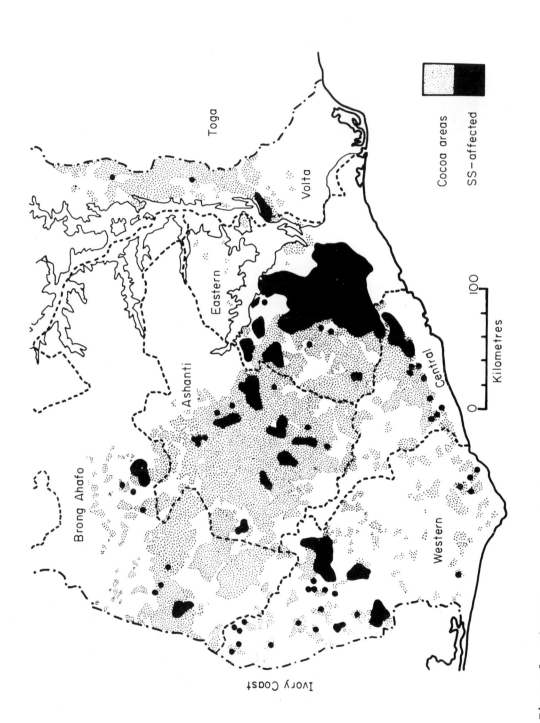

Figure 1. The main cocoa-producing areas of Ghana and the areas worst affected by swollen shoot disease.

trees were reported, 45 million in the Eastern Region alone.

THE ERADICATION CAMPAIGN (1946-62)

Infected trees were being eradicated as early as 1936, but the first intensive "cutting-out" campaign was launched officially in 1946. In 1947 compulsory powers were granted to the Department of Agriculture to remove infected trees. This caused serious social and political unrest, as farmers opposed eradication, especially of symptomless, apparently healthy "contact" trees next to those with symptoms. The campaign had to be suspended pending arbitration by "three scientists of international repute" (Berkeley et al., 1948).

The cutting-out campaign was intensified in 1949 and 1554 km² of the worst affected part of the Eastern Region was designated the "Special Area" around which it was intended to create a disease-free belt, the "cordon sanitaire" of Ross & Broatch (1951), that could be widened to prevent spread to the rest of the country where infection was only sporadic (Walker, 1950).

Compulsory cutting-out was replaced in 1951 by a voluntary system and control measures were discontinued in the abandoned areas of devastation and mass infection in the Eastern Region. In 1952, compulsory treatment was resumed, but until 1961, only areas with the best soils wore treated in the Special Area (Quartey-Papafio, 1961).

By 1961 over 104 million infected trees had been removed (Fig. 2) and the disease was considered to be well under control except in the devastated area of the Eastern Region, where it had been contained (Quartey-Papafio, 1961). During this period, about 80% of all Department of Agriculture expenditure on cocoa was being spent on swollen shoot control (Hammond, 1957).

Compensation for trees removed, and grants for replanting, were paid to farmers in different ways, ranging from payments per area of farm treated and replanted to payments per tree cut and per tree replanted.

ERADICATION SINCE 1962

The official Cocoa Division organization responsible for disease control was disbanded in July 1962 and eradication virtually ceased until the Division was partially re-established in 1964. Instead of monetary compensation, the "plant-as-you-cut" scheme was introduced in 1969. Treated farms were replanted and maintained for three years by Cocoa Division staff before being returned to the owner.

The work done under this scheme has been very limited. Areas cleared of infected trees have usually been small and scattered and they are almost invariably surrounded by infected trees on the same farm or neighbouring farms from which virus spreads to the new plantings.

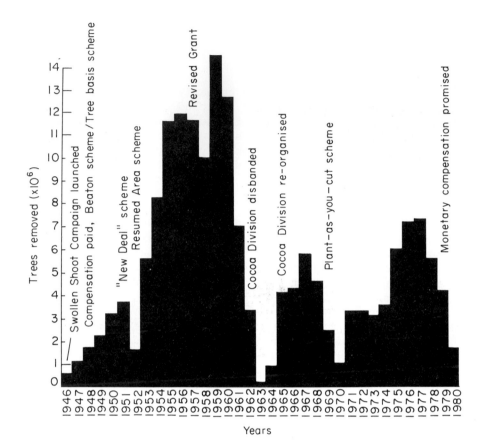

Figure 2. Number of cocoa trees destroyed during the swollen shoot eradication campaign 1946-1980.

Replanting and maintaining the new farms have been expensive and not always done satisfactorily. Furthermore, the scheme effectively discourages farmer involvement. Consequently, only a small proportion of the treated outbreaks has been replanted successfully. The scheme has thus been ineffective as an alternative to monetary compensation and as a means of rehabilitation.

The plant-as-you-cut scheme still operates and by December 1980 over 15 860 ha had been replanted in the Eastern Region, accounting for 84% of all farms replanted under the scheme. Some additional replanting is done independently by farmers using seed and seedlings supplied by Cocoa Division.

These replanting schemes are being supplemented by two important Cocoa Rehabilitation Projects started in 1970 and 1975. The aim is to replant 31 788 ha of land and rehabilitate 20 648 ha of existing cocoa, much of it in parts of Eastern Region that are devastated by swollen shoot.

THE CURRENT STATE

An intensive country-wide survey of infection was started in 1969 (Gyasi, 1978) and completed in October, 1979. It showed that swollen shoot is still mainly concentrated in the Eastern Region, where about 96% of all known infected trees occur (Table 1). Within this region there has been considerable increase in the number of infected trees in the former "area of scattered outbreaks" where the disease was previously reported to have been controlled. Furthermore, the virus is spreading south-westward into adjacent districts of the Central Region previously reported to be little affected.

Table 1. Summary of eradication measures and replanting in different regions of Ghana

Region	Treatments 1946-1961*		1969-79 survey and revision	
	Trees removed† (x10^6)	Replanted (ha)	Infected trees (x10^6)	Untreated¶ (x10^6)
Eastern	98.14 (69.5)	21 605	59.50	38.68
Ashanti/ Brong Ahafo	1.59 (38.3)	662	1.34	0.02
Central/ Western	5.24 (76.3)	370	1.41	0.56
Volta	0.13 (31.0)	14	0.02	0.00
Total	105.10 (69.3)	22 651	62.27	39.26

* Data from Quartey-Papafio (1961)
† Figures in brackets indicate the percentage of trees that were removed during the initial treatment of outbreaks
¶ Estimated number of infected trees still untreated in April 1981

The 1969-79 survey had obvious limitations; it took 10 years and was completed at different times in different districts so that the number of infected trees must have increased considerably due to further spread. Moreover, there were gross overestimates because the infected trees in large outbreaks were not counted individually and the whole area was assumed to be totally affected.

AN ASSESSMENT OF THE ERADICATION PROGRAMME

The most recent survey indicates that swollen shoot is now more prevalent than ever despite the eradication of over 179 million trees since 1946 (Fig. 2) including *c*. 165 million in the Eastern Region. Clearly control has not been achieved and there are now over 39 million affected trees awaiting removal, including over 38 million in the Eastern Region (Table 1). This unsatisfactory state is due to various factors, some of which are epidemiological whereas others are social, political or administrative.

The late discovery of the disease

Swollen shoot disease must have been present in the Eastern Region of Ghana long before it was discovered in 1936. Indeed, as early as 1932, farmers had seen patches of dying cocoa, one of which is said to have been of *c*. 0.4 ha. This suggests that infection must have started some years earlier, perhaps soon after cocoa was first planted in the district *c*. 1907 (Dale, 1962).

Thus, swollen shoot had already reached epidemic proportions in the Eastern Region when first discovered. The limited cutting-out measures that were adopted in the early years were totally inadequate (Thorold, 1975). Furthermore, the emphasis during the 1939-45 war was on food production, and intensive control operations could not start until 1948.

Discontinuity in the eradication campaign

In eradication schemes of this type, it is necessary to pursue the campaign adequately and continuously. Clearly, this has not been so in Ghana. Discontinuities caused by the war, farmer opposition, political intervention, lack of trained manpower and resources or finance have characterized the eradication programme and facilitated the build-up of inoculum.

Whilst the compulsory programme has been abandoned, some trees have been removed voluntarily by farmers and others have been removed by Cocoa Division with the farmers' consent. However, the trees removed were usually in such an advanced stage of infection that they were virtually moribund and no longer important sources of infection.

Compensation

The monetary compensation for trees cut and grants for replanting ceased when the eradication campaign was abandoned in 1962. Since then farmers have taken insufficient interest in disease control and many replanted farms, especially small ones, have been neglected. An official statement in 1979 that the payment of compensation *might* resume has led many farmers to refuse treatment until compensation is forthcoming. This explains the recent dramatic decline in number of trees removed (Fig. 2).

Eradication procedure

Infected trees should be removed as soon as possible after infection occurs for eradication measures to be effective, otherwise there is an opportunity for local spread to neighbouring trees and for "jump spread" when mealybugs are blown to new areas. In recent years it has seldom been possible to remove trees promptly even after they have developed symptoms as the resources available are inadequate and some-times the farmer's consent is not readily given. Consequently, infected trees are left for months or even years after discovery. Furthermore, it is now standard practice to delay cutting trees bearing pods until after harvest and this virtually limits control operations to December-June.

The removal of cocoa trees with symptoms is seldom sufficient to eradicate virus from an area and adjoining "contact" trees must also be removed because symptoms are easily overlooked and others are latently infected. This led Posnette (1943) to suggest that control could be achieved most economically by removing trees with symptoms and an adjacent ring of symptomless "contact" trees. It has seldom been possible to do this in Ghana due to opposition from the farmers or lack of resources, and Hammond (1957) estimated that virus had been eradica-ted from only 2% of all farms treated. The limitations of the cutting-out policy adopted in Ghana are also indicated by the large number of infected trees removed on reinspection and retreatment of outbreaks after the initial treatment has been done (Table 1).

With the inducement of higher compensation grants, Quartey-Papafio (1961) at one stage envisaged removing more than one ring of "contact" trees and any remaining apparently healthy trees on farms where cutting-out had reduced the tree population to less than 124/ha. However, it has not always been possible to persuade Ghanaian farmers to remove even the first ring of contact trees. The problem has been exacerbated in recent years by the decreasing profitability of cocoa farming, and Cocoa Division has not always had the staff or facilities to ensure the regular reinspection of farms being treated. The lack of resources for field operations and the transfer of labour to replanting and maintain-ing new farms are some of the factors limiting the effectiveness of reinspection and retreatment.

At present, examining trees for symptoms at regular intervals is the only practical method of detecting infection in the field. Consequently, the success of the eradication programme depends largely on the efficiency of the inspections. Disease symptoms tend to be least con-spicuous during the dry season when the condition of the trees deteriorates and leaves and branches are shed or damaged, especially by capsids. In recent years, capsid control in Ghana has been totally inadequate and the extensive damage encountered has reduced the effectiveness of routine inspections for swollen shoot infection.

Logistics and practical difficulties

Any assessment of the eradication campaign in Ghana must take account
of the exceptional difficulties encountered and the vast scale of the
undertaking. At the outset, the cocoa-buying stations were known but
not the location, size or ownership of individual farms. Many farms
are small and in inaccessible areas of difficult forest country far
from the nearest roads. Existing maps, roads and personnel were
totally inadequate, and many survey and cutting-out teams had to be
recruited, trained, housed and provided with transport. The work
proved to be difficult, slow and tedious in extremely trying conditions,
with hot, humid weather for much of the year. The enormous resources and
funds required to maintain and supervise operations on an adequate scale
have meant that other ways of improving agricultural efficiency have had
to be neglected. Moreover, operations have at times been greatly
curtailed, as at present, because of economic difficulties. Other
difficulties of the eradication procedure are discussed by Thresh
(1958a).

Latent and missed infections

In view of the many difficulties experienced in the field, it is hardly
surprising that the survey teams fail to identify some infected trees
or even whole outbreaks, which are eventually discovered belatedly when
many trees are affected.

   Infections which are missed also include recently infected trees
which have yet to produce symptoms ("latent infections"). Such trees
present the greatest problem, because the latent period is very
variable and ranges from a few weeks to more than 2 years (Posnette,
1947). Latent infections can be widely distributed around outbreaks
(Thresh & Lister, 1960) and they are likely to support a larger mealybug population than trees in more severe stages of infection (Cornwell,
1956). Thresh (1958b) and Owusu (1972) have shown that mealybugs can
acquire virus from trees during the latter part of the latent period.

   A coppicing technique (Thresh & Lister, 1960) has been used to study
the number and distribution of latent and missed infections around
natural disease outbreaks in Ghana (Legg *et al.*, 1972). Preliminary
examination of the results shows that inapparent infection revealed by
coppicing is extensive, ranging from 2 to 48% of the trees within 30 m
of the outbreak area. Such infection is often irregularly and
unpredictably distributed. Furthermore, many of the outbreaks coppiced
were not isolated ones as originally assumed. These trials have confirmed field experience, indicating that the eradication of outbreaks
in heavily infected areas is difficult to achieve unless unacceptably
drastic measures are adopted.

Wild hosts

Swollen shoot virus originated in wild hosts and these can still
initiate new outbreaks, as in the Western Region of Ghana (Posnette,

1981). However, wild hosts are relatively unimportant now that so much infected cocoa remains and the main spread of the virus is within and between cocoa plantings.

Reinfection of replanted cocoa

It has long been known that cocoa can be successfully re-established in devastated farms, yet much of the replanted cocoa on farmers' plots in the Eastern Region is quickly reinfected. This rapid reinfection is due to the smallness of many farms and their close proximity to abundant sources of infection. Hammond (1957) observed that half of the 16 million infected trees cut out in the Eastern Region were recently planted and not yet bearing. More recently Owusu (1978) has again warned of the danger of replanting in the Eastern Region next to, or intermingled with, old infected trees. The rapid reinfection of replanted farms generates frustration and some farmers become disillusioned and refuse any further treatment.

THE FUTURE

At present there is no alternative to eradication as the basic method for controlling swollen shoot disease. However, the voluntary measures adopted since 1962 have serious limitations and there is an urgent need to return to the former system of compulsory cutting-out. Even if this is done immediately there is an enormous backlog of work that will take at least 2 years to complete in order to protect existing plantings and any others that are made. This estimate is based on the deployment of 730 senior or supervisory staff and 16 000 field assistants and labourers. They will have to be supported by administrators, instructors and those responsible for paying compensation to farmers. Almost all the personnel required are already available within Cocoa Division but they are now largely involved in replanting farms treated under the plant-as-you-cut scheme. However, replanting could again become the responsibility of the farmers as it was in the earlier stages of the eradication campaign.

It is recognized that, at least for the immediate future, many of the treated or presently abandoned cocoa farms are unlikely to be replanted with cocoa and will be used to grow food and other crops. This trend towards more diversification of agriculture will increase the separation between cocoa plantings and so restrict virus spread and facilitate control. However, it will be important to ensure that infected farms are not abandoned to become dangerous foci of infection. There are obvious advantages in planting large contiguous areas away from all known sources of infection, but further research is required to determine the critical size of planting and the minimum isolation distance to avoid serious risk of infection.

Ultimately the planting of cocoa types resistant to, or tolerant of, infection is likely to facilitate rehabilitation but there are problems. Tolerance showed early promise (Posnette & Todd, 1951), yet the avail-

able types have not performed consistently under Ghanaian field
conditions (Legg et al., 1980), and it appears more feasible to exploit
resistance. Cocoa types much less susceptible than those now being
distributed to farmers have recently become available (Legg & Lockwood,
1981) and they are being assessed for use in the Eastern Region.
These types are not immune and they will have to be deployed judicious-
ly if they are to be exploited to best advantage.

   Experience has shown that no single measure will solve the swollen
shoot problem and several measures will have to be taken together. Thus
it may be necessary to consider other means of restricting virus spread.
There may be scope, for example, for a perimeter row or rows of banana,
coffee, citrus, cola, oil palm or some other crop immune to swollen
shoot that would provide an effective barrier and greatly decrease the
risk of reinfection. This approach has long been advocated but has
never been evaluated properly.

Acknowledgements

Acknowledgements are due to Cocoa Production Division for unpublished
data, and to Messrs P.S. Hammond and D.F. Edwards, formerly of the
Cocoa Division, but presently members of the World Bank Cocoa Team in
Ghana, for helpful discussions. This paper is published with the
permission of the Director, Cocoa Research Institute of Ghana.

REFERENCES

Berkeley G.H., Carter W. & Van Slogteren E.L. (1948) Report on the
     commission of enquiry into the swollen-shoot disease of cacao in
     the Gold Coast. *Colonial* 236, 1-10.
Cornwell P.B. (1956) Some aspects of mealybug behaviour in relation
     to the efficiency of measures for the control of virus disease of
     cacao in the Gold Coast. *Bulletin of Entomological Research* 47,
     137-66.
Dale W.T. (1962) Diseases and pests of cocoa. A. Virus diseases.
     Agriculture and Land Use in Ghana (Ed. by J.B. Wills), pp. 286-316.
     Oxford University Press, London.
Gyasi E.K. (1978) *Interim Report on Intensive Cocoa Survey (Country-
     wide) for February 1970-December 1976*, pp. 1-58. Cocoa Production
     Division, Ministry of Cocoa Affairs, Accra.
Hammond P.S. (1957) Notes on the progress of pest and disease control
     in Ghana. *Report of the 1957 Cocoa Conference*, pp. 110-8. The
     Cocoa, Chocolate and Confectionary Alliance, London.
Legg J.T. (1979) The campaign to control the spread of cocoa swollen
     shoot virus in Ghana. *Plant Health. The Scientific Basis for
     Administrative Control of Plant Diseases and Pests* (Ed. by D.L.
     Ebbels & J.E. King), pp. 285-93. Blackwell Scientific Publications,
     Oxford.
Legg J.T., Owusu G.K. & Lovi N.K. (1972) Virus research. Joint CRIG/
     Cocoa Production Division coppicing experiments to assess latent

CSSV infection. *Report, Cocoa Research Institute, Ghana, for 1969-70*, pp. 64-5.

Legg J.T., Kenten R.H. & Lockwood G. (1980) Tolerance trials with young cocoa trees infected with cocoa swollen-shoot virus. *Annals of Applied Biology* 95, 197-207.

Legg J.T. & Lockwood G. (1981) Resistance of cocoa to swollen-shoot virus in Ghana. I. Field trials. *Annals of Applied Biology* 97, 75-89.

Owusu G.K. (1972) Virus research. Acquisition of swollen shoot virus by mealybugs from cocoa plants during the period of latent infection. *Report, Cocoa Research Institute, Ghana, for 1969-70*, pp. 60-1.

Owusu G.K. (1978) Performance of the Eastern Region Cocoa Project, Disease control: swollen shoot disease. *Report on the Evaluation of the Ghana Government/World Bank Cocoa Rehabilitation Projects in the Eastern and Ashanti Regions*, pp. 87-8. Ghana Government/World Bank Cocoa Project Evaluation Committee, Accra.

Partiot M., Amefia Y.K., Djiekpor E.K. & Bakar K.A. (1978) Le "swollen shoot" du cacaoyer au Togo: Inventaire préliminaire et première estimation des pertes causées par la maladie. *Café Cacao Thé* 22, 217-28.

Posnette A.F. (1940) Transmission of swollen-shoot diseases of cacao. *Tropical Agriculture* 17, 98.

Posnette A.F. (1943) Control measures against swollen shoot virus disease of cacao. *Tropical Agriculture* 20, 116-23.

Posnette A.F. (1947) Virus diseases of cacao in West Africa. I. Cacao viruses 1A, 1B, 1C and 1D. *Annals of Applied Biology* 34, 388-401.

Posnette A.F. (1981) The role of wild hosts in cocoa swollen shoot disease. *Pests, Pathogens and Vegetation* (Ed. by J.M. Thresh), pp. 71-8. Pitman, London.

Posnette A.F. & Todd J.M. (1951) Virus diseases of cacao in West Africa. VIII. The search for virus-resistant cacao. *Annals of Applied Biology* 38, 385-400.

Quartey-Papafio E. (1961) Notes on the progress of swollen shoot disease control in Ghana. *Report of the 1961 Cocoa Conference*, pp. 176-8. The Cocoa, Chocolate & Confectionary Alliance, London.

Ross S.D. & Broatch J.D. (1951) A review of the swollen shoot control campaign in the Gold Coast. *Report of the 1951 Cocoa Conference*, pp. 92-7. The Cocoa, Chocolate & Confectionary Alliance, London.

Thorold C.A. (1975) Virus diseases. *Diseases of Cocoa*. Clarendon Press, Oxford.

Thresh J.M. (1958a) The control of cacao swollen shoot disease in West Africa. A review of the present situation. *Technical Bulletin, West African Cocoa Research Institute* No. 4, 36 pp.

Thresh J.M. (1958b) Virus research. The availability of cacao swollen shoot virus to mealybugs feeding on infected trees. *Report of the West African Cocoa Research Institute for 1956-57*, pp. 78-81.

Thresh J.M. & Lister R.M. (1960) Coppicing experiments on the spread and control of cocoa swollen shoot disease in Nigeria. *Annals of Applied Biology* 48, 65-74.

Walker R.E. (1950) Disease control and rehabilitation in the special area. *Report of the 1950 Cocoa Conference*, pp. 8-11. The Cocoa, Chocolate & Confectionary Alliance, London.

The possible role of honeybees in long-distance spread of prunus necrotic ringspot virus from California into Washington sweet cherry orchards

G.I. MINK

Washington State University, Irrigated Agriculture Research
and Extension Center, Prosser, WA 99350, USA

INTRODUCTION

Orchard conditions

Management practices in orchards of sweet cherry (*Prunus avium* L.)
facilitate the spread of pollen-borne viruses. All commercial sweet
cherry cultivars are self-incompatible and pollinator cultivars are
planted as every third tree in every third row. To promote adequate
cross pollination during the short flowering season most growers rent
hives of honeybees. These hives are placed throughout orchards at the
rate of 2.5 hives/ha.

Since 1961 Washington state has had a voluntary fruit tree virus
certification programme for nursery trees which receives strong support
from commercial nurserymen and growers. Consequently, most of the
cherry orchards planted during the past 20 years have been established
with trees free from the two important pollen-borne viruses of cherry
(Nyland *et al.*, 1974), prunus necrotic ringspot virus (NRSV) and prune
dwarf virus (PDV). However, in recent years the incidence of cherry
rugose mosaic disease, caused by strains of NRSV (Nyland *et al.*, 1974),
has increased dramatically (Mink, 1980). In orchards 15-20 years old
the disease can usually be observed first in one tree or a few widely
scattered trees. Observations by growers suggest that nearly all of
these initial infections have appeared since 1971. In orchards where
rugose mosaic has become a significant problem, patches of diseased trees
have developed around the initial infection sites despite attempts by
growers to control spread by removing diseased trees.

Inter-state bee traffic

In 1970 Washington beekeepers began transporting approximately 50 000
beehives to California annually between December and February to pollinate almond and some cherry orchards. The pollination season begins in
southern Californian almond orchards in mid-February and ends in mid-
March. In northern California the almond and sweet cherry pollination
season extends from mid-March to early April. When the California

Plumb R.T. & Thresh J.M. (1983) *Plant Virus Epidemiology.*
Blackwell Scientific Publications, Oxford.

season ends all Washington-based hives are returned to central Washington and prepared for the local pollination season which begins in cherry orchards in early to mid-April. Because cherry trees throughout central Washington flower over a relatively short period, the same hives are rarely, if ever, used to pollinate two Washington cherry orchards in one season.

Virus isolates

Studies by Nyland & Lowe (1964) and Mink (1980) indicate that rugose mosaic strains of NRSV can be differentiated from ordinary strains (Nyland et al., 1974) by host reaction and serological tests. In 1979 NRSV was isolated from several almond trees in central California which expressed symptoms of almond calico and almond virus bud failure (Nyland et al., 1974). These isolates were symptomatologically and serologically similar to NRSV isolates obtained from rugose mosaic-diseased cherry trees in Washington. Subsequent studies were initiated to determine if commercial honeybees spread rugose mosaic strains of NRSV between California and Washington. Initial tests to establish the presence of NRSV in pollen stored in commercial bee hives are reported here.

MATERIALS AND METHODS

Pollen sampling procedures

To determine if bees from a given hive had collected and stored pollen from NRSV-infected trees, co-operating beekeepers removed freshly gathered pollen from individual cells of combs alongside brood combs. The pollen plugs were stored in vials at $-20°C$ until tested by enzyme-linked immunosorbent assay (ELISA) and by mechanical inoculation to Chenopodium quinoa. Pollen used as healthy and infected controls was collected by hand from known healthy and NRSV-infected cherry trees.

In 1980 pollen samples were taken a week before the cherries flowered (7 April) and again during full bloom (18 April) from five beehives that had remained in Washington throughout the 1979/80 winter and spring. All samples were tested concurrently by ELISA.

On 4-8 April 1980, pollen samples were taken from 13 hives as they entered Washington from California. The hives were selected at random by Mr. James Bach, Chief Apiary Inspector, Plant Industries Division, Washington State Department of Agriculture, from several loads of hives that arrived during that period. These hives originated from four different orchards in northern California.

In 1981 a co-operating beekeeper placed 40 preselected hives in a large (144 ha) almond orchard near Bakersfield in southern California. This represented the approximate southern limit of the Washington-based bees. ELISA tests done in January with dormant flower buds taken from this orchard indicated that over 90% of the trees were infected with

NRSV. The beekeeper inserted one frame containing empty combs into each of the 40 hives before flowering in early February. On 23 February approximately 0.5 g of freshly gathered almond pollen was taken from each marked frame and sent by air to Prosser, WA, where the samples arrived and were tested on 26 February.

ELISA

The NRSV antiserum was made against isolate G by R.W. Fulton (1957) and was prepared for ELISA by the methods of Clark & Adams (1977). Anti-virus globulin was conjugated to alkaline phosphatase type VII (Sigma Chemical Company, St. Louis, MO). Tests were performed in Gilford poly-styrene cuvette strips (1 cm path length) and absorbance at 405 nm was recorded with an EIA automatic analyser PR-50 (Gilford Instrument Laboratory, Oberlin, OH). Pollen samples were triturated 1:10 (W/V) in phosphate-buffered saline, pH 7.4, containing 0.5 ml/l Tween 20, 20 g/l polyvinylpyrrolidone (mol. wt. 44 000) and 2µg/l ovalbumin. All test conditions were as described earlier (Mink, 1980).

Infectivity assays

Pollen samples were triturated in a mortar with 0.01 M neutral phosphate buffer and mechanically inoculated on carborundum-dusted leaves of *C. quinoa* using two plants per test.

RESULTS

1980

Little, if any, NRSV was detected in the samples taken from Washington hives approximately one week before cherry trees in the area began to bloom (Table 1). These samples consisted mainly of pollen from local weeds and possibly a few apricot trees near the hives. However, virus was detected in all samples taken from each of the five hives on 18 April. At that time the hives had been 7 days in a cherry orchard which was in full bloom on the day before sampling.

In 1980 NRSV was detected by ELISA in five of six hives that entered Washington from California cherry orchards (Table 2). However, infectivity was detected only in samples taken from one hive (CA-5). Little, if any, virus was detected in hives from the prune or almond orchard. Symptomatologically the virus isolated from CA-5 resembled the ordinary strain of NRSV (Mink, 1980). However, in agar gel double-diffusion tests, isolate CA-5 produced a precipitin line which appeared to fuse with lines produced by several NRSV isolates obtained locally from rugose mosaic-diseased cherry trees.

Table 1. Results of ELISA indexing for prunus necrotic ringspot virus in pollen taken from Washington bee hives in 1980 before and during the cherry blossom period.

Hive tested	Collection date	
	Prebloom (7 April)	Full bloom (18 April)
WA-1	.00*	1.12
WA-2	.16	1.02
WA-3	.00	.71
WA-4	.00	.79
WA-5	.00	.53
CRM pollen†		.85
Healthy pollen¶		.00

* Absorbance at 405 nm (average of 3 tests)
  Pollen hand-collected on 16 April   †from cherry tree with rugose mosaic   ¶from healthy cherry tree

1981

In 1981 NRSV was detected by ELISA in all pollen samples taken from 40 hives sampled in a California almond orchard containing many infected trees (Table 3). The results indicate that samples from each hive contained 0.1-1.0 µg of detectable virus/100 mg of pollen. These values were similar to those obtained with dormant flower buds taken from trees in the same orchard. Individual pollen samples taken from a particular hive produced similar absorbance values (1.43-2.16) indicating that virtually all pollen stored in these hives came from NRSV-infected trees.

Although NRSV was present in all pollen samples taken while the hives were in California, when these samples were assayed for infectivity 3 days later in Washington, NRSV was transmitted to *C. quinoa* from only 20 of 40 hives (Table 3). These hives arrived in Washington on 9 March. Pollen samples taken from all hives on 9-14 March still contained virus detectable by ELISA. However, NRSV was transmitted from only two of the 40 hives (2 and 34).

To prevent further access to the California almond pollen by bees after the hives arrived in Washington each marked frame was covered with a plastic screen. Samples from these frames were tested periodically between 9 March and mid-April when cherry trees in Washington began to flower. However, no infectivity was detected in any of the 40 hives during this period, indicating that NRSV associated with California almond pollen was totally inactivated within 2-3 weeks of collection.

Table 2. ELISA and *Chenopodium quinoa* indexing tests for prunus
necrotic ringspot virus in pollen taken from beehives entering
Washington from Californian orchards on 4-8 April, 1980.

Hive	Californian orchard	A405nm	Infectivity on *C. quinoa*
CA-1	Cherry	.73	−
CA-2	"	.36	−
CA-3	Cherry/Prune	.35	−
CA-4	"	.13	−
CA-5	"	2.12	+
CA-6	"	1.58	−
CA-7	Prune	.05	−
CA-8	"	.03	−
CA-9	"	.04	−
CA-10	"	.05	−
CA-11	Almond	.08	−
CA-12	"	.16	−
CA-13	"	.08	−
CRM pollen		.72	+
Healthy pollen		.01	−

In 1981 the flowering periods for northern California and central
Washington overlapped. In at least one instance a load of hives was
removed from a California almond and cherry orchard at 1700 hours and
was placed in a flowering cherry orchard in Washington early the
following morning. One of these hives was selected at random on arrival
and placed in a field cage containing an uninfected cherry tree of the
cultivar Bing in full bloom. This tree set 120 fruit of which 75
reached full maturity, suggesting that some bees had carried viable
cherry pollen from California to Washington. Low levels of both NRSV
and PDV were detected by ELISA in stored pollen samples taken from this
hive. By mid-July 1981 there was no indication that either virus had
been transmitted to the caged tree. However, PDV was detected in five
seeds.

DISCUSSION

Pollen collected and stored by honeybees is eventually consumed by bee
larvae as a source of protein (Oertel, 1967). Furthermore, we have been
unable to germinate bee-stored almond or cherry pollen in dilute sugar
solutions under conditions in which fresh hand-collected pollen
germinated profusely. Therefore, it seems unlikely that the pollen
deposited in hives will play any direct role in the introduction and
spread of rugose mosaic. However, we have demonstrated that NRSV can

Table 3. Results of ELISA (A) and *Chenopodium quinoa* (B) indexing tests for prunus necrotic ringspot virus in pollen* taken from 40 beehives located in a Californian almond orchard.

Hive no.	Tested by A	B	Hive no.	Tested by A	B	Hive no.	Tested by A	B	Hive no.	Tested by A	B
1	1.73†	+¶	11	1.01	-	21	2.34	-	31	1.89	+
2	1.68	+	12	1.60	-	22	2.13	+	32	2.21	-
3	1.84	+	13	1.65	-	23	1.44	-	33	1.96	-
4	1.48	+	14	1.60	-	24	1.96	-	34	1.98	+
5	1.70	+	15	1.67	+	25	1.90	+	35	1.81	-
6	1.88	-	16	1.73	+	26	1.86	-	36	2.36	+
7	1.54	-	17	1.60	+	27	1.76	-	37	2.00	+
8	1.41	-	18	1.68	-	28	1.82	+	38	1.82	-
9	1.96	+	19	1.75	+	29	2.29	+	39	2.46	+
10	.66	-	20	1.78	+	30	2.06	-	40	2.53	+
V1§	1.93	+	V1	1.67		V1	2.20		V1	2.27	+
V2§	.70	+	V2	.71	+	V2	.85	+	V2	.83	+

* Pollen samples (0.1 g) were triturated in 2 ml of grinding buffer and tested in triplicate
† Absorbance at 405 nm; average of 3 replications
¶ Infectivity detected by sap inoculation on two *C. quinoa* plants
§ V1 and V2 = purified NRSV diluted to 1 µg/ml and 0.1 µg/ml, respectively, included in each plate as positive controls.

be detected readily by ELISA in stored pollen. In unpublished experiments I have also detected PDV by ELISA in stored cherry pollen but not in stored almond pollen. Consequently it appears that bee-stored pollen can be used to determine whether or not bees from a particular hive have foraged on trees infected with either NRSV or PDV. This will be useful in selecting hives for future experiments.

In tests now under way we have found that honeybees emerging from the hive after depositing their pollen loads still have small, but detectable, amounts of pollen adhering to their bodies. Infectivity has not yet been associated with pollen taken from a few bees selected at random for assay. However, on current evidence, it seems that some bees which forage on NRSV-infected trees may carry virus-contaminated pollen grains on their bodies for several days. Although infectivity of NRSV associated with bee-gathered pollen decreases with time, I have demonstrated here that NRSV can remain infectious in some conditions for up to 14 days. If fertilization is necessary to transmit NRSV from pollen to cherry trees, then the ability of honeybees to function as vectors will depend primarily on conditions affecting pollen viability. However, if NRSV is transmitted mechanically to cherry trees by abrading contaminated pollen onto flower parts, then honeybees could function as vectors for several days after foraging on diseased trees even when the pollen is no longer viable.

Honeybees exposed to NRSV-infected trees in California have not yet been shown to transmit NRSV to Washington cherry trees. However, the preliminary evidence suggests that many of the conditions necessary for this to occur do exist and over 50 000 hives return to Washington each year from California. A substantial number of these contain bees which are potential vectors of NRSV, yet relatively few initial infection sites occur in local orchards which can be attributed to this type of spread. This suggests that unusual conditions must prevail if bees originating in California are to initiate such infections.

Acknowledgement

This work was supported in part by funds provided by the Washington Tree Fruit Research Commission.

Washington State University College of Agriculture Scientific Paper No. 5929, Project 1719.

REFERENCES

Clark M.F. & Adams A.N. (1977) Characteristics of the microplate method of enzyme-linked immunosorbent assay for the detection of plant viruses. *Journal of General Virology* 34, 475-84.

Fulton R.W. (1957) Comparative host ranges of certain mechanically transmitted viruses of *Prunus*. *Phytopathology* 47, 215-20.

Mink G.I. (1980) Identification of rugose mosaic-diseased cherry trees by enzyme-linked immunosorbent assay. *Plant Disease* 64, 691-4.

Nyland G. & Lowe S.K. (1964) The relation of cherry rugose mosaic and almond calico viruses to *Prunus* ringspot virus. (Abstract) *Phytopathology* 54, 1435-6.

Nyland G., Gilmer R.M. & Moore J.D. (1974) "Prunus" ringspot group. *Virus diseases and Noninfectious Disorders of Stone Fruits in North America* USDA Agricultural Handbook 437, pp. 104-32. US Government Printing Office, Washington DC.

Oertel E. (1967) Nectar and pollen plants. *Beekeeping in the United States* USDA Agricultural Handbook 335, pp. 10-6. US Government Printing Office, Washington DC.

Epidemiology and control of groundnut bud necrosis and other diseases of legume crops in India caused by tomato spotted wilt virus

D.V.R. REDDY, P.W. AMIN, D. McDONALD &
A.M. GHANEKAR
Groundnut Improvement Program, International Crops Research Institute
for the Semi-Arid Tropics (ICRISAT),
Patancheru P.O., A.P. 502 324, India

INTRODUCTION

Tomato spotted wilt virus (TSWV) was first reported in India in tomato in 1964 (Todd et al., 1975). The occurrence of TSWV on a legume in India was first recorded in 1968 (Reddy et al., 1968). The "bud necrosis disease" of groundnut, caused by TSWV, is now considered to be one of the most damaging groundnut diseases in India (Ghanekar et al., 1979a; Reddy, 1980). Bud necrosis is likely to have been present in India for some time although it has only recently become economically important. TSWV has also been reported on groundnuts in Brazil (Costa, 1941), the United States of America (Halliwell & Philley, 1974), South Africa (Klesser, 1966) and Australia (Helms et al., 1961). This chapter considers the epidemiology and control of bud necrosis and gives a brief account of other economically important diseases of legumes in India caused by TSWV.

OCCURRENCE AND DISTRIBUTION OF LEGUME DISEASES CAUSED BY TSWV IN INDIA

Our surveys show that bud necrosis is widely distributed in the main groundnut-growing regions of India and that it is endemic in the states of Andhra Pradesh and Tamilnadu. Extensive infection has also been seen in parts of the states of Maharashtra, Gujarat, Rajasthan and western Uttar Pradesh. The greater incidence of bud necrosis in recent years may be related to the expansion of irrigation projects which has led to continuous cropping of groundnuts and other hosts of TSWV. Until recently most groundnuts were grown in the rainy season but increased demand has caused an expansion of the post-rainy season, irrigated crop.

The economically important leaf curl diseases of green and black gram (Vigna radiata) (Nene, 1972) have recently been shown to be caused by TSWV (Ghanekar et al., 1979b) and field trials showed that TSWV can also cause economically important diseases of pea (cultivar Bonneville), broad bean (cultivar Local), cowpea (cultivar C-152) and soyabean (cultivar Bragg).

Plumb R.T. & Thresh J.M. (1983) Plant Virus Epidemiology.
Blackwell Scientific Publications, Oxford.

ROLE OF THRIPS IN TSWV TRANSMISSION

The isolate of TSWV infecting groundnut is transmitted by two thrips species, *Frankliniella schultzei* and *Scirtothrips dorsalis*. The former is the most efficient vector and is chiefly responsible for disease spread in the field (Amin *et al.*, 1981). Nearly 50% of individuals of laboratory-bred *F. schultzei* transmitted TSWV when tested in the laboratory. *Thrips tabaci* had been presumed to transmit TSWV in tomato crops (Todd *et al.*, 1975) but was not found on groundnuts in our survey. In addition, our attempts to transmit the TSWV isolate that causes bud necrosis from groundnut to groundnut and from groundnut to the highly susceptible green and black gram and cowpea by *T. tabaci* were unsuccessful.

Larvae of *F. schultzei* require a minimum acquisition access time of 30 min but the frequency of transmission increases up to an optimum acquisition time of 48 h. Adults cannot acquire the virus, but can transmit it if it is acquired when they are larvae. The incubation period is 8-9 days, which is the time required for the thrips to complete the larval and pupal instars and become adults. The virus is retained in adults throughout their life, irrespective of acquisition time. Males and females transmit the virus equally efficiently.

Hosts of thrips and TSWV

F. schultzei infests many different host plants, including crops, ornamentals and weeds many of which are also susceptible to TSWV (Table 1). Crop plants such as green and black gram are highly susceptible to both the virus and the vector, and cropping with these short-duration legumes helps to perpetuate virus and vector. Tomato, egg plant, groundnut and ornamentals such as zinnia and chrysanthemum which are susceptible to both TSWV and *F. schultzei* are commonly grown during the summer season. *Ageratum conyzoides* and *Cassia tora*, two common weeds in groundnut fields, harbour numerous *F. schultzei* and more than 50% are infected with TSWV. These weeds are usually abundant soon after monsoon showers and are likely to provide sources of inoculum.

SEED TRANSMISSION

Groundnut plants infected when young (50-60 days after sowing) produce shrivelled seeds of which only 30% germinate. However, when plants at least 10 weeks old were infected, 85% of the seed they produced germinated normally. The virus was not detected in seedlings raised from seed collected from plants infected when young (2800 seeds) or when older (1600 seeds). Infective virus was recovered from the testa of immature and freshly harvested mature seeds, but virus could only be detected serologically in the testas of dried seeds. Neither serologically detectable nor infective virus was recovered from the cotyledons and embryos of freshly harvested or dried seeds. Freshly harvested mature seeds with testas containing infective virus, failed to transmit TSWV when grown on (A.M. Ghanekar & R. Rajeshwari, unpublished data).

Table 1. Some hosts of *Frankliniella schultzei* and tomato spotted wilt virus (TSWV).

Papilionaceae:	*Arachis hypogaea* (C) (T) (V), *Canavalia gladiata* (C) (T) (V), *Crotalaria juncea* (C) (T) (V), *Desmodium triflorum* (W) (T) (V), *Glycine max* (C) (T) (V), *Pisum sativum* (C) (T) (V), *Tephrosia purpurea* (W) (T), *Vicia faba* (C) (T) (V), *Vigna mungo* (C) (T) (V), *V. radiata* (C) (T) (V), *V. unguiculata* (C) (T) (V).
Compositae:	*Acanthospermum hispidum* (W) (T) (V), *Ageratum conyzoides* (W) (T) (V), *Carthamus tinctorius* (C) (T), *Chrysanthemum indicum* (O) (T), *Cosmos bipinstatus* (O) (T) (V), *Helianthus annuus* (C) (T), *Lagasca mollis* (W) (T) (V), *Tridax procumbens* (W) (T), *Xanthium strumorium* (W) (T) (V), *Zinnia elegans* (O) (T) (V).
Solanaceae:	*Lycopersicon esculentum* (C) (T) (V), *Nicotiana tabacum* (C) (T), *Solanum melongena* (C) (T) (V), *S. tuberosum* (C) (T) (V).
Caesalpiniceae:	*Cassia tora* (W) (T) (V), *C. obtusifolia* (W) (T) (V), *C. occidentalis* (W) (T).
Liliaceae:	*Allium cepa* (C) (T).
Convolvulaceae:	*Datura stramonium* (W) (T).
Labiateae:	*Leucas aspara* (W) (T).
Papaveraceae:	*Papaver* sp. (W) (T).
Zygophyllaceae:	*Tribulus triticus* (W) (T).
Asclepidaceae:	*Calotropis gigantica* (W) (T) (V).
Amaranthaceae:	*Celosia argentea* (W) (T).

C = Crop   V = Host of TSWV
O = Ornamental   T = Host of *F. schultzei*
W = Weed

INCIDENCE OF BUD NECROSIS

Figs. 1 & 2 show the population fluctuations of *F. schultzei*, the principal vector of TSWV, on groundnuts in the rainy (June-September) and post-rainy (January-March) seasons in Hyderabad. The first

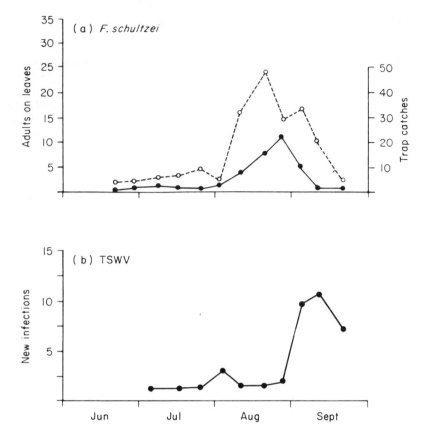

Figure 1.   (a) Adult *F. schultzei* collected during June-September,
            1979, on the ICRISAT farm. In suction traps O---O
            and on ten terminal leaves ●——●.
          (b) Percentage of crop stand first showing symptoms of
            tomato spotted wilt virus.

invasion of *F. schultzei* occurred in the first week of July to weeds
such as *C. tora* and *A. conyzoides*, which emerge soon after the first
rain. The *F. schultzei* populations were low until the second week
of July but subsequently increased rapidly to reach a maximum in
the last week of August and early September. By mid-September
populations on the crop declined sharply (Fig. 1a). The maximum
disease incidence, which ranged from 50 to 100%, occurred in the
rainy season, 2-3 weeks after the maximum number of *F. schultzei*
was recorded (Table 2) whereas in the post-rainy season crop in the
Hyderabad region disease incidence was only 20-30%. In the post-rainy
season crop, maximum numbers of *F. schultzei* occurred in January and
February and most new infections of TSWV were in February (Fig. 2).
F. schultzei populations declined sharply from March to July when they

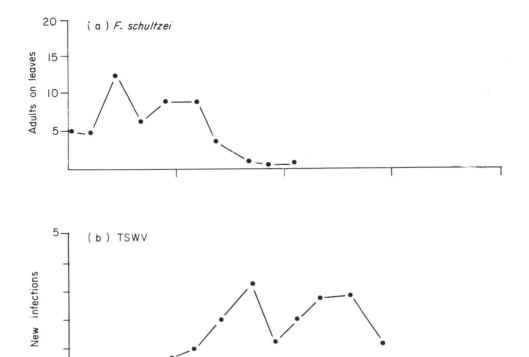

Figure 2. (a) Adult *F. schultzei* counted on plants during January-
April, 1980, on the ICRISAT farm.
(b) Percentage of crop stand first showing symptoms of
tomato spotted wilt virus.

were found on the flowers of summer crops such as chrysanthemum,
ziñnia, tomato, egg plant, onions and weeds such as *Calotropis
gigantica, C. tora* and *Tribulus triticus*.

The number of *F. schultzei* caught in suction traps and recorded on
young terminal buds followed similar patterns (Fig. 1a). Factors that
contribute to the massive build-up and migration of *F. schultzei* are
unknown, although from initial observations it seems that temperature
and wind speed may be important. Seventeen of the 22 mass flights
recorded from June 1980 to April 1981 occurred when the wind velocity
at 3 m above the crop at 14.00 h was less than 10 km/h, and 5 flights
occurred at wind speeds of 10-15 km/h. Few thrips were caught when the
wind speed exceeded 15 km/h and none when wind speed was greater than
20 km/h. Twenty of 22 mass flights occurred when the maximum tempera-

Table 2. Effect of sowing date on the incidence of bud necrosis in groundnuts on the ICRISAT farm.

Sowing date	Disease incidence (%)		Yield (kg/ha)	
	At 50 days	At 100 days	Dry pods	Seeds
15 June 1979	14.1	59.2	917	668
10 July 1979	82.3	99.3	121	49
S.E.	±1.77	±0.76	±44	±30.6

tures were 20°-35°C and none occurred above 40°C. Körting (1930) and Johansson (1946) have also stressed the influence of temperature on mass flights of thrips. Relative humidity differed greatly (13-86%) when mass flights occurred, indicating that it may be less important than wind speed and temperature. Körting (1930) also showed that relative humidity and atmospheric pressure did not affect thrips migration.

Most infections by TSWV were recorded shortly after mass invasions by *F. schultzei* and this may indicate that primary infection is more important than secondary spread. However, limited secondary spread within the groundnut crop by the progeny of migrant thrips may occur.

CONTROL OF BUD NECROSIS DISEASE

Husbandry

Since high incidence of bud necrosis disease in Hyderabad is primarily associated with infestation by immigrant thrips during August-September and January-February, sowing dates could be expected to affect disease incidence and subsequent crop loss. Groundnuts sown in mid-June, with the onset of the rains, were infected less and yielded more than crops sown later (Table 2). Disease incidence was lower in the November-sown than in the December-sown crop but overall disease incidence in post-rainy season crops was much lower than in rainy season crops and different sowing dates did not give yield advantages.

An increase in plant density decreased the proportion of infected plants, although the number of infected plants per unit area was unaffected (Table 3).

Table 3. Effect of plant spacing on the incidence of bud necrosis in groundnuts on the ICRISAT farm.

Spacing (cm)	Post-rainy season 1978-79		Post-rainy season 1979-80	
	Disease %	Yield (kg/ha)	Disease %	Yield (kg/ha)
37.5 x 5.0	10	3493	7	2153
37.5 x 15.0	23	2855	16	1524
75.0 x 5.0	20	2289	9	1570
75.0 x 15.0	40	1745	18	917
150.0 x 5.0	23	1270	11	740
150.0 x 15.0	43	777	21	409
SE: Inter-row spacing	±110.7 (11 D.F.)		±80.4 (13 D.F.)	
SE: Intra-row spacing	±47.7 (6 D.F.)		±44.2 (9 D.F.)	

Insecticides

Carbofuran (1 kg a.i./ha) applied to the soil at planting protected the crop from disease for 3 weeks and had no effect upon the subsequent disease incidence, but gave a small, uneconomic increase in the yield. The incidence of bud necrosis was reduced by weekly sprays of systemic insecticides such as dimethoate (450 ml a.i./ha). However, the yield increase did not cover the cost of the insecticide applications and also such excessive use of pesticides cannot be recommended on environmental grounds. Non-phytotoxic mineral oil emulsions such as JMS oil and Sunoco 7E oil at 2% concentration, alone or in combination with dimethoate, also failed to decrease disease incidence. Insecticidal protection could be expected to be more economical if the applications were timed to coincide with the maximum immigration of the vector. However, because of the expertise and equipment required, monitoring vector numbers is impractical.

Resistance

Several germplasm lines of *Arachis hypogaea* have been screened under natural conditions where the disease incidence in standard cultivars exceeded 60%. No line was resistant or immune to the virus although some cultivars such as NC Ac2242 and NC Ac2575 have some resistance to the vector. These vector-resistant lines have either thick, leathery leaves or more phenolic compounds in the leaves than susceptible cultivars.

However, Robut 33-1, a high-yielding cultivar, although susceptible when mechanically inoculated in the laboratory, was much less infected than most other cultivars when grown in both rainy and post-rainy seasons (Table 4). The basis for this lower disease incidence in Robut 33-1 is being investigated.

Table 4. The incidence of bud necrosis in groundnut cultivars TMV2 and Robut 33-1 in rainy and post-rainy seasons on the ICRISAT farm.

Cultivar	Disease incidence* (%) at maturity				
	Rainy season† crop			Post-rainy season¶ crop	
	1978	1979	1980	1978/79	1979/80
TMV2	87	100	94	48	34
Robut 33-1	34	50	35	28	20

* = Each cultivar was grown in four 100 m² plots at 15 x 75 cm spacing.
† = June-October   ¶ = November-April

Some wild *Arachis* spp., including *A. chacocnse* (Collection Number 10602) and *A. pusilla* (Collection Number 12922), were not infected despite repeated mechanical sap inoculations in laboratory tests. Further tests using grafts and transmission by viruliferous adults of *F. schultzei* are required before it can be determined if these species are genetically resistant to TSWV.

CONCLUSIONS

Bud necrosis disease, caused by TSWV, is now one of the most important diseases of groundnut in India.

Early planting, increased plant density and use of the cultivar Robut 33-1 appear to minimize losses due to bud necrosis in Hyderabad, but need to be investigated in different regions of India before they can be generally recommended. Eliminating primary sources of infection does not appear to be feasible and roguing of infected plants is unlikely to be effective in controlling virus spread, as most infection is from sources outside the crop. The effects on disease spread of inter-cropping groundnut with quick-growing cereals such as pearl millet are now being investigated. Groundnuts planted with green gram or black gram showed a very high incidence of bud necrosis and these plants are not recommended for inter-cropping because they are seriously affected by TSWV.

The most effective method of controlling TSWV would be to grow resistant cultivars. However, attempts to locate sources of high resistance in *A. hypogaea* have so far been unsuccessful. The only known source of resistance to TSWV appears to be in wild *Arachis* spp. and these are now being tested.

ICRISAT Conference Paper CP-46

Acknowledgements are due to Dr. R.W. Gibbons, Program Leader, Groundnut Improvement Program, ICRISAT, for many valuable suggestions.

REFERENCES

Amin P.W., Reddy D.V.R., Ghanekar A.M. & Reddy M.S. (1981) Transmission of tomato spotted wilt virus, the causal agent of bud necrosis disease of peanut (*Arachis hypogaea* L.) by *Scirtothrips dorsalis* Hood and *Frankliniella schultzei* (Trybom) (Thysanoptera, Thripidae). *Plant Disease* <u>65</u>, 663-5.

Costa A.S. (1941) Una molestia de virus do amendoim (*Arachis hypogaea* L.). A mancha anular. *Biologico* <u>7</u>, 249-51.

Ghanekar A.M., Reddy D.V.R., Iizuka N., Amin P.W. & Gibbons R.W. (1979a) Bud necrosis of groundnut (*Arachis hypogaea*) in India caused by tomato spotted wilt virus. *Annals of Applied Biology* <u>93</u>, 173-9.

Ghanekar A.M., Reddy D.V.R. & Amin P.W. (1979b) Leaf curl disease of mung and urd beans caused by tomato spotted wilt virus. (Abstract) *Indian Phytopathology* <u>32</u>, 163.

Halliwell R.S. & Philley G. (1974) Spotted wilt of peanut in Texas. *Plant Disease Reporter* <u>58</u>, 23-5.

Helms K., Grylls N.E. & Purss G.S. (1961) Peanut plants in Queensland infected with tomato spotted wilt virus. *Australian Journal of Agricultural Research* <u>12</u>, 239-46.

Johansson E. (1946) Studier och försok rörande de pa gras och sadesslag levande tripsarnas biologi och skadegörelse. II. Tripsarnas frekvers och sprid ning i Jämförelse med andra sugande insekters samt deras fröskadegorande betydelse. *Meddelanden från Statens Växtskyddsanstalt Stockholm.* <u>46</u>, 1-59.

Klesser P.J. (1966) Tomato spotted wilt virus on *Arachis hypogaea*. *South African Journal of Agricultural Science* <u>9</u>, 731-6.

Körting A. (1930) Beitrag zur Kenntnis der Lebensgewohnheiten und der phytopathogen Bedeutung einiger an Getreide lebender Thysanopteran. *Zeitschrift für angewandte Entomologie* <u>16</u>, 451-512.

Nene Y.L. (1972) Leaf curl. *A Survey of Viral Diseases of Pulse Crops in Uttar Pradesh*, pp. 142-153. University Press, Pantnagar.

Reddy D.V.R. (1980) International aspects of groundnut virus research. *Proceedings of the International Workshop on Groundnuts, 13-17*

102   D.V.R. REDDY, P.W. AMIN, D. McDONALD & A.M. GHANEKAR

October, 1980, Patancheru, A.P., India (Ed. by R.W. Gibbons),
pp. 203-10. International Crops Research Institute for the Semi-
Arid Tropics (ICRISAT), Patancheru.

Reddy M., Reddy D.V.R. & Appa Rao A. (1968) A new record of virus
disease on peanut. *Plant Disease Reporter* 52, 494-5.

Todd J.M., Ponniah S. & Subramanyam C.P. (1975) First record of tomato
spotted wilt virus from Nilgiris in India. *Madras Agricultural
Journal* 62, 162-3.

Epidemiology of beetle-borne viruses of grain legumes in Central America

RODRIGO GÁMEZ* & RAÚL A. MORENO†
*Centro de Investigación en Biología Celular y Molecular, Universidad de
Costa Rica, Ciudad Universitaria, Costa Rica
†Centro Agronómico Tropical de Investigación y Enseñanza,
Turrialba, Costa Rica

INTRODUCTION

Common bean (*Phaseolus vulgaris*) is the most important grain legume in
the American tropics. It originated in southern Mexico, Central and
South America and has been cultivated in these areas since pre-Columbian
times. Through centuries of cultivation, a great range of genotypes
adapted to the varied ecological conditions of Mesoamerica has arisen as
a result of natural selection and limited human intervention. In Central
American countries, particularly in Guatemala, there is remarkable
diversity in disease resistance, seed colour, seed size, plant archi-
tecture, yield and other characteristics (Echandi, 1975). Cowpea (*Vigna
unguiculata*), which is African in origin, was introduced to the area in
post-colonial times and possibly on several different occasions. Only
in the last decade has this crop been cultivated in restricted areas in
the warm lowland tropics, where the climate is unsuitable for beans. It
is an increasingly important crop and a potential substitute for beans
in these areas. A few cultivars have been selected for their yield and
adaptation to local conditions.

Leaf-feeding beetles, of which numerous species are present in Central
America, are abundant and important pests of bean, cowpea and other crops.
Their importance as virus vectors and the identity of the viruses they
transmit has only recently been studied in this area (Gámez, 1973, 1980).

This chapter reviews the information available on the ecology of
beetle vectors, on their relationship with virus spread and on the contrasting epidemiological characteristics of beetle-borne viruses of bean
and cowpea in Central America.

THE VIRUSES

Identity

Beetle-transmitted viruses are the largest group of insect-borne viruses
of legumes in Central America (Gámez, 1980), and include members of the

Plumb R.T. & Thresh J.M. (1983) *Plant Virus Epidemiology.*
Blackwell Scientific Publication, Oxford.

como, bromo and sobemovirus groups (Table 1). Isolates of cowpea severe
mosaic virus (CPSMV) (De Jager, 1979) from Central America are serologically indistinguishable (Fulton & Scott, 1977), but distinct strains
of bean rugose mosaic virus (BRMV) have been reported in Costa Rica,
El Salvador and Guatemala (Gámez, 1972, 1982; Cartín & Gámez, 1972;
Gálvez et al., 1977). Southern bean mosaic virus (SBMV) occurs as
severe and mild strains in bean in Costa Rica (Murillo, 1967). Strains
of bean curly dwarf virus (BCDV) (Meiners et al., 1977), bean yellow
stipple virus (BYSV) (Gámez, 1976) and bean mild mottle virus (BMMV)
(Waterworth et al., 1977) have not been distinguished.

Table 1. Beetle-borne viruses and beetle vectors in Central America.

Virus group and virus	Vector	References
Comovirus group:- Cowpea severe mosaic	*Cerotoma ruficornis rogersi* *Cerotoma atrofasciata* *Diabrotica balteata* *Diabrotica adelpha* *Gynandrobrotica variabilis* *Epilachna varivestis* *Systena sp.*	Díaz, 1972; González et al., 1975; Valverde et al., 1978.
Bean rugose mosaic	*Cerotoma ruficornis rogersi* *Diabrotica balteata* *Diabrotica adelpha* *Diabrotica undecimpunctata*	Gámez, 1972, 1982; Cartín & Gámez, 1973.
Bean curly dwarf	*Epilachna varivestis*	Meiners et al., 1977.
Bromovirus group:- Bean yellow stipple	*Cerotoma ruficornis rogersi* *Diabrotica balteata* *Diabrotica adelpha*	Gámez, 1976.
Sobemovirus group:- Southern bean mosaic	*Diabrotica adelpha*	Murillo, 1967
Ungrouped:- Bean mild mosaic	*Diabrotica undecimpunctata* *Epilachna varivestis*	Waterworth et al., 1977.

Transmission

A review of the transmission of viruses by beetles has been published (Fulton *et al.*, 1980). The varied virus-vector relationships indicate that more complex biological phenomena are involved than mere surface contamination of the insects' mouth parts.

The time that beetles retain the ability to transmit virus depends on the virus, the beetle, host plant species and environment. *Cerotoma ruficornis rogersi* retains BRMV much longer (9 days) than *Diabrotica* spp. (3 days) (Gámez, 1972; Cartín & Gámez, 1973) and the proportion of individual insects which transmit BRMV following an acquisition feed varies also with the vector species. *C. ruficornis rogersi* (80% transmission) is the most efficient vector compared with *D. balteata* (20%) and *D. adelpha* (30%). Similarly, *C. ruficornis rogersi* (90%) is a more efficient vector of CPSMV than other species (19-56%) (Valverde *et al.*, 1978). Only very low rates of transmission (30% or less) of BYSV are obtained with *C. ruficornis rogersi* and *D. balteata* (Gámez, 1976).

Distribution

BRMV occurs in Costa Rica, Guatemala and El Salvador (Gámez, 1972, 1982; Cartín & Gámez, 1973; Granillo *et al.*, 1975), Honduras and Nicaragua.

Knowledge of the distribution of other bean viruses is limited. BYSV has been reported in Costa Rica (Gámez, 1976), BCDV in El Salvador (Meiners *et al.*, 1977) and Costa Rica (H. Hobbs, unpublished data) and BMMV in El Salvador (Waterworth *et al.*, 1977), but all of these viruses may occur elsewhere in Central America.

CPSMV is widely distributed and has been isolated in all countries and from many different localities in Central America (Díaz, 1972; González *et al.*, 1975), the Caribbean and South America (Fulton *et al.*, 1980). The distribution of these viruses appears to follow closely the distribution of the beetle vectors, and beetles and viruses seem to be more prevalent in the basal and premontane altitudinal zones than at higher altitudes (Holdridge, 1978).

THE BEETLES

Identity, distribution and hosts

Several beetle species have been recognized as virus vectors in Central America (Table 1), most of them within the family Chrysomelidae and only one (*Systena* sp.) in the Coccinellidae. Many other beetle species exist in the area and some may be virus vectors.

The known species of vectors have been recorded in all countries from Guatemala to Panama. Chrysomelid beetles usually have a limited host range on which they feed and reproduce, and this may influence which viruses they transmit (Fulton *et al.*, 1980). Naturally these beetles occur on various secondary herbs, mostly in the families Cucurbitaceae, Compositae, Rosaceae and Leguminoseae (Risch, 1979). In an ecological sense beetles can be "specialist" or "generalist" herbivores. Of the known virus vectors, *C. ruficornis rogersi* is a relative specialist; larvae eat only bean and other legume roots but the adults, while preferring bean vegetation, also feed on squash. The two most abundant generalist species, *D. balteata* and *D. adelpha* can reproduce on the roots of bean and squash, but prefer maize; the adults eat both squash and bean leaves of all ages as well as very young maize shoots (Risch, 1979).

Field populations and dispersal

Traditional farming systems in Central America rely largely on polycultures. These involve planting two or more crops simultaneously and sufficiently close together to interact (Risch, 1979). One reason why maize, bean and squash have been grown together for millenia in Mesoamerica appears to be that this decreases insect pest damage and virus spread in the legume crop. Studies in Turrialba, Costa Rica (Risch, 1979), have shown that, whenever a polyculture involving bean contained at least one non-host plant, usually maize, beetle numbers were significantly reduced, compared with bean monocultures. This pattern was generally observed 40-60 days after planting for each of the beetle species studied and continued until the end of the season (Risch, 1979). Similar results were obtained when cowpea monocultures and cowpea-maize dicultures were compared (González, 1978; Valverde, 1978; Valverde *et al.*, 1982a).

Many, complex factors determine the differences in beetle numbers between monocultures and polycultures, but the most important is the pattern of beetle movement. In polycultures with maize, beetles avoid feeding on host plants shaded by maize and the maize stalks impede flight. Beetles landing on non-hosts remain for significantly less time than on hosts, resulting in increased rates of emigration from polycultures (Risch, 1979). This effect on beetle movement influences the rate and pattern of virus spread.

EPIDEMIOLOGY

Incidence

The incidence of beetle-borne viruses in bean and cowpea is very different. Although five different viruses occur in bean in Central America, their incidence normally ranges from 0% to *c.* 5% in monocultures, while they are only sporadic in traditional maize-bean polycultures. The incidence of CPSMV is much greater and from 1977 to 81 ranged from 16 to 100% in experimental and commerical cowpea fields in Turrialba and elsewhere in Costa Rica, even where cowpeas were a new crop (González *et al.*, 1975; González, 1978; Valverde, 1978; Valverde *et al.*, 1982a). The incidence of BYSV in cowpea is much less, ranging from 1% or less to, exceptionally, 30% (González, 1978; Valverde, 1978). Great differences in incidence of viruses in bean and cowpea have even been observed in adjacent fields under identical conditions. The reasons for this are not known but bean, as an indigenous species, may have evolved resistance to local pathogens, whereas the introduced cowpea has not.

Rainfall seems to influence the incidence of CPSMV and, in two different trials in Turrialba, the percentage of infected plants was 92 and 24 for the rainy (May-November) and dry (December-April) cropping seasons, respectively. There was a positive correlation between total populations of beetles and the proportion of CPSMV-infected plants (Valverde, 1978; Valverde et al., 1982a). There was no obvious effect of rainfall on the incidence of BYSV.

In cowpea monocultures and maize-cowpea dicultures in Turrialba during the wet and dry seasons of 1975/6 and 1977/8, beetle populations and the incidence of CPSMV and BYSV were consistently smaller in poly-cultures than in dicultures in 1975/6 but not in 1977/8. This may have been because beetle populations (0-30/trap) in 1977/8 were larger than in 1975/6 (0-15/trap) (González, 1978; Valverde, 1978; Valverde et al., 1982a). However, in polycultures when the detrimental effect of maize on beetle numbers is seen (40-80 days after sowing) (Risch, 1979) most virus spread has occurred.

TEMPORAL AND SPATIAL PATTERNS OF VIRUS SPREAD IN COWPEAS

Rate of increase

The rate of increase of CPSMV follows a typical sigmoid curve (González, 1978; Valverde, 1978; Araujo & Moreno, 1979). The exponential phase of this curve begins c. 4 weeks after sowing and reaches a plateau at c. 7 weeks. Populations of C. ruficornis rogersi and D. balteata increased 2 weeks after planting, which probably initiates the exponential phase of spread. Daily rates of increase of CPSMV ranging from 0.16 to 0.24 units have been calculated (Valverde, 1978; Araujo & Moreno, 1979) using the "compound interest" equation of Vanderplank (1963). In one field trial in the dry season, the small beetle population during early crop growth was associated with unusually heavy rains and probably accounted for the reduced rate (0.03/units) of disease increase.

There is a correlation between beetle damage, beetle populations and rates of disease spread. As shown in Fig. 1, the feeding damage of the insects on cowpea during the early stages of growth appears to correlate closely with the final incidence of CPSMV (Shannon & King, 1980). There is also a close correlation between virus incidence and numbers of C. ruficornis rogersi and D. balteata trapped in yellow pans (González, 1978; Araujo & Moreno, 1979) or screen cages (Valverde, 1978) (Fig. 2).

The percentage of viruliferous individuals of C. ruficornis rogersi collected in the field at different intervals after planting varies with disease progress (Fig. 3); the number of viruliferous D. balteata remains consistently small (Valverde, 1978; Araujo & Moreno, 1979).

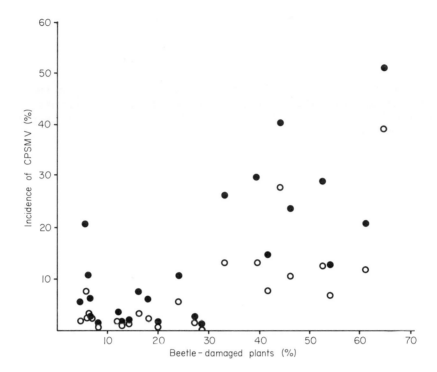

Figure 1. Relation between chrysomelid damage 26 days after planting
and the incidence of CPSMV after 39 days (O) and 46 days (●)
(From Shannon & King, 1980).

Pattern of spread and field sources

Seed transmission of CPSMV has not been recorded and the virus is considered to spread into cowpea plantings from outside sources. As the
beetle populations increase, particularly in monocultures, secondary
spread within the crop becomes increasingly prevalent (González, 1978;
Valverde, 1978). A grouping of infected plants can be detected by
"doublet count" analyses (Vanderplank, 1946).

Many species of wild legumes have been tested as possible perennating
hosts of viruses in Turrialba. Only the weed species *Vigna vexillata*
was found to be naturally infected with CPSMV and appears to be a good
host for *C. ruficornis rogersi* and *D. balteata* (Valverde, 1978;
Valverde *et al.*, 1982a). During the early stages of crop growth at one
site the incidence of CPSMV tended to decrease with increasing distance
from the virus source, which was an infected cowpea field. As the crop
matured this effect disappeared, possibly due to the increasing
importance of secondary spread (H. Hobbs, unpublished information).

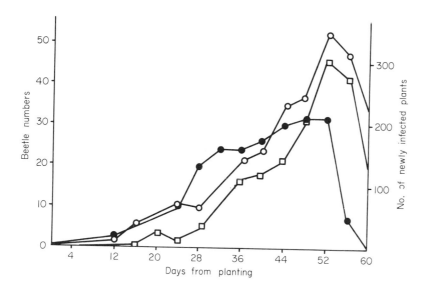

Figure 2. Relation between incidence of CPSMV (□——□) and populations of the beetle vectors *Diabrotica balteata* (O——O) and *Cerotoma
ruficornis rogersi* (●——●) in cowpea monocultures (From Valverde *et
al.*, 1982a).

EFFECT ON YIELD AND CONTROL

In field experiments at Turrialba CPSMV-infected cowpea plants were
categorized according to their growth stage at infection. Yield
reductions of 56-90% were recorded and early infections had the greatest
effect (Valverde *et al.*, 1982b). Similar losses have been reported in
other tropical regions.

Early attempts to control CPSMV by controlling vectors with insecticides were unsuccessful. During the rainy season in Turrialba, no
significant reductions in CPSMV incidence or increases in dry grain
yields were obtained when carbaryl was applied between one and three
times a week from crop emergence. Carbofuran applied at planting
decreased early infection but not the final virus incidence (Shannon &
King, 1980).

Polyculture may be a satisfactory alternative means of control. In
the experiments described earlier, maize and beans were planted simultaneously, and there was initially, a more rapid increase of disease in
the diculture (González, 1978; Valverde, 1978). This was because young
maize plants in the inter-spaces of cowpea rows attracted more *D.
balteata* than were attracted to the monoculture. Although *D. balteata*
is a less efficient vector of CPSMV than *C. ruficornis rogersi* this
could be counteracted by larger populations. If cowpeas are inter-
planted as maize approaches maturity, as in the traditional maize-bean

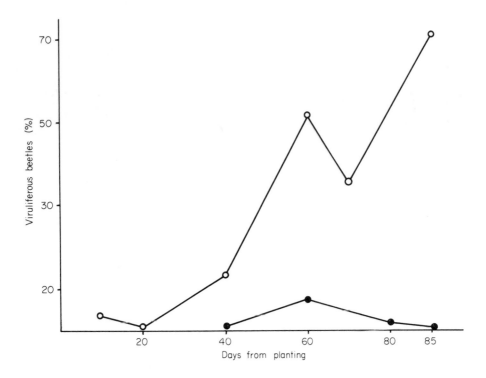

Figure 3. Percentage of *Cerotoma ruficornis rogersi* (O——O) and *Diabrotica balteata* (●——●) collected in cowpea fields and carrying CPSMV (From Valverde, 1978).

system, the deterrent effect of maize on beetle movement and reproduction could result in a more consistent reduction in virus spread. CPSMV can also be decreased by inter-planting with taller crops such as cassava (*Manihot esculenta*) and plantain (*Musa* sp.). These reduce the total populations of *D. balteata* and *C. ruficornis rogersi* compared with cowpea monoculture probably because of the strong shading effect (Araujo & Moreno, 1979).

In experiments on the effect of barrier crops on the spread of CPSMV (Shannon & King, 1980) cassava and bean decreased both final virus incidence and rate of increase (Fig. 4).

Cowpea cultivars differ in their susceptibility to CPSMV (González, 1978). The use of tolerant cultivars and appropriate polycultures or barrier crops seem to be promising means of controlling CPSMV.

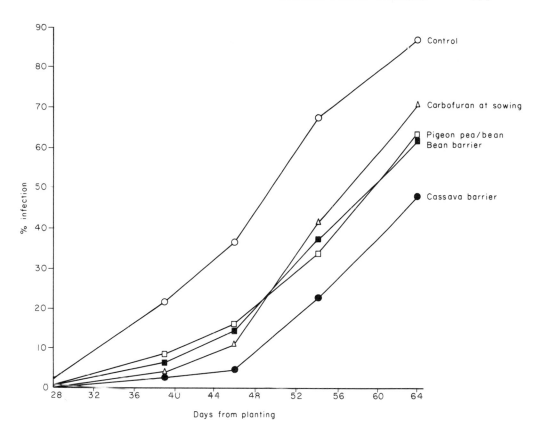

Figure 4. The spread of CPSMV in cowpea as a monoculture (control) (O——O), with carbofuran applied to the seed at sowing (Δ——Δ) and with 50 cm barriers of common beans (■——■), weeds/pigeon pea (□——□) or cassava (●——●) (From Shannon & King, 1980).

Acknowledgements

This work was supported in part by grants from the Consejo Nacional de Investigaciones Científicas y Technológicas (CONICIT) and the Vicerrectoría de Investigación de la Universidad de Costa Rica. Rodrigo Gámez is a research fellow of CONICIT. We are grateful to C. González, R. Valverde, E. Araujo, H. Hobbs and A. King for their interest and participation in this study, and to J. Wolley for his comments on the manuscript.

REFERENCES

Araujo E. & Moreno R. (1979) Disseminação de doenças foliares do
    feijão macassar (Vigna unguiculata (L.) Walp.) en diferentes sistemas
    de cultivos. I. Viroses. Fitopatología Brasileira 4, 281-91.
Cartín F. & Gámez R. (1972) Transmission of two new strains of bean
    rugose mosaic virus by chrysomelid beetles. Abstracts, XIII Annual
    Meeting, Caribbean Division, American Phytopathological Society, p. 12.
De Jager D.P. (1979) Cowpea severe mosaic virus. CMI/AAB Descriptions
    of Plant Viruses No. 209, 4 pp.
Díaz A.J. (1972) Estudio y caracterización de un mosaico del frijol de
    costa (Vigna sinensis) en El Salvador. Phytopathology 62, 754
    (Abstract).
Echandi E. (1975) Bean (Phaseolus vulgaris) diseases in the tropical
    Americas. Tropical Diseases of Legumes (Ed. by J. Bird & K.
    Maramorosch), pp. 165-6. Academic Press, New York.
Fulton J.P. & Scott H.A. (1977) Bean rugose mosaic and related viruses
    and their transmission by beetles. Fitopatología Brasileira 2, 9-16.
Fulton J.P., Scott H.A. & Gámez R. (1980) Beetles. Vectors of Plant
    Pathogens (Ed. by K.F. Harris & K. Maramorosch), pp. 115-32.
    Academic Press, New York.
Gálvez G.E., Cardenas M., Kitajima E.W., Díaz-Chávez A.J. & Nieto M.P.
    (1977) Purification, serology, electron microscopy and properties
    of the ampollado strain of bean rugose mosaic virus. Turrialba 27,
    343-50.
Gámez R. (1972) Los virus del frijol en Centro América. II. Algunas
    propiedades y transmisión por crisomélidos del virus del mosaico
    rugoso del frijol. Turrialba 22, 249-57.
Gámez R. (1973) Goals and means for protecting Phaseolus vulgaris in
    the tropics. Potential of Field Beans and Other Food Legumes in
    Latin America. Series Seminars, Centro Internacional de Agricultura
    Tropical, Cali, Colombia No. 2E, pp. 233-6.
Gámez R. (1976) Los virus del frijol en Centro América. IV. Algunas
    propiedades y transmisión por insectos crisomélidos del virus del
    moteado amarillo del frijol. Turrialba 26, 160-6.
Gámez R. (1980) Virus transmitidos por crisomélidos. Problemas de
    Producción de Frijol (Ed. by H.F. Schwartz & G. Gálvez), pp. 239-
    59. Centro Internacional de Agricultura Tropical, Cali.
Gámez R. (1982) Bean rugose mosaic virus. CMI/AAB Descriptions of
    Plant Viruses No. 246, 4 pp.
González C.E. (1978) Identidad, transmisión por insectos crisomélidos
    y epifitiología de virus del frijol de costa (Vigna unguiculata (L)
    Walp.) en Costa Rica. Thesis, Universidad de Costa Rica.
González C.E., Moreno R. & Gámez R. (1975) Identidad, incidencia y
    distribución de virus del frijol de costa (Vigna sinensis) en Costa
    Rica. Abstracts, Annual Meeting, Caribbean Division, American
    Phytopathological Society, pp. 44-5. Centro Internacional de
    Agricultura Tropical, Cali.
Granillo C.R., Díaz A., Anaya M. & Jiménez G.E. (1975) A new virus
    disease of beans transmitted by chrysomelid beetles. Tropical
    Diseases of Legumes (Ed. by J. Bird & K. Maramorosch), pp. 115-7.
    Academic Press, New York.

Holdridge L.R. (1978) *Ecología Basada en Zonas de Vida*. Instituto Interamericano de Ciencias Agrícolas, San José, Costa Rica.

Meiners J.P., Waterworth B.E., Lawson R.H. & Smith F.F. (1977) Curly dwarf mosaic disease of beans from El Salvador. *Phytopathology* 67, 163-8.

Murillo J.I. (1967) Estudio sobre dos aislamientos virosos del frijol en Costa Rica. *XIII Reunión Anual, Programa Cooperativo Centroamericano para el Mejoramiento de Cultivos Alimenticios*, pp. 52-5. San José, Costa Rica.

Risch S.J. (1979) Effect of plant diversity on the population dynamics of several species of beetle pests in monocultures and polycultures of corn, beans and squash in Costa Rica. *PhD Thesis, Michigan University*.

Shannon P. & King A. (1980) Cropping systems entomology, Costa Rica. *Progress Report 1978-1979*. Centro Agronómico Tropical de Investigación y Enseñanza, Costa Rica/Overseas Development Agency, UK.

Valverde R. (1978) Epifitiología e importancia agronómica del virus del mosaico del frijol de costa (*Vigna unguiculata* (L.) Walp.) *Thesis, Universidad de Costa Rica*.

Valverde R., Moreno R. & Gámez R. (1978) Beetle vectors of cowpea mosaic virus in Costa Rica. *Turrialba* 28, 90-1.

Valverde R., Moreno R. & Gámez R. (1982a) Incidence and some ecological aspects of cowpea severe mosaic virus in two cropping systems in Costa Rica. *Turrialba* 32, 29-32.

Valverde R., Moreno R. & Gámez R. (1982b) Yield reductions in cowpea (*Vigna unguiculata*) infected with cowpea mosaic virus in Costa Rica. 32, 89-90.

Vanderplank J.E. (1946) A method for estimating the number of random groups of adjacent diseased plants in homogeneous fields. *Transcriptions of the Royal Society of South Africa* 31, 269-78.

Waterworth H.E., Meiners J.P., Lawson R.H. & Smith F.F. (1977) Purification and properties of a virus from El Salvador that causes mild mosaic in bean cultivars. *Phytopathology* 67, 169-73.

The biology of the principal aphid virus vectors

V.F. EASTOP
British Museum (Natural History), Cromwell Road,
London SW7 5BD, UK

INTRODUCTION

Aphids transmit some of the most important viruses of crop plants. More than 200 species have been shown experimentally to transmit viruses and undoubtedly many others as yet untested have this ability. Many aphid species are pests in their own right, damaging the plant by removing sap or injecting toxic saliva, but this paper summarizes the biologies of only 25 species of particular importance as virus vectors. They were selected either because they transmit a number of different viruses, or because of the damage caused by a particular virus to an important crop. Only brief biological information concerning virus transmission in the field is included as a book could be written on any one economically important species. For example, a description of the rose aphid, *Macrosiphum rosae*, by Mordvilko (1919) and an account of the ecology of *Myzus persicae* by van Emden *et al.* (1968) each occupy 94 pages.

In this chapter a brief summary is given of the biology of each species, and where possible, a recent "key" reference.

TAXONOMY

Most of the 228 species of aphids recorded as virus vectors belong to the subfamily Aphidinae, which includes the genera *Aphis, Myzus* and *Macrosiphum* (Eastop, 1977). The Aphidinae have undergone adaptive radiation since the Cretaceous period in association with the radiation of herbaceous angiosperms. Many of the other aphid subfamilies are associated with trees. At first sight Aphidinae appear pre-eminent as virus vectors (Table 1) and they are probably most important in the field, but not necessarily because of any intrinsic superiority as vectors. The viruses of herbs have in general been more intensively studied than those of trees. Consequently, tree-feeding aphids have more often been tested as vectors of viruses infecting herbaceous plants than herb-feeding aphids have been tested as vectors of viruses infecting tree species.

Plumb R.T. & Thresh J.M. (1983) *Plant Virus Epidemiology.*
Blackwell Scientific Publications, Oxford.

Table 1. Vector species of Aphidoidea by families and subfamilies.

Family and subfamily	World species	No. of vector species /No. tested
Aphididae		
Lachninae	347	1/4
Chaitophorinae	141	6/14
Drepanosiphinae	446	14/20
Aphidinae	2229	200/236
Greenideinae	127	0/0
Phloemyzinae	1	0/0
Anoeciinae	32	0/3
Hormaphidinae	171	0/0
Pemphiginae	226	4/9
Adelgidae	57	1/1
Phylloxeridae	69	1/1
Total	3846	227/288

Aphidinae are probably more important than the other subfamilies because their winged forms fly in large numbers through young crops. Aphidinae commonly outnumber the total of all other subfamilies in trap catches by a ratio of 3:1 (Johnson & Eastop, 1951; Eastop, 1958a) and the ratio may exceed 20:1 (Eastop, 1957; Hughes *et al.*, 1964). Similarly, within the subfamily Aphidinae, a few species may be so abundant that even if relatively inefficient, they would be the principal vectors. For instance, the large flights of *Brachycaudus helichrysi* in western Europe and of *Aphis citricola* (Raccah, this volume) in the Mediterranean may result in considerable virus spread in crops they do not colonize.

All the parthenogenetic individuals of many species of the Drepanosiphinae (= Callaphididae, Callipterinae) are winged, but few are common in arable crops. The genus *Therioaphis* is an exception as apterous viviparae are commonly produced, several species feed on herbaceous legumes and vagrant alatae are often found in other crops. The other two main subfamilies, Lachninae and Pemphiginae, may only produce one or two generations of alatae each year and although large numbers of pemphigine sexuparae may fly in the autumn, this is likely to be too late in the growing season to be of much importance in virus transmission.

BIOLOGY

Most aphids have similar life-cycles. Winter is spent as eggs that hatch to produce parthenogenetic females, known as fundatrices. Those

of most subfamilies are apterous, have short legs, antennae, etc. and
are very fecund. Their progeny are also parthenogenetic females that
may be either winged or wingless, depending largely on the species but
sometimes partly on environment. Commonly a succession of wingless
parthenogenetic generations is terminated by the production of many
winged parthenogenetic females which found new colonies of apterous
parthenogenetic females. In autumn males and sexually reproducing
females are produced, which mate to give rise to the overwintering eggs.
For most species of aphids the whole cycle occurs on the same species of
plant, or on a closely related group of plant species. However, about
10% of aphid species alternate between different hosts. The sexual
generation produces overwintering eggs, usually on a woody primary host
from which the spring migrants fly to a secondary host, usually a herb,
on which parthenogenetic generations occur in the summer. Ten of the 25
virus vectors discussed in this paper alternate between primary and
secondary hosts in at least part of their geographic range.

Commonly each aphid has only one species of primary host plant, and
when there are two or more species they are closely related. Notable
exceptions are *Aphis citricola* with *Citrus* (Rutaceae) and *Spiraea*
(Rosaceae); *A. fabae* with *Euonymus* (Celastraceae), *Viburnum* (Caprifoliaceae) and *Philadelphus* (Philadelphaceae), and *A. gossypii* with *Rhamnus*
(Rhamnaceae) and *Catalpa* (Bignoniaceae). Each of these three exceptions
may have different underlying causes. The populations called *A.
gossypii* on Bignoniaceae and Rhamnaceae may be different species. The
A. fabae group is difficult taxonomically and may have a history of
hybridization. *A. citricola* is similar to *A. pomi* and *Spiraea* is
probably the original primary host, *Citrus* having been acquired second-
arily.

In a few species, particularly in the Arctic, there are only three
generations each year, the fundatrices being the grandparents of the
sexuales, which are often produced during the long photoperiods around
midsummer. In many aphids, however, the production of sexuales is
triggered by the lengthening nights in late summer and autumn, provided
that the temperature also falls below a critical threshold.

In temperate climates, most species of aphid have a single, annual,
sexual generation resulting in the production of cold-resistant, overwintering eggs. In some species a final autumn generation consists
entirely of sexuales, but in other species part of the population may
continue to reproduce parthenogenetically through the winter, e.g.
M. persicae (Blackman, 1972). Where the winter is sufficiently mild,
as in the maritime climate of the Atlantic seaboard of western Europe,
parthenogenetically reproducing populations of such aphids may persist
on plants for as long as foliage is available. In such circumstances
many of the alatae leaving virus-infected biennials or perennials, such
as beet seed crops or grasses, in spring may carry virus. Cockbain &
Heathcote (1965) showed that *Myzus persicae* was a better vector from
the beet on which they developed of persistently transmitted than of
non-persistently transmitted viruses.

Some aphid species and populations have completely lost the ability to produce sexuales and, consequently, overwintering eggs. Aphids with a variable life cycle, such as *M. persicae*, may produce a few males in autumn which can contribute to the "gene pool", overwintering as sexually produced eggs. The ability to overwinter parthenogenetically can thus be inherited through the sexual phase. This ensures that both types of overwintering behaviour can appear in subsequent generations, even after a very cold winter which would kill individuals attempting to overwinter parthenogenetically, or if the winter was not sufficiently cold to break diapause of the eggs. Such alternative strategies are of particular importance in variable, cool temperate climates such as that of Britain (Blackman, 1976). Some agricultural practices may facilitate parthenogenetic overwintering. Heathcote & Cockbain (1966) investigated the role as virus vectors of aphids overwintering parthenogenetically on mangel roots stored in clamps. Heathcote et al. (1965) considered weeds as a source of beet viruses, and the role of weeds in the biology of pests including aphids is the subject of a recent book (Thresh, 1981).

In the tropics, reproduction may be entirely parthenogenetic and parthenogenetically reproducing populations can be found throughout the year on plants in suitable condition. Polyphagous aphid species and those of Gramineae are more likely to find overwintering hosts in a suitable condition than aphids specific to particular annuals, deciduous shrubs or trees.

In the dry tropics, aphids are seldom found during the hottest season. When temperatures increase suddenly aphids may be killed on the plants, but more often their fecundity declines to very low levels at high temperatures. Each autumn the area is repopulated, but to what extent this represents recolonization and to what extent increase from a few resident individuals that survived in particularly favoured conditions is not known, and is likely to vary between places and between seasons. The uncertainty could perhaps be resolved by electrophoretic characterization of populations of a few common species in successive years. Extended irrigation creates crop protection problems in many areas as virus-infected weeds and crop plants growing as "volunteers" on the shaded banks of irrigation ditches survive the dry season to provide local sources of virus, and also support populations of aphid vectors. Populations remaining on crops and weeds throughout the year will be more exposed to pesticides than those moving to wild hosts. Thus more of each year's population is likely to have originated from individuals that survived applications the previous year. Pesticide resistance is thus likely to develop faster in areas of large-scale irrigation than where no irrigation is practised. Many other aspects of aphid biology which affect virus transmission are considered in Harris & Maramorosch (1977).

The geographical origin of aphid pests can be deduced in a number of ways. In the case of the more host-specific aphids, the natural distribution of their host plants is likely to indicate their geographical origin. During the last 200 years the introduction of some pests has been well-documented, e.g. the introduction to North America of the palaearctic, *Acyrthosiphon pisum*, *Schizaphis graminum* and *Therioaphis trifolii*, and the introduction to Europe from America of *Eriosoma lanigerum*. However, in the last case this is not proof that the aphid originated in America, as it could previously have been introduced to America from the Far East. A number of pests, e.g. *Acyrthosiphon kondoi*, *A. pisum*, *Aphis citricola*, *Hysteroneura setariae*, *Sitobion avenae* and *S. fragariae* have extended their range in recent years. This is well-documented in the taxonomic and economic literature and trap catch records, and it is also evident from museum collections.

Often only a small part of the gene pool of a species in its region of origin is represented in the introduced population (the "Founder effect"). The introduced genotype(s) are frequently anholocyclic, i.e. able to survive the winter without a sexual phase. For instance, *Macrosiphum euphorbiae* until recently overwintered only parthenogeneti-cally in Europe, although a sexual generation on rose has long been known from North America. The absence of *M. euphorbiae* from European collections before 1917 and its abundance in collections after that date also indicate that it is an introduction (Eastop, 1968b). The sexuales of *Toxoptera* spp. are known only from the Far East and the males of *Rhopalosiphum maidis* are more common in the Far East than elsewhere, suggesting an oriental origin for those aphids. Other circumstantial evidence of the origin of a pest aphid may be provided by the distribu-tion of its close relatives. In some aphid genera the species living on wild plants may occur only in particular geographical regions, e.g. *Lipaphis* spp. in the palaearctic region and *Myzus* spp. in the palae-arctic and northern oriental region. The *Aphis fabae* group appears to be of Old World origin while the *A. helianthi* group occupies a similar niche in the New World. The genus *Acyrthosiphon* is predominantly palaearctic: the few American species are atypical in morphology and/or biology and, except perhaps for the circumpolar species, may be wrongly placed generically.

Aphids transferred from one continent to another may not only lack an appreciable part of their genetic variation but may also be intro-duced without their natural enemies. Hence the deliberate introduction of natural enemies from climatically comparable parts of the pest's natural geographical range is a common method used in attempts at biological control.

There are also records of introduced aphids flourishing on native plants and of introduced plants suffering from native aphids. For instance *Nasonovia ribisnigri* colonizes the sticky flower heads of *Lactuca* and related Compositae in Europe but will also colonize the sticky flower heads of introduced species of *Nicotiana* and *Petunia* (Solanaceae) and *Martynia* (Martyniaceae). The resources devoted by the latter genera to producing sticky hairs may have been at the expense of chemical defences, making the plants susceptible to any insect capable of overcoming the physical defences.

APHID VIRUS VECTORS

Acyrthosiphon pisum: The pea aphid.   Commonwealth Institute of
                         Entomology (CIE) map A23 (1952).

A. pisum is recorded as a vector of more than 30 viruses including non-
persistently transmitted viruses of beans, peas, beet, clover, cucurbits
narcissi and crucifers and persistently transmitted viruses of peas and
beans.   *A. pisum* is a complex of populations of unknown inter-fertility.
Colonies occur on many herbaceous Papilionaceae and also on *Sarothamnus*
and a few leguminous shrubs.   Host alternation in the strict sense does
not occur, but at least some populations can colonize members of several
genera of Papilionaceae.   Both apterous and alate males are known and
some populations probably overwinter parthenogenetically in mild
climates.   The pea aphid is probably of European origin, but is of
greater economic importance in North America and some of the other
regions to which it has been introduced, even though only part of the
"gene pool" of the species has yet been transferred.   Green, pink and
yellow populations occur in Europe, some with evident host preferences,
but only green forms are recorded from America.   Some of the earlier
records of *A. pisum* from Australia were based on misidentified
Macrosiphum euphorbiae, but a member of the *A. pisum* group occurs on
Cytisus in Tasmania.   *A. pisum* is not known from the Australian main-
land.   Lucerne-feeding *A. pisum* was first recorded in New Zealand in
1977.   Blackman (1981) discusses the genetical implications of the
races and strains described from various regions and Eastop (1971)
refers to part of the enormous literature on the biology of *A. pisum*.
Only one record is known of resistance to insecticides, to organophosphorus in North America (Anon., 1967).

Amphorophora agathonica: The American large red raspberry aphid.

A. agathonica is a vector of several non-persistently transmitted
viruses including raspberry (black) necrosis and rubus yellow net
viruses.   It remains throughout the year and overwinters as eggs on
Rubus idaeus ssp. *strigosus* and raspberry cultivars probably derived
from it.   It is a North American species, widespread in the central and
northern United States of America and southern Canada, where it was
earlier confused with, and referred to as, the European species *A. rubi*.
Kennedy & Schaeffers (1974a,b) and Daubeny (1978a,b) describe the
biology of *A. agathonica* in New York, and its behaviour on resistant
raspberry cultivars.

Amphorophora idaei: The large raspberry aphid.

A. idaei is the only known vector of raspberry (black) necrosis virus
in Europe.   The aphid is known only from *Rubus idaeus sensu stricto*
and cultivars derived from it, in central and northern Europe where it
overwinters as eggs.   Until recently it was often confused with *A. rubi*
which lives on brambles in Europe.   Blackman (1981) discussed the
genetics of host resistance and Rautapää (1967) considered the biology
of *A. idaei*.

Aphis citricola: The green citrus aphid, Spiraea aphid. CIE map A256
                     (1969).

A. citricola, better known until recently as *A. spiraecola*, is a vector
of several non-persistently transmitted viruses, including bean common
mosaic, beet mosaic, at least one strain of citrus tristeza, papaw distortion mosaic, plum pox, potato virus Y and tobacco etch viruses. *A.
citricola* damages the young growth of *Citrus* and particularly the loose-
skinned cultivars of orange. It also occurs on apple (where it has been
confused with *A. pomi*), pear, *Cydonia, Rhaphiolepis, Viburnum, Eupator-
ium* and some other Compositae, and sporadically on the young growth of
many other shrubs and herbaceous perennials. *A. citricola* overwinters
as eggs on both *Citrus* and *Spiraea* and is probably of Far Eastern origin.
It is permanently parthenogenetic in South-East Asia and other tropical
areas to which it has been introduced. It has been in North America at
least since 1907, and was introduced more recently to the Mediterranean
region (*c*. 1939), Africa (1961), Australia (1926) and New Zealand (1931).
Barbagallo (1967), Micieli de Biaze (1970) and Heinze (in Kranz *et al.*,
1977) present accounts of *A. citricola*. Asakawa (1975) records this
species as resistant to insecticides in Japan.

Aphis craccivora: The groundnut (peanut), cowpea or black legume aphid,
                      CIE map A99 (1959).

A. craccivora transmits about 20 viruses non-persistently, including
broad bean mosaic, bean common mosaic, beet mosaic, cabbage black ring
spot, cowpea mosaic, cucumber mosaic, alfalfa (lucerne) mosaic, papaw
distortion mosaic and pea mosaic viruses, and the persistently transmitted subterranean clover stunt, groundnut rosette, groundnut mottle,
and millet red leaf viruses. *A. craccivora* is widespread in warm temperate, subtropical and tropical regions. It lives mainly on Leguminosae
but, especially under drought conditions, will colonize irrigated crops
or succulent members of other families. Over much of the world its
reproduction is entirely parthenogenetic but it does reproduce sexually
in central Europe. A population with an unusual karyotype is known
from lupin in Iran and populations with differing abilities as virus
vectors are recorded from Africa. Hoffmann (1968, 1972) gives an account
of the aphid in Europe and Müller (1977), Heinze (in Kranz *et al.*, 1977)
and Eastop (1966) refer to accounts from other regions.

Aphis fabae: The black bean aphid, dolphin. CIE map A174 (1963).

A. fabae transmits about 35 viruses. Those transmitted non-persistently
include potato virus Y and viruses causing mosaics of beet, cucumber,
pea and soyabean. The persistently transmitted viruses include beet
yellow net and potato leaf roll.

  A. fabae feeds on diverse herbaceous plants during the summer, including broad bean and sugarbeet and it can overwinter as eggs on *Euonymus
europaeus* (Celastraceae), *Viburnum* (Caprifoliaceae) and *Philadelphus*
(Philadelphaceae). *A. fabae* is the name applied to a group of closely
related species, the taxonomy of which is still confused. The occur-

rence of primary hosts in three distantly related families may indicate
earlier hybridization. Host-alternation between *E. europaeus* and broad
bean is well documented, but many other plants are colonized and
Chenopodium is a favoured weed host in mid and late summer. Resistance
to dimethoate is reported from Czechoslovakia (Hůrkova & Hlináková,
1977). Eastop (in Kranz *et al.*, 1977) refers to recent accounts of *A.
fabae*.

Aphis gossypii: The melon and cotton aphid. CIE map A18 (revised 1968).

A. gossypii is the vector of about 50 non-persistently transmitted
viruses of abaca, banana, *Phaseolus* beans, beet, Cruciferae, cowpea,
cucurbits, papaw, pea, pepper, potato, radish, soyabean, strawberry,
sugarcane, sweet potato, tobacco and tomato. *A. gossypii* is also
recorded as a vector of seven persistently transmitted viruses including
cotton anthocyanosis, a strain of groundnut rosette, and lily rosette.
The name *A. gossypii* is currently applied to aphids from many hosts and
may include several taxonomic entities. Large populations develop on
cotton, hibiscus, cucurbits, on many other herbs and on the young growth
of shrubs. Members of this complex can overwinter as eggs on both
Rhamnaceae and Bignoniaceae. There are also records of overwintering
eggs on *Citrus* in the Orient. The populations overwintering on Rhamnaceae are probably European in origin, but an oriental or North
American origin is possible for those overwintering on *Catalpa*. The
taxonomic confusion in the group precludes a precise account of the
biology, but permanently parthenogenetic populations with a wide host
range occur in the tropics. Members of the complex are now found in all
except the coldest regions of the world. Börner & Heinze (1957) give
many references to *A. gossypii* and Thomas (1968), Tamaki & Allen (1969)
and Eastop (in Kranz *et al.*, 1977) have also published accounts. The
records of insecticide resistance in North and South America, Germany
and China are summarized in Georghiou (in press).

Aphis nasturtii: The buckthorn-potato aphid.

A. nasturtii is recorded as a vector of nine non-persistently transmitted viruses, including beet mosaic and cabbage black ringspot and
potato viruses A, Y and aucuba mosaic, and the persistently transmitted
potato leaf roll virus. *A. nasturtii* occurs on potato, watercress,
Polygonum and *Rumex* and overwinters as eggs on *Rhamnus cathartica*. The
summer host range is probably larger than indicated but confusion with
similar species is possible. *A. nasturtii* is European in origin, but
has been introduced to North America and probably to the Kenya highlands.
It probably occurs naturally in the Middle East and in the foothills of
the Himalayas, but again there may be confusion with similar species.
Shands & Simpson (1971) describe the biology of *A. nasturtii*.

Aulacorthum solani: The glasshouse-potato aphid. CIE map 86 (1958).

A. solani transmits about 40 viruses including both persistently and
non-persistently transmitted viruses of beet and potato. *A. solani* is
one of the most polyphagous aphids, and shows little preference for

particular plant taxa. It occurs mostly on herbs in cool and often
moist places, or on the young growth of shrubs. It can overwinter as
eggs on many different species. In mild and warm climates most populations reproduce parthenogenetically. *A. solani* is probably of European
origin but is now more widespread in agricultural areas than indicated
in the 1958 map, occurring in all but the hottest and driest places.
Wave *et al.* (1965) and Müller *et al.* (1973) have given accounts of *A.
solani*. The report of insecticide resistance from the United Kingdom
needs confirmation.

Brevicoryne brassicae: The cabbage aphid. CIE map A37 (1954).

B. brassicae transmits about 15 viruses non-persistently, including
cabbage black ringspot, cauliflower mosaic and watercress mosaic viruses.
It is also recorded as transmitting cauliflower mosaic virus in a persistent manner (Chalfant & Chapman, 1962). *B. brassicae* feeds on many
genera in the tribe Brassiceae and on some other genera of Cruciferae.
It sometimes also occurs on other plants that contain mustard oils such
as *Tropaeolum* (Tropaeolaceae). *B. brassicae* is probably of palaearctic
origin, and has been known in England since mediaeval times. It is now
widespread in agricultural areas where it overwinters as eggs on
crucifers, except in warm climates where it is permanently parthenogenetic. Dunn & Kempton (1972) found evidence of genetic variation in host-
colonizing abilities. Daiber (1970, 1971a,b) gives accounts of *B.
brassicae* in South Africa. A biotype has been reported feeding on
previously aphid-resistant rape in New Zealand (Lammerink, 1968).

Cavariella aegopodii: The willow-carrot aphid.

C. aegopodii transmits seven viruses non-persistently, parsnip yellow
fleck virus which is semi-persistently transmitted and four persistently
transmitted viruses of Umbelliferae including carrot mottle, carrot red
leaf and parsnip mottle viruses. *C. aegopodii* occurs on many different
Umbelliferae in the summer. Permanently parthenogenetic populations may
occur in some places but *C. aegopodii* overwinters as eggs on *Salix* spp.
in Europe, southern Australia and New Zealand. It is probably of
palaearctic origin, having been introduced to North and South America,
Australia and New Zealand, and also occurs in the more temperate parts
of Africa and Asia. Its importance as a vector of carrot viruses in
Australia may have been partly due to the absence of parasites. Börner
& Heinze (1957) give references to *C. aegopodii*.

Chaetosiphon fragaefolii: The strawberry aphid.

C. fragaefolii transmits four strawberry viruses persistently and two
non-persistently. The species occurs only on cultivated strawberries
in Europe, temperate Africa and Asia, Australia and New Zealand. It
probably originated in North America where related taxa, referred to as
C. thomasi, occur on strawberries, *Potentilla* and *Rosa*. The *C.
fragaefolii* known from cultivated strawberries elsewhere may represent
only part of the genome of the more variable species occurring in North
America. This is suggested by the occurrence of sexuales which fail to

produce fertile eggs in Europe, where overwintering is parthenogenetic. Resistance to endosulfan developed during 1960-1964 in Washington State (Shands, 1967). Börner & Heinze (1957) give references to *C. fragaefolii*.

Hyperomyzus lactucae: The currant-sowthistle aphid.

H. lactucae transmits 10 viruses non-persistently and lettuce necrotic yellows virus persistently. It lives only on *Sonchus* species in the summer and overwinters as eggs on *Ribes nigrum* and occasionally on other *Ribes* spp. It is probably permanently parthenogenetic on *Sonchus* in warmer regions and at low latitudes where *Ribes* is not grown (Martin, this volume). It is palaearctic in origin but is now widespread in Europe, Middle East, African highlands, Himalayas, Australia, New Zealand, North and South America, and Japan. Lettuce necrotic yellows virus may originally have been transmitted to *Sonchus* by *H. carduellinus* which also colonizes Australian and South-East Asian species belonging to several other genera of Liguliflorae.

Lipaphis erysimi: The American turnip aphid. CIE map 203 (1965).

L. erysimi was formerly known as *Rhopalosiphum pseudobrassicae*. It transmits 10 viruses non-persistently including cabbage black ringspot and cauliflower mosaic viruses. *Lipaphis* is a small, palaearctic genus associated with Cruciferae. In cool climates *L. erysimi* lives on wild members of a few genera of *Cruciferae*, including *Capsella*, *Cardamine*, *Erysimum* and *Sisymbrium* on which it overwinters as eggs. In warmer places cultivated brassicas, radish and stocks (*Matthiola*) are colonized by permanently parthenogenetic populations. Some variation in karyotype, has been observed (Blackman, 1980). As with several other crucifer-feeding insects, it also colonizes members of other families containing mustard oil, e.g. *Gynandropsis gynandra* (= *pentaphylla*) of the Capparaceae (= Capparidaceae, including Cleomaceae). Börner & Heinze (1957), Lal (in Kranz *et al.* 1977), and Eastop (1966) refer to the principal accounts of this aphid which is regularly caught in traps in most of the warmer parts of the world where it is a pest of mustard and other crucifers.

Macrosiphum euphorbiae: The potato aphid. CIE map A44 (1954).

M. euphorbiae (= *solanifolii*), transmits more than 50 viruses, including non-persistently transmitted viruses of beans, beet, crucifers, clover, lucerne, pea, papaw, pepper, potato, sweet potato, tobacco, chrysanthemum, dahlia, daffodil, freesia, tulip, and persistently transmitted viruses of beet, pea, potato and tomato. *M. euphorbiae* may overwinter as eggs on rose, but other populations are permanently parthenogenetic on herbs. It is a polyphagous aphid (Eastop, 1981) occurring on members of several families of both monocotyledonous and dicotyledonous plants. The species probably originated in America, and it was unknown from Europe before 1917, and probably only part of the genome has yet been introduced to Europe and other temperate and subtropical regions. Before 1973 only parthenogenetic overwintering was known from England but the eggs of a new, holocyclic genotype may have been introduced on rose around 1973 when another North American aphid, *Wahlgreniella nervata*, was also found for the first time overwintering on rose in England. The distribution is now wider than the 1954 map indicates and includes most places where temperate and subtropical crops are grown. Heinze (in Kranz *et al.*, 1977) gives a recent account of *M. euphorbiae* and Börner & Heinze (1957) refer to many of the accounts of this aphid (as *M. solani* Kittel). The report of dimethoate resistance in the United Kingdom requires confirmation.

Metopolophium dirhodum: The rose-grain aphid.

M. dirhodum transmits barley yellow dwarf virus persistently. It over-winters as eggs on rose and the summer forms occur on many Gramineae including oat, barley and wheat. The species can also overwinter parthenogenetically on Gramineae; *Bromus carinatus* and *B. catharticus* are particularly favoured hosts. *M. dirhodum* is probably palaearctic in origin and is widespread in Europe and the Middle East, as far east as Himachal Pradesh. It now also occurs in Africa and North and South America. Some outbreaks of *M. dirhodum* in South America and South Africa have wrongly been attributed to the "green bug", *Schizaphis graminum*. Henderson & Perry (1978), Jones (1980) and Watt (1979) des-cribe aspects of the biology of *M. dirhodum*.

Myzus persicae: The peach-potato aphid. CIE map 45 (revised 1979).

M. persicae transmits about 120 plant viruses, including many non-persistently transmitted viruses and about a dozen that are persistently transmitted, including beet mild yellowing, beet yellow net, pea enation mosaic, pea leaf roll, potato leaf roll, radish yellows and tobacco vein distorting and yellow vein banding viruses.

M. persicae occurs on diverse herbaceous plants in late spring and early summer, but can be difficult to find in mid and late summer, except on irrigated crops. It overwinters as eggs on peach, (*Prunus persica*) and, more rarely, on a few other *Prunus* spp. Reproduction is entirely parthenogenetic in many parts of the world where peach is absent, and where the climate permits active stages to survive the coldest season. In hot regions (e.g. the Middle East) *M. persicae* is often common in the "winter" but absent during the hottest season. It is assumed that each autumn these areas are recolonized, often from afar, by populations surviving in cooler habitats at high altitudes. However, local survival in moist, shaded positions may be more important than currently suspected.

M. persicae is probably of western oriental origin and has been widely distributed by man. It is uncertain how long *M. persicae* has been in South America. The worldwide distribution of 9 aphid-trans-mitted viruses of potato, which is a South American member of the Solanaceae, is surprising as no native South American aphid is known from Solanaceae. At least some of the aphid-transmitted viruses of potato may have originated in a different plant species in Europe.

Several distinctive karyotypes of *M. persicae* are recorded from Chile and California (Blackman, 1980), so the aphid may have been present in the Americas before the European discovery of North America in the 16th century. However, these karyotypes may have been overlooked in the Old World or may have originated recently in the New. Morphometrically the American aphids of unusual karyotype cannot be differentiated from European specimens. Any population of *M. persicae* occurring in South America prior to the introduction of peach is likely to have been entirely parthenogenetic as there is no evidence that *M. persicae* reproduces sexually on native South American species, although it uses *Prunus nigra* as a primary host in North America. If the aphid-transmitted viruses of potato evolved in potato then they did so relatively recently, before the sexually reproducing palaearctic aphid populations diverged morphologically from the South American populations. Alternatively the viruses relied for transmission on the small native aphid fauna none of which is known from Solanaceae.

The ecology of *Myzus persicae* is discussed in van Emden *et al.* (1969) Heinze (in Kranz *et al.*, 1977) gives a more recent brief account and resistance to insecticides is summarized in Georghiou (in press).

Pentalonia nigronervosa: The banana aphid. CIE map A42 (1968).

P. nigronervosa transmits abaca bunchy top and banana bunchy top viruses persistently, banana mosaic virus non-persistently, and the unclassified cardamom katte virus. *P. nigronervosa* occurs on monocotyledonous plants of several families, and is principally found on *Heliconia* and *Musa* (Musaceae), *Alocasia, Colocasia, Caladium* (Araceae), *Costus* and *Zingiber* (Zingiberidaceae) and *Palisota* (Commelinaceae). Reproduction is largely, if not entirely, parthenogenetic although sexuales occur. The distribution is almost coextensive with the cultivation of banana. Ilharco (1968) presents an account of *P. nigronervosa*.

Rhopalosiphum maidis: The cereal-leaf aphid. CIE map A67 (1956).

R. maidis transmits seven viruses non-persistently, including abaca mosaic, banana mosaic, cucumber mosaic, onion yellow dwarf, sugarcane mosaic and watermelon mosaic viruses, and also the persistently transmitted barley yellow dwarf and maize leaf fleck viruses. Although it may have a sexual phase (on Rosaceae?) in eastern Asia where males are common and where it probably originated, *R. maidis* is permanently parthenogenetic on many genera of Gramineae over most of its range in the warm temperate, subtropical and tropical regions. A number of "biotypes" have been recognized by their differing host plant preferences and Blackman (1981) has discussed the genetical implications. Populations with structurally different karyotypes occur (Blackman, 1980, and personal communication). Eastop (in Kranz *et al.* 1977) gives a brief account of *R. maidis*.

Rhopalosiphum padi: The bird-cherry aphid. CIE map A288 (1971).

R. padi transmits, non persistently, four viruses of dicotyledonous and

three of monocotyledonous plants which it does not colonize. The
species also transmits barley yellow dwarf and maize leaf fleck viruses
persistently. *R. padi* overwinters as eggs on *Prunus padus*, and more
rarely on other *Prunus* spp., and occurs during the summer on many
Gramineae including oat, barley, wheat and maize. In warm regions
reproduction is entirely parthenogenetic on Gramineae, and occasionally
on other monocotyledonous plants including members of the Cyperaceae,
Typhaceae and even Iridaceae. It is probably originally of palaearctic
origin but is now one of the most widely distributed aphids, being known
from both circumpolar regions and also from tropical mountains.
Specimens from low altitudes in the tropics are often green, while those
developing at low temperatures are a dark olive-brown. Resistance to an
organophosphorus insecticide has been recorded from Portugal (Anon.,
1974). There are accounts of the biology of *R. padi* in central England
(Rogerson, 1947; Leather & Dixon, 1981) and Finland (Rautapää, 1976,
1980). Baker & Turner (1919) considered the summer forms of *R. padi* in
Virginia, USA., under the name *R. prunifoliae*, but the data concerning
specimens on apple may well apply to other species.

Schizaphis graminum: The green bug. CIE map A173 (1963).

S. graminum transmits barley yellow dwarf and millet red leaf viruses
persistently and sugarcane mosaic virus non-persistently. *S. graminum*
occurs on barley, wheat, sorghum and maize and on a number of wild
grasses including *Arundo*, *Eleusine*, *Panicum* and *Phalaris*. The eggs
produced by sexuales in the autumn are laid on Gramineae, principally
Poa pratensis, but over much of the world reproduction is entirely
parthenogenetic. *S. graminum* is of palaearctic origin although now
widespread in the warmer wheat-growing areas of the world. Two
structurally distinct populations with different host plant preferences
were introduced to North America around 1882 and 1968. The history and
genetical implications for agriculture have been discussed by Blackman
(1981). Resistance to some organophosphorus insecticides in the Mid-
West of the USA was recorded by Peters *et al*. (1975). Barbulescu
(1975) describes its biology in Rumania and Heinze (in Kranz *et al*.
(1977) summarizes recent accounts of *S. graminum*.

Sitobion avenae (= *granarium*): The grain aphid (English grain aphid
                                   in the USA)

S. avenae transmits barley yellow dwarf and radish yellows viruses
persistently and bean yellow mosaic and pea mosaic viruses non-persistently. *S. avenae* lives on Gramineae and occasionally on other monocotyledonous plants of Cyperaceae, Juncaceae and Iridaceae with no host
alternation. In continental climates eggs are laid on Gramineae, but
in maritime regions much of the population overwinters parthenogenetically. *S. avenae* is probably of palaearctic origin and is widespread
from Scandinavia to North Africa and from Spain through Turkey to
Iran, Pakistan, Kashmir and Uttar Pradesh. It occurs in Ethiopia and
has recently been introduced to the Kenya highlands, Zimbabwe and South
Africa. It is widespread in both North and South America. Records
from Australasia and from eastern Asia mostly apply to other species,

particularly *S. miscanthi*, and early records from Africa were also based on misidentifications of other species. Rautapää (1976, 1980), Vickerman & Wratten (1979) and Lowe (1981) have given recent accounts of different aspects of its biology. The relative susceptibilities of wheat cultivars have been discussed by Kay *et al.* (1981).

Sitobion miscanthi: The grain aphid (in Australia).

S. miscanthi transmits barley yellow dwarf and millet red leaf viruses persistently. *S. miscanthi* occurs on many Gramineae, including oat, barley, wheat, bamboo and a number of dicotyledenous plants including *Capsella bursa-pastoris* and *Polygonum hydropiper*. It may alternate from *Akebia* in eastern Asia or *S. akebiae* may be a distinct species alternating to Gramineae. *S. miscanthi* is widespread in eastern Asia, including Japan, China, northern India, Nepal, Pakistan, Malaya, Australia, New Zealand, Fiji, Tahiti, Tonga and Hawaii. Early records of "*Macrosiphum granarium*" from Australia and South-East Asia usually apply to *S. miscanthi*. As the numerous *Sitobion* spp. of the Old World tropics have not yet been satisfactorily distinguished from each other, details of their biologies and geographical distribution are equally uncertain. Miyazaki (1971) writes that *S. akebiae* "may produce hibernating eggs on various kinds of plants". David (1976) gives an account of *S. miscanthi* in India.

Toxoptera citricida: The black citrus aphid. CIE map A132 (1961).

T. citricida transmits citrus tristeza, citrus vein enation, abaca mosaic and pea mosaic viruses. It normally lives only on Rutaceae, and is a native of eastern Asia but now occurs in many warm parts of the world. *T. citricida* is widespread in subtropical and warm temperate South America, including Brazil, Uruguay, Argentina, Chile and Peru, and also occurs in Guyana, Surinam and Trinidad. It is also widespread in Africa south of the Sahara, in South-East Asia, Australia and New Zealand. *T. citricida* is, however, absent from North America, the Mediterranean area and the Middle East. Any extension of the range of *T. citricida* in the Caribbean, Central America, or to the Middle East would create a serious threat to the important citrus industries in California, Florida and the Mediterranean. *T. citricida* tolerates cool conditions, occurring up to about 2000 m in the Kenya highlands. It overwinters as eggs in Japan but is permanently parthenogenetic over most of its range. *T. citricida* is toxic to many aphid predators. Tao & Chu (1971) give details of its biology in Taiwan and Heinze (in Kranz *et al.*, 1977) summarizes its world wide importance.

EPILOGUE

Doubts are sometimes cast upon the reality of species. This has arisen from the appreciable gene flow assumed to occur between some populations of plants commonly regarded as distinct species and partly from the instability observed in the relationships between some cereals and their rusts and mildews. Similar variation or instability is manifest by biotypes of insects which develop resistance to pesticides or the ability to colonize previously resistant crop cultivars. There is, however, little evidence for appreciable gene flow between insect species. The nervous systems of animals have led to complex behaviour, courtship rituals etc. which tend to reduce the flow of genes even between closely related species. Taxonomic problems in entomology are more often concerned with the difficulty of finding characters to distinguish biologically distinct species than with any blurring of the boundaries between species caused by gene flow. Clinal variation is likely in widespread species as their distribution is limited by many parameters. Genotypes well adapted to the northern part of a range may be absent from the southern part. Habitats are patchy and populations long isolated from the main gene pool of a species may be regarded as subspecies. In practice it may either be difficult to recognize these subspecies or, if they are easily recognizable, it may be correspondingly difficult to assign them to the correct species. In either case this is a problem of interpreting the observed differences rather than of unstable taxa. There is a reality and stability in the species of most insects and of most vascular plants for periods of time that are significant in relation to agriculture. If this were not so, then host specificity would cease to have any meaning, or perhaps could not exist. In the absence of stability of species there would be no reason to identify either virus vectors or crops!

REFERENCES

Anon. (1967) *Report of the third session of the FAO Working Party of Experts on Resistance of Pests to Pesticides, Rome, 11-18 September 1967.* 20 pp. Food & Agriculture Organization, Rome.

Anon. (1974) *Cereal aphids.* Ministry of Agriculture, Fisheries and Food, Advisory leaflet 586 (revised by K.S. George & R.T. Plumb), 7pp.

Asakawa M. (1975) Insecticide resistance in agricultural insect pests of Japan. *Japan Pesticide Information* 23, 5-8.

Baker A.C. & Turner W.F. (1919) Apple-grain *Aphis. Journal of Agricultural Research* 28, 311-24.

Barbagallo S. (1967) Contributo alla conoscenza degli afide degli agrumi. I. *Aphis spiraecola* Patch. *Bolletino del Laboratorio di Entomologia Agraria "Fillipo Silvestri" di Portici* 24, 49-83.

Barbulescu A. (1975) Duration and number of generations of the cereal greenbug (*Schizaphis graminum* Rond.), *Analele Institutului de Cercetări pentru Cereale și Plante Tehnice, Fundulea* 40, 221-30 (in Romanian).

Blackman R.L. (1972) The inheritance of life-cycle differences in *Myzus persicae* (Sulz.) (Hem., Aphididae). *Bulletin of Entomological Research* 62, 281-94.

Blackman R.L. (1976) The puzzle of the adaptable aphid. *Spectrum* 144, 2-4.

Blackman R.L. (1980) Chromosome numbers in the Aphididae and their taxonomic significance. *Systematic Entomology* 5, 7-25.

Blackman R.L. (1981) Aphid genetics and host plant resistance. *Bulletin SROP* IV/1, 13-9.

130   V.F. EASTOP

Börner C. & Heinze K. (1957) Aphidina - Aphidoidea, *Handbuch der Pflanzenkrankheiten* 5(2) Homoptera II (Ed. by P. Sorauer), pp.1-402. Paul Parey, Berlin.

Chalfant R.B. & Chapman R.K. (1962) Transmission of cabbage viruses A and B by cabbage aphid and the green peach aphid. *Journal of Economic Entomology* 55, 584-90.

Cockbain A.J. & Heathcote G.D. (1965) Transmission of sugar beet viruses in relation to the feeding, probing and flight activity of alate aphids. *International Congress of Entomology, London* 12, 521-3

Daiber C.C. (1970) Cabbage aphids in South Africa: the influence of temperature on their biology. *Phytophylactica* 2, 149-56.

Daiber C.C. (1971) Cabbage aphids in South Africa: their field populations during the year. *Phytophylactica* 3, 15-28.

Daiber C.C. (1971) Cabbage aphids in South Africa: their parasites, predators and disease. *Phytophylactica* 3, 137-46.

Daubeny H.A. (1978a) Chilcotin red raspberry. *Canadian Journal of Plant Science* 58, 279-82.

Daubeny H.A. (1978b) Skeena red raspberry. *Canadian Journal of Plant Science* 58, 565-8.

David S.K. (1976) A taxonomic review of *Macrosiphum* (Homoptera: Aphididae) in India. *Oriental Insects* 9, 461-93.

Dunn J.A. & Kempton D.P.H. (1972) Resistance to attack by *Brevicoryne brassicae* among plants of Brussels Sprouts. *Annals of Applied Biology* 72, 1-11.

Eastop V.F. (1957) The periodicity of aphid flight in East Africa. *Bulletin of Entomological Research* 48, 305-10.

Eastop V.F. (1958a) Flight periodicity of some aphids and psyllids in Nigeria. *Entomologist's Monthly Magazine* 94, 32-3.

Eastop V.F. (1958b) The history of *Macrosiphum euphorbiae* (Thomas) in Europe. *Entomologist* 91, 198-201.

Eastop V.F. (1966) A taxonomic study of Australian Aphidoidea (Homoptera). *Australian Journal of Zoology* 14, 399-592.

Eastop V.F. (1971) Keys for the identification of *Acyrthosiphon* (Hemiptera : Aphididae). *Bulletin of the British Museum (Natural History) Entomology* 26, 3-115.

Eastop V.F. (1977) Worldwide importance of aphids as virus vectors. *Aphids as Virus Vectors* (Ed. by K.F. Harris & K. Maramorosch), pp. 3-62. Academic Press, New York.

Eastop V.F. (1981) The wild hosts of aphid pests. *Pests, Pathogens and Vegetation* (Ed. by J.M. Thresh), pp. 285-309. Pitman, London.

van Emden H.F., Eastop V.F., Hughes R.D. & Way M.J. (1969) The ecology of *Myzus persicae*. *Annual Review of Entomology* 14, 197-270.

Georghiou G.P. (in press) *The Occurrence of Resistance to Pesticides in Arthropods*. Food and Agriculture Organisation, Rome.

Harris K.F. & Maramorosch K. (Eds.) (1977) *Aphids as Virus Vectors*. Academic Press, New York.

Heathcote G.D. & Cockbain A.J. (1966) Aphids from mangold clamps and their importance as vectors of beet viruses. *Annals of Applied Biology* 57, 321-36.

Heathcote G.D., Dunning R.A. & Wolfe M.D. (1965) Aphids on sugar beet and some weeds in England, and notes on weeds as a source of beet viruses. *Plant Pathology* 14, 1-10.

Henderson I.F. & Perry J.N. (1978) Some factors affecting the build-up
of cereal aphid infestations in winter wheat. *Annals of Applied
Biology* 89, 177-83.

Hoffman H. (1968) Beiträge zur Kenntnis der Biologie und Taxonomie der
schwarzglänzenden, langsiphonigen *Aphis*-Arten Mitteleuropas. *Archiv
der Freunde der Naturgeschichte in Mecklenburg* 14, 129-65.

Hoffman H. (1972) Biologische und taxonomische Untersuchungen an fünf
weniger bekannten *Aphis*-Arten von Leguminosen. *Zeitschrift für
angewandte Zoologie* 59, 289-332.

Hughes R.D., Casimir M., O'Loughlin G.T. & Martyn E.J. (1964) A survey
of aphids flying over eastern Australia in 1961. *Australian Journal
of Zoology* 12, 174-200.

Hůrkova J. & Hlináková M. (1977) Toxicity of some current aphicides to
the black bean aphid *Aphis fabae* (Homoptera, Aphididae). *Acta
Entomologica Bohemoslavaca* 74, 349-50.

Ilharco F.A. (1968) *Pentalonia nigronervosa* Coquerel na ilha da Madeira.
Bocagiana 17, 1-26.

Johnson C.G. & Eastop V.F. (1951) Aphids captured in a Rothamsted
suction trap, 5ft. above ground level from June to November, 1947.
Proceedings of the Royal Entomological Society of London (A) 26,
17-24.

Jones M.G. (1980) Observations on primary and secondary parasites of
cereal aphids. *Entomologist's Monthly Magazine* 115, 61-71.

Jones R.A.C. (1981) The ecology of viruses infecting wild and culti-
vated potatoes in the Andean region of South America. *Pests,
Pathogens and Vegetation* (Ed. by J.M. Thresh), pp. 89-107. Pitman,
London.

Kay D.J., Wratten S.D. & Stokes S. (1981) Effects of vernalisation and
aphid culture history on the relative susceptibilities of wheat
cultivars to aphids. *Annals of Applied Biology* 99, 71-5.

Kennedy G.G. (1974) Evidence for non-preference and antibiosis in
aphid-resistant red raspberry cultivars. *Environmental Entomology*
3, 773-7.

Kennedy G.G. & Schaefers G.A. (1974) The distribution and seasonal
history of *Amphorophora agathonica* Hottes on Latham red raspberry.
Annals of the Entomological Society of America 67, 356-8.

Kranz J., Schmutterer H. & Koch W. (1977) *Diseases, Pests and Weeds in
Tropical Crops*. Paul Parey.

Lammerink J. (1968) A new biotype of cabbage aphid (*Brevicoryne
brassicae* (L.)) on aphid resistant rape (*Brassica napus* L.). *New
Zealand Journal of Agricultural Research* 11, 341-4.

Leather S.R. & Dixon A.F.G. (1981) Growth, survival and reproduction
of the bird-cherry aphid, *Rhopalosiphum padi*, on its primary host.
Annals of Applied Biology 99, 115-8.

Lowe H.J.B. (1981) Resistance and susceptibility to colour forms of
the aphid *Sitobion avenae* in spring and winter wheats *(Triticum
aestivum)*. *Annals of Applied Biology* 99, 87-98.

Micieli de Biaze L. (1970) Notizie sull' *Aphis spiraecola* Patch
(Homoptera, Aphididae). *Bolletino del Laboratorio di Entomologia
Agraria "Filippo Silvestri" di Portici* 28, 194-203.

Miyazaki M. (1971) A revision of the tribe Macrosiphini of Japan
(Homoptera : Aphididae, Aphidinae). *Insecta Matsumurana* 34, 1-247.

132   V.F. EASTOP

Mordvilko A. (1919) *Fauna of Russia and Neighbouring Countries.*
   I. Insecta. Hemiptera. Zoological Museum of the Academy of Sciences
   Petrograd (in Russian).
Müller F.P. (1977) Vergleich einer tropischen mit einer mitteleuropä-
   ischen Population von *Aphis craccivora* Koch. *Deutsche Entomologische*
   Zeitschrift 24, 251-60.
Müller F.P., Hinz B. & Möller F.W. (1973) Übertragung des Enationen-
   virus der Erbse durch verschiedene Unterarten und Biotypen der
   Grünfleckigen Kartoffelblattlaus *Aulacorthum solani* (Kaltenbach).
   Zentralblatt für Bakteriologie II 128, 72-80.
Peters D.C., Wood E.A. Jr. & Starks K.J. (1975) Insecticide resistance
   in selections of the greenbug. *Journal of Economic Entomology* 68,
   339-40.
Rautapää J. (1967) The bionomics of the raspberry aphids *Aphis idaei*
   v.d. Goot and *Amphorophora rubi* (Kalt.) (Hom., Aphididae). *Annales*
   Agriculturae Fenniae 15, 272-93.
Rautapää J. (1976) Population dynamics of cereal aphids and method of
   predicting population trends. *Annales Agriculturae Fenniae* 15,
   272-93.
Rautapää J. (1980) Light reactions of cereal aphids (Homoptera,
   Aphididae). *Annales Entomologici Fennici* 46, 1-12.
Rogerson J.P. (1947) The oat bird-cherry aphis, *Rhopalosiphum padi* L.,
   and comparison with *R. crataegellum*, Theo. (Hemiptera, Aphididae).
   Bulletin of Entomological Research 38, 157-76.
Shands W.A. & Simpson G.W. (1971) Seasonal history of the buckthorn
   aphid. Suitability of alder-leaved buckthorn as a primary host in
   northeastern Maine. *Technical Bulletin, University of Maine* 51,
   24 pp.
Shands C.H. Jr. (1967) Resistance of the strawberry aphid to endosul-
   phan in south western Washington. *Journal of Economic Entomology*
   60, 968-70.
Tamaki G. & Allen W.W. (1969) Competition and other factors influencing
   the population dynamics of *Aphis gossypii* and *Macrosiphoniella*
   sanborni on greenhouse chrysanthemums. *Hilgardia* 39, 447-505.
Tao C-C. & Chiu S-C. (1971) Biological Control of Citrus, Vegetable
   and Tobacco Aphids. *Special Publication of the Taiwan Agricultural*
   Research Institute, Taipei 10, 110 pp.
Thomas K.H. (1968) Die Blattläuse aus der engeren Verwandtschaft von
   Aphis gossypii Glover und *A. frangulae* Kaltenbach unter besondere
   Berücksichtigung ihres Vorkommens an Kartoffel. *Entomologisches*
   Abhandlungen aus dem Staatliches Museum für Tierkunde in Dresden 35,
   337-89.
Thresh J.M. (Ed.) (1981) *Pests, Pathogens and Vegetation.* Pitman, Londo
Vickerman G.P. & Wratten S.D. (1979) The biology and pest status of
   cereal aphids (Hemiptera : Aphididae) in Europe: a review.
   Bulletin of Entomological Research 69, 1-32.
Watt A.D. (1979) The effect of cereal growth stages on the reproduc-
   tive activity of *Sitobion avenae* and *Metopolophium dirhodum*. *Annals*
   of Applied Biology 91, 147-57.
Wave H.E., Shands W.A. & Simpson G.W. (1965) Biology of the Foxglove
   Aphid in the Northeastern United States. *United States Department*
   of Agriculture Technical Bulletin 1338, 1-40.

EURAPHID: synoptic monitoring for migrant vector aphids

L.R. TAYLOR
Rothamsted Experimental Station, Harpenden,
Herts AL5 2JQ, UK

INTRODUCTION

The first entomological problem in virus epidemiology is to identify the
vectors and the second is to assess their relative importance in field
transmission and the need for their control. Aphid-borne viruses
present a special problem in that the appearances, distribution and
behaviour of aphid vectors are more erratic than those of most other
insects and their relative importance fluctuates widely from place to
place, from year to year (Fig. 1), and often from day to day. This
dynamic flexibility raises doubts about the validity of some aspects
of the prevailing experimental approach.

PARAMETERS

Continued agricultural interest in aphids as crop pests and as virus
vectors has generated a vast amount of information on their taxonomy,
biochemistry, physiology, behaviour and ecology which provides the basis
for modelling crop infestation and infection. Some models are simple
visual or conceptual images for considering specific crop problems.
Others are complex numerical systems involving sophisticated functional
relationships for projecting population processes over long periods of
time and in different environments. There is, as yet, no complete
behavioural dynamics for any single aphid species in its whole environment.

Models for forecasting that are simple empirical projections from
past measurements, such as the Watson-Hurst regressions for weather-
dependent virus yellows of sugarbeet (Hurst 1965; Watson 1966; Watson
et al., 1975), may conceal complex systems within their superficially
simple regression relationships (Fig. 2) (Taylor, 1977a). They are,
however, internally consistent in that the data are specific to the
model. Constructed models, involving complex behavioural and ecological functional equations that are derived from individuals and integrated to simulate population growth and redistribution, are less
internally consistent. They synthesize concepts based on information

Plumb R.T. & Thresh J.M. (1983) *Plant Virus Epidemiology.*
Blackwell Scientific Publications, Oxford.

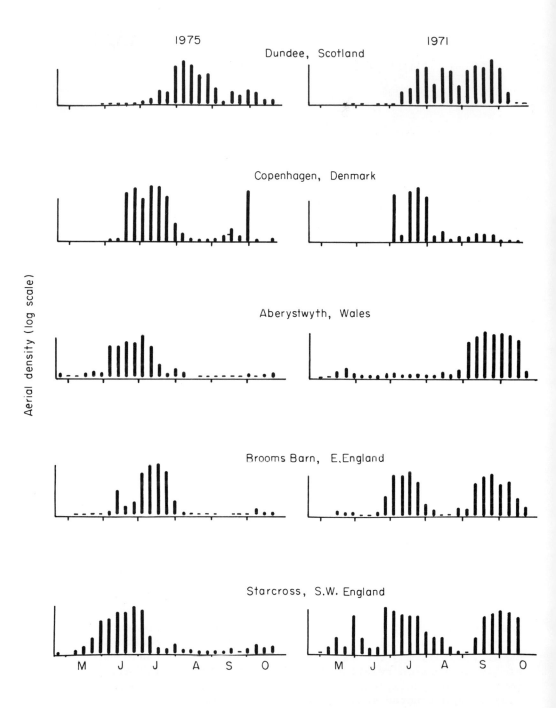

Figure 1. The timing, size and geographical distribution of *Rhopalosiphum padi* aerial population density at different sites in different years (from Taylor & French, 1981).

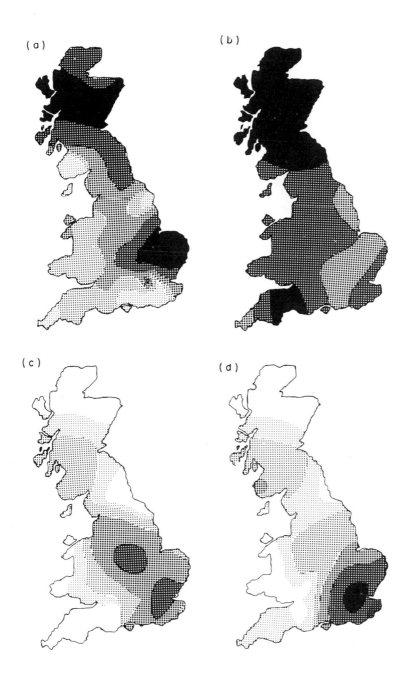

Figure 2. Aerial density maps for *Myzus persicae* migrations:
(a) summer 1972, (b) autumn 1972, (c) spring 1973, (d) difference
map, spring 1973/autumn 1972, i.e. overwintering survival. High
density and maximum survival heavily shaded. (For details see
Taylor, 1977a).

UNIVERSITY OF GREENWICH LIBRARY

from various species, morphs and clones, grouped together largely by inference because the experimental evidence is too time-consuming to produce or too detailed to have been obtained for many individuals or species. The problem is compounded in virus epidemiological or infection models because the geographical source of the virus is rarely known and several aphid species may be involved. Also, aphid behaviour during transmission adds more uncertainty because it is influenced by the recent history of the individual vector concerned and this is not known. Usually the only directly recorded features of the model are the spatial distribution on plants, in fields or regionally, of some stage in vector development - egg, nymph or adult - and disease incidence assessed by sampling. These spatial samples should be repeated sequentially to give some measure of the dynamic, temporal succession leading to the annual timing of the crucial agricultural event, the date when some predetermined economic threshold of damage is exceeded. But this is rarely done.

Temporal or physiological growth and reproductive functions may have been measured for the particular aphid species on a particular crop host at some time, but rarely over the whole range of field conditions and usually with a limited range of cultured aphids and hosts. Nor do we yet have any widely accepted standard function for the relationships involved; even the fundamental temperature x development response curve is not yet resolved in a form acceptable to all modellers. Simulation is an alternative to formally constructed models, but projection beyond the measured range of physical and biological environments remains speculative.

The spatial/behavioural processes of redistribution on crops or alternate hosts and the functional relationships are also not yet formalized. It is seldom known exactly what to measure, or how; but current indications are that random diffusion is not appropriate and complex specific functional relationships exist between population density and distance that are intractable mathematically (R.A.J. Taylor, 1980, 1981).

For instance, the time taken for an aphid to inoculate a given plant with virus depends on factors other than the transmission characteristics determined in the laboratory. It depends on the feeding behaviour of the individual aphid on the carefully chosen part of the individual plant encountered in the field. This, in turn, depends on the nature of the aphid clone, but more particularly on the individual's recent history; on those social, nutritional and meteorological experiences of its parent that determined its morph potential, and on its own development of that potential. These resulted in the individual's morph development in accordance with a graduated scale of migratoriness far more sophisticated than is suggested by the crude terms "alate" and "apterous". Moreover, its own social, nutritional and meteorological experience will decide if and how long it flies, where and when it lands, and if and how long it stays to feed on the plants it finds.

Some of the behavioural components in the migratory infection flight

are known in outline for a few species but most is known about *Aphis
fabae* which is considered an unimportant vector. The components cannot
be known in a given circumstance for a particular individual because
each individual's experience is unique. It is impossible, therefore, to
forecast from laboratory experiments what will happen in the field.
Even a statistical distribution of probabilities will be too speculative
to have much value if projected for a generation into the future, or
when an unknown number of species may be involved. This point is also
important when seeking to validate, by laboratory tests, *post factum*
analysis of samples from natural field infections.

At least 317 aphid species are known to fly over Britain (Taylor *et
al*., 1981), an unknown number of which may be field vectors. Thus the
time taken to build any realistic model of all the potential infection
probabilities, based on laboratory measurements, could exceed the
commercial life of a given crop cultivar. It seems likely that cultivars can affect the aerial morph behaviour of common aphid species (see
below). Thus, at least for non-persistently transmitted viruses, the
whole balance of a laboratory-based probability model could be upset
faster than it could be rebuilt, unless it is capable of being continuously updated by new functional relationships and parameters. Continuous
measurement of field parameters is required to make this possible.

APHID POPULATION INSTABILITY

The problem of modelling populations is almost equally great when
considering aphid pests as it is when they are virus vectors, and sometimes the difficulties are more easily recognized without the additional
complication of virus ecology. For example, migrant greenbugs,
Schizaphis graminum, which are virus vectors and also directly damaging
pests of cereals, have been sampled continuously at Bushland, Texas, for
28 years. From 1953 to 1967 migration occurred each spring and autumn;
from 1968 to 1970 migration occurred only in late summer; from 1971 to
1975 it occurred in spring and summer, and from 1976 to 1980 it was too
small to record (Daniels, 1981). These changes in timing and intensity
of migration corresponded to changes in the aphid biotypes that affected
population distribution on wheat, grasses and sorghum, and had serious
agricultural repercussions (Harvey & Hackerott, 1969a, 1969b). The
lifespan of the new biotypes and their reproduction in the laboratory
responded differently to temperature (Daniels & Chedester, 1981). These
changes appear to have parallels in the differential reproductive
functions of insecticide-resistant biotypes (Eggers-Schumacher, 1981).

Sudden adaptive changes in behaviour and physiology are readily
recognized in the simplified environment of the Great Plains of North
America. They are usually obscured in the complex, mosaic environment
of maritime Europe where, like the distribution of insecticide-resistant
population elements that could equally well be called ecotypes, their
appearance is geographically patchy. Hence the cycles of migration of
European aphids are less predictable than the formal description of life
histories implies. They reflect the recent success, or failure, in
population survival and growth, in different years and places (Fig. 1).

Seasonal cycles of distribution, as of abundance, are greatly affected by the distributions of crops (Fig. 2). Concentration of the summer migration of *Myzus persicae* in eastern England reflects the large populations of many pest species on agricultural crops. Concentration of spring migration in southern England reflects the increased opportunity for overwintering survival there of many, but not all, anholocyclic species (Taylor *et al.*, 1982). The seasonal distribution of holocyclic species such as the damson-hop aphid *Phorodon humuli* more closely reflects the distribution of their winter hosts (Fig. 3). But not all distributions are so host-dominated, and this is most clearly demonstrated in non-pest species such as the elder aphid *Aphis sambuci*, where there is little agricultural manipulation of the spatial and temporal environment (Fig. 4).

SAMPLING FOR FORECASTING

The practical issue arises of how best to measure the basic spatial distributions of aphids; what is ideal? Disregarding whether or not they are infective, the ideal would be regular and frequent samples of vector populations at all stages and also records of the changing distribution of current host plants, so as to assess the likelihood of damaging populations arising. Information is also required on the timing of crop invasion and the infestation thresholds that justify control measures. These measurements could be obtained by regular host plant inspection, within crops and elsewhere. The recognition of levels of infectivity is a separate issue (see below).

As an example, Way *et al.* (1977) inspect spindle, *Euonymus europaeus*, the overwintering host of the bean aphid, *Aphis fabae*. Eggs are counted in autumn; inspections in spring assess overwintering and estimate likely time of emigration to beans; summer inspections of the bean crop determine dates of arrival and estimate dates for reaching treatment thresholds. The method is successful and spray warnings are issued that are highly successful in anticipating what would have been damaging infestations, in-so-far as this can be measured.

This is a direct damage problem. It involves a single, holocyclic pest species with, effectively, only one overwintering host. This host is a tree with a restricted and mapped distribution, so that the same trees can be sampled annually using a standard procedure of known efficiency (Cammell *et al.*, 1978). The aphid is, for all practical purposes, entirely holocyclic in Great Britain and so the egg sample is easy to collect and count, and usually yields a reliable population estimate. However, considerable labour is involved in sampling widely scattered hosts and may be required for other, more immediately pressing local problems, especially at times of *A. fabae* recession. If sampling is ever discontinued, any subsequent resurgence of *A. fabae* would find the system unprepared and the new outbreak would have to be tackled *de novo*, as has often happened in the past. Meanwhile the summer phase of the life cycle is incompletely understood. The summer migration from beans and the subsequent reproduction on diverse herbaceous hosts

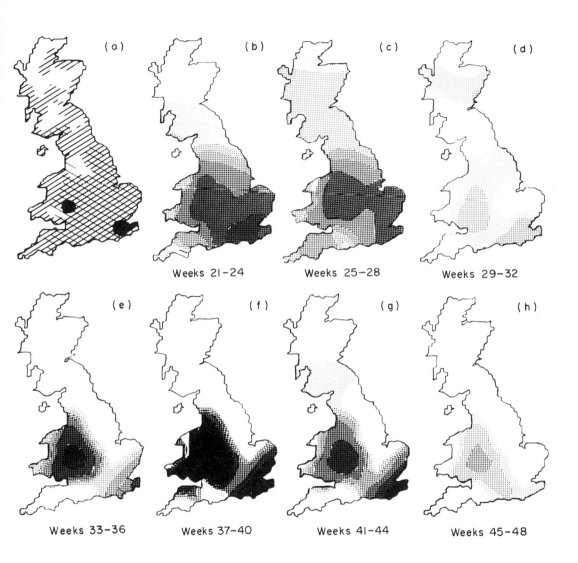

Figure 3. (a) The distribution of *Prunus spinosa* (including damson) ▨ , wild hop ◰ and hop gardens ■ . (b-h) 4-weekly maps of aerial density of damson hop aphid at 12.2 m during 1973. (b,c) The first migration from *Prunus* is diffuse, extending to the north of Scotland. (d) There is a clear break between migrations. (e,f,g,h) The second migration, dispersing from the two main hop-growing areas, lasts longer. (For further details see Taylor, Woiwod & Taylor, 1979.)

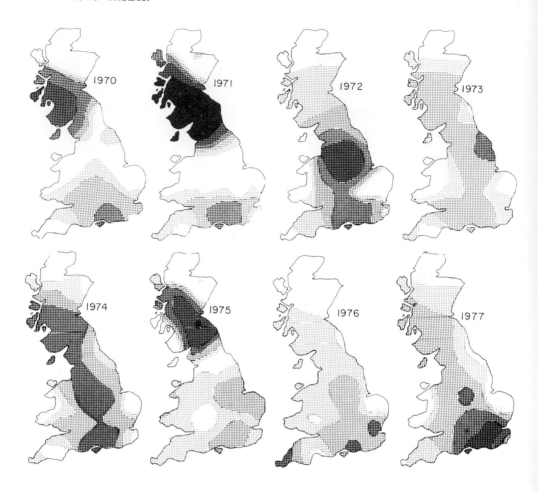

Figure 4. Aerial density of *Aphis sambuci* (elder aphid) 1970-77 (from Taylor & Taylor, 1979).

that produces the return migration of sexuparae to the winter host in late autumn, remain a mystery. Sampling for this stage is impossible to standardize because the herbaceous hosts are widely distributed and highly variable.

Nevertheless, the *A. fabae* survey remains an exemplary exercise on a national scale. It facilitates crop protection, can be assessed for efficiency and ultimately for economic return, and is applicable to any kind of subsequent control treatment. There are also the essential components of geographical continuity, sequential seasonal replication and measureable efficiency. The system most nearly approaches the ideal for host-plant sample monitoring.

AERIAL SAMPLING

The efficiency of host-plant sampling for *A. fabae* can be matched by continuous aerial monitoring (Way *et al.*, 1981). Aerial sampling has the limitation that it does not assess apterae, but it has the advantage of sampling the migrants produced in each of the three seasonal population cycles, irrespective of the botanical or geographical variability of the hosts. Trapping is equally effective whether the aphid is holocyclic, anholocyclic, or an unknown mixture of each. It also provides a daily sample of those migrants that have flown from hosts, instead of an estimate of the potentially migrant alatae. This can be important when weather restricts flight; violent rain that damages the teneral alatae can cause the failure of migration from an apparently large vector source, as can temperatures below the threshold for take-off, because sufficient delay may cancel the migratory drive (Halgren & Taylor, 1968; Dry & Taylor, 1970). Also host-sampling programmes for species that may overwinter both holo- and anholocycly, depending on weather, need constant revision because the expectations for holocyclic populations can be subverted by an early invasion of anholocyclic individuals from outside the host-sampling area (Hille Ris Lambers, 1980).

The Insect Survey at Rothamsted (Taylor, 1973) was developed to overcome some of the host-sampling problems by means of continuous synoptic aerial monitoring. It is not regarded as ideal but is feasible (Taylor, 1974b) and deals continuously, efficiently and simultaneously with all migrant species, including those not yet recognized as pests or vectors. Consequently, the incentive to sample does not fluctuate with crop economics, pest populations or virus incidence. Aerial monitoring differs from crop sampling in providing an opportunity to increase understanding of long-term fluctuations in those stages of the seasonal cycle that are away from crops and therefore inaccessible. It is also possible to look back into the past demographic history of those species not yet suspected of being vectors because they do not appear frequently in crop samples (van Harten, 1981). Monitoring does not produce instant solutions to pest or virus problems, although it does produce hard evidence for sound scientific interpretation.

The Insect Survey at Rothamsted was devised in 1959 to integrate insect migration into population dynamics (Taylor, 1979). It was not expected to provide automatic, instant, crop invasion forecasting, although deposition rates can closely reflect aerial density when measured appropriately (Taylor, 1977b; Light, 1980). The first survey suction trap was designed, built and tested at Rothamsted in 1963 and began operation in 1964, sampling at a height of 12.2 m (Taylor & Palmer, 1972). A transect was established down eastern Great Britain by 1966 with another across southern England by 1970. Traps sampled the aerial insect population over Great Britain effectively by 1975, excluding north-western Scotland (Taylor *et al.*, 1981). Traps were started in the Netherlands in 1968 (van Harten, 1981) and there were three in 1982; in Denmark in 1971 (Philipsen, 1977); in N. Ireland in 1976; in France in 1977 (Robert & Choppin de Janvry, 1977; Bouchery,

1979) and there will be ten in 1982; in Belgium in 1980 (Latteur, 1980)
(two in 1982). These were integrated into a voluntary co-operative
venture, EURAPHID, in 1980 (Taylor & French, 1981). Other sites are
believed to be operating, or initiated, in Poland, Hungary, Czechoslovakia, Switzerland and Italy. Similar traps have been operating in
Sweden since 1978 (Wiktelius, 1980), and they are expected to be
standardized to the Rothamsted model in the near future.

The sampling system provides a measure of the aerial density of
aphids in flight at 12.2 m. At this height in the air, the fine scale
distribution of individuals is random, unlike their distribution on
wild host plants and crops. Weekly means of daily samples differ little
over considerable distances. Large-scale spatial trends in population
density over, say, western Europe can thus be mapped effectively with a
program such as SYMAP V (Laboratory of Computer Graphics, Harvard),
using samples from a modest number of sites (Fig. 5). Against the background of these standard maps, systematic field samples can provide
fine geographical detail, but only for species that remain on crops.

The sampling system has now produced mean seasonal timing of flights
and geographical distributions for most migrant aphid species as base-
line data for expectations and as indicators for current deviations
(Taylor et al., 1981; Heie et al., 1981). The data are already used
extensively in risk projection (Bardner et al., 1981). They present a
possibility of continuously improving aphid life-cycle information
applicable to the currently predominant strains, clones or biotypes,
despite their ecological flexibility. These data become increasingly
valuable as they accumulate and make possible the retrospective confirmation of new concepts and the development of new hypotheses of crop
infestation. This is vital in a subject where it takes two decades to
test any hypothesis. Aerial sampling does not compete with crop
sampling but provides a continuous, objective record by which even
spasmodic crop sampling can be related to a continuously developing
model.

There are now nearly two decades of continuous data from the first
operational site at Rothamsted (Taylor 1974a, 1974b) and seventeen
sites have completed one decade. With about 35 sites now operating in
Western Europe, the prospect for sound temporal and spatial distribution data about aphid populations in the future seems hopeful. Using
supplementary RADAR to follow dense concentrations and with suitable
meteorological data, large migrations can sometimes be tracked and
anticipated (Cochrane, 1980). When live samples are also tested for
virus (Plumb, 1976), the essential quantitative background becomes
available for integrating isolated items of information to make crop
infestation and infection an intelligible process (Plumb, 1981).

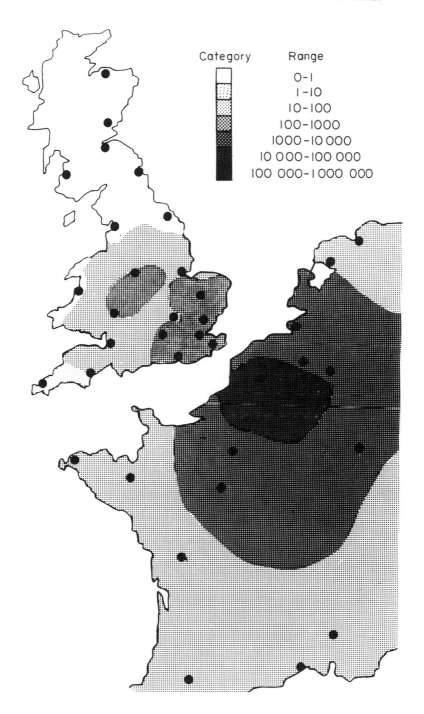

Figure 5. The distribution of sampling sites in Western Europe and aerial density of *Metopolophium dirhodum* in June, 1979 (adapted from Dewar, Woiwod & Choppin de Janvry, 1980).

REFERENCES

Bardner R., French R.A. & Dupuch M.J. (1981) Agricultural benefits of
    the Rothamsted Aphid Bulletin. *Rothamsted Report for 1980, Part 2,*
    21-39.
Bouchery Y. (1979) Pour l'étude des vols de pucerons une méthode de
    capture par piège à aspiration. *Phytoma* 304, 39-40.
Cammell M.E., Way M.J. & Heathcote G.D. (1978) Distribution of eggs of
    the black bean aphid, *Aphis fabae*, Scop., on the spindle bush
    Euonymus europaeus L., with reference to forecasting infestations
    of the aphid on field beans. *Plant Pathology* 27, 68-76.
Cochrane J. (1980) Meteorological aspects of the numbers and distribu-
    tion of the rose-grain aphid, *Metopolophium dirhodum* (Wlk.) over
    south-east England in July 1979. *Plant Pathology* 29, 1-8.
Daniels N.E. (1981) Migration of greenbugs in the Texas Panhandle in
    relation to their biotypes. *Texas Agricultural Experiment Station
    Miscellaneous Publications* 1487, 1-4.
Daniels N.E. & Chedester L.D. (1981) Biological and small grain
    resistance experiments with biotype E greenbug. *Texas Agricultural
    Experiment Station Progress Report* 3870, 1-7.
Dewar A.M., Woiwod I.P. & Choppin de Janvry E. (1980) Aerial migrations
    of the rose-grain aphid, *Metopolophium dirhodum* (Wlk.), over
    Europe in 1979. *Plant Pathology* 29, 101-9.
Dry W.W. & Taylor L.R. (1970) Light and temperature thresholds for
    take-off by aphids. *Journal of Animal Ecology* 39, 493-504.
Eggers-Schumacher H.A. (1981) Potential of multiplication in insecti-
    cide resistant and susceptible strains of *Myzus persicae* (Sulz.)
    M.Sc. Thesis, London University.
Halgren L.A. & Taylor L.R. (1968) Factors affecting flight responses
    of alienicolae of *Aphis fabae* Scop., and *Schizaphis graminum* Rodani
    (Homoptera : Aphididae). *Journal of Animal Ecology* 37, 583-93.
van Harten A. (1981) Bladluizen, virusziekten en rooidata bij
    pootaardappelen. *Gewasbescherming* 12, 57-71.
Harvey T.L. & Hackerott H.L. (1969a) Recognition of a greenbug biotype
    injurious to sorghum. *Journal of Economic Entomology* 62, 776-9.
Harvey T.L. & Hackerott H.L. (1969b) Plant resistance to a greenbug
    biotype injurious to sorghum. *Journal of Economic Entomology* 62,
    1271-4.
Heie O.E., Philipsen H. & Taylor L.R. (1981) Synoptic monitoring for
    migrant insect pests in Great Britain and Western Europe. II. The
    species of alate aphids sampled at 12.2 m by Rothamsted Insect Survey
    suction trap at Tåstrup, Denmark, between 1971 and 1976. *Rothamsted
    Report for 1980, Part 2,* 105-14.
Hille Ris Lambers D. (1980) Integrated control of aphid-borne viruses
    of potatoes. *Integrated Control of Insect Pests in the Netherlands*
    (Ed. by A.K. Minks & P. Gruys), pp. 59-66. Centre for Agricultural
    Publishing and Documentation, Wageningen.
Hurst G.W. (1965) Forecasting the severity of sugarbeet yellows. *Plant
    Pathology* 14, 47-53.
Latteur G. (1980) Les pucerons des céréales. *Rapport d'Activité 1980,
    Centre de Recherches Agronomiques de l'Etat, Gembloux,* pp. 49-50.

Philipsen H. (1977) Fangst af bladlus i luftplankton. Sugefaelde på Højbakkegård som lod i et muligt varslingssystem. *Ugeskrift for Agronomer, Hortonomer, Forstkandidater og Licentiater* No. 6, 99-103.

Plumb R.T. (1976) Barley yellow dwarf virus in aphids caught in suction traps, 1969-73. *Annals of Applied Biology* 83, 53-9.

Plumb R.T. (1981) Aphid-borne virus diseases of cereals. *Euraphid, Rothamsted 1980* (Ed. by L.R. Taylor), pp. 18-21. Rothamsted Experimental Station, Harpenden.

Robert Y. & Choppin de Janvry E. (1977) Sur l'intérêt d'implanter en France un réseau de piègeage pour améliorer la lutte contre les pucerons. *Bulletin Technique d'Information* No. 323, 559-68.

Taylor L.R. (1973) Monitor surveying for migrant insect pests. *Outlook on Agriculture* 7, 109-16.

Taylor L.R. (1974a) Insect migration, flight periodicity and the boundary layer. *Journal of Animal Ecology* 43, 225-38.

Taylor L.R. (1974b) Monitoring change in the distribution and abundance of insects. *Rothamsted Report for 1973, Part 2*, 202-39.

Taylor L.R. (1977a) Migration and the spatial dynamics of an aphid, *Myzus persicae*. *Journal of Animal Ecology* 46, 411-23.

Taylor L.R. (1977b) Aphid forecasting and the Rothamsted Insect Survey *Journal of the Royal Agricultural Society of England* 138, 75-97.

Taylor L.R. (1979) The Rothamsted Insect Survey - an approach to the theory and practice of synoptic pest forecasting in agriculture. *Movements of Highly Mobile Insects: Concepts and Methodology* (Ed. by R.L. Rabb & G.G. Kennedy), pp. 147-85. North Carolina State University, Raleigh.

Taylor L.R. & French R.A. (1981) Synoptic aerial monitoring as a basis for an aphid forecasting system in Europe. *Euraphid, Rothamsted 1980* (Ed. by L.R. Taylor), pp. 9-12. Rothamsted Experimental Station, Harpenden.

Taylor L.R., French R.A., Woiwod I.P., Dupuch M.J. & Nicklen J. (1981) Synoptic monitoring for migrant insect pests in Great Britain and Western Europe. I. Establishing expected values for species content, population stability and phenology of aphids and moths. *Rothamsted Report for 1980, Part 2*, 41-104.

Taylor L.R. & Palmer J.M.P. (1972) Aerial sampling. *Aphid Technology* (Ed. by H.F. van Emden), pp. 189-234. Academic Press, London.

Taylor L.R., Woiwod I.P., Tatchell M., Dupuch M.J. & Nicklen J. (1982) Synoptic monitoring for migrant insect pests in Western Europe. III. Distribution of seasonal migrations and relative commonness of species. *Rothamsted Report for 1981, Part 2*, 23-121.

Taylor L.R., Woiwod I.P. & Taylor R.A.J. (1979) The migratory ambit of the hop aphid and its significance in aphid population dynamics. *Journal of Animal Ecology* 48, 955-72.

Taylor R.A.J. (1980) A family of regression equations describing the density distribution of dispersing organisms. *Nature, London* 286, 53-5.

Taylor R.A.J. (1981) The behavioural basis of redistribution. 1. The Δ-model concept. *Journal of Animal Ecology* 50, 573-86.

Taylor R.A.J. & Taylor L.R. (1979) A behavioural model for the evolu-
    tion of spatial dynamics. *Population Dynamics* (Ed. by R.M. Anderson,
    B.D. Turner & L.R. Taylor), pp. 1-27. Blackwell Scientific
    Publications, Oxford.
Watson M.A. (1966) The relation of annual incidence of beet yellowing
    viruses in sugar beet to variations in weather. *Plant Pathology* 15,
    145-52.
Watson M.A., Heathcote G.D., Lauckner F.B. & Sowray P.A. (1975) The
    use of weather data and counts of aphids in the field to predict
    the incidence of yellowing viruses of sugar-beet crops in England in
    relation to the use of insecticides. *Annals of Applied Biology* 81,
    181-98.
Way M.J., Cammell M.E., Alford D.V., Gould H.J., Graham C.W., Lane A.,
    Light W.I. St. G., Rayner J.M., Heathcote G.D., Fletcher K.E. &
    Seal K. (1977) Use of forecasting in chemical control of black bean
    aphid, *Aphis fabae* Scop., on spring-sown field beans, *Vicia faba* L.
    Plant Pathology 26, 1-7.
Way M.J., Cammell M.E., Taylor L.R. & Woiwod I.P. (1981) The use of
    egg counts and suction trap samples to forecast the infestation of
    spring-sown field beans *Vicia faba*, by the black bean aphid, *Aphis
    fabae*. *Annals of Applied Biology* 98, 21-34.
Wiktelius S. (1980) Vindspridning av insekter. *Forskning och Framsteg,
    1980*, No. 1, 1-9.

Monitoring insect vector populations and the detection of viruses in vectors

B. RACCAH

Division of Virology, Agricultural Research Organization,
The Volcani Center, Bet Dagan, Israel

INTRODUCTION

Many viruses are dependent on vectors for entry into host cells and for
transfer to new hosts and new sites. Therefore studies of plant virus
epidemiology frequently involve monitoring vectors so as to interpret
the pattern and sequence of virus spread. The majority of insect
vectors are airborne at some stage in their life cycle, thus monitoring
systems are based on sampling insects in flight or on landing. Yellow
water pans or impaction traps have been used extensively for some years
and these traps and their uses are described elsewhere (Taylor & Palmer,
1972). In many cases, trap catches have been correlated with virus
spread and the efficiency with which insects transmit viruses has been
assessed in transmission experiments in the laboratory or glasshouse.
In some of these experiments the insect species tested were those
occurring most frequently in traps (Shaunak & Pitre, 1971). However,
the importance of vectors is not due solely to the numbers occurring
and their effectiveness in transmitting virus. It is also influenced
by epidemiological and ecological factors, such as the host preferences
of vectors, the distribution of virus in the host plant and meteorological conditions. Therefore, in order to determine the actual virus-
transmitting potential of naturally occurring vectors, they have to be
trapped live in the field and tested for infectivity.

The most appropriate method depends on the objectives of the study,
and the following features are important in considering the spread of
insect-borne viruses:-

1. assessing changes in numbers of airborne vectors or those
   landing in crops,
2. discriminating between vector and non-vector species,
3. determining capacity to transmit,
4. determining the viruses present in the crop and their rate of
   spread.

More than one trapping method is required to assess all these parameters and, in this review, particular attention is given to methods of

Plumb R.T. & Thresh J.M. (1983) *Plant Virus Epidemiology.*
Blackwell Scientific Publications, Oxford.

determining the relative importance of vectors.

The recent development of sensitive serological techniques for detecting plant viruses, such as enzyme-linked immunosorbent assay (ELISA) (Clark & Adams, 1977) and serologically specific electron microscopy (SSEM) (Derrick, 1973), has provided new methods of detecting viruses in insect vectors. We were able to detect cucumber mosaic virus (CMV) in the melon aphid, *Aphis gossypii*, using ELISA (Gera *et al.*, 1978), and SSEM has been used to visualize potato leafroll virus (PLRV) in the green peach aphid *Myzus persicae* (Roberts & Harrison, 1979). The advantages of these techniques in epidemiological studies are obvious and the prospects and limitations of their use are discussed in this chapter. Ultimately, it is anticipated that monitoring will make forecasting of epidemics possible. The success of non-conventional control measures against viruses (Loebenstein & Raccah, 1980) depends on timely application, and vector monitoring will also be helpful in the choice and timing of control measures.

MONITORING ALIGHTING APHID VECTORS

Although Moericke water traps (Moericke, 1951) are simple to operate, certain aphid species are attracted differentially (Eastop, 1955), and this may cause bias when assessing their relative importance. Moreover these traps not only record landing but also attract aphids in flight. Therefore Irwin (1980) designed a modified water trap designed specifically to monitor aphids landing on soyabean. The trap consists of a perspex receptacle containing a horizontal, coloured (green) tile (HCT). The colour was chosen to resemble the characteristics of soyabean vegetation based on reflection spectrophotemetry. Instead of a preservative or sticky tanglefoot, Irwin & Goodman (1981) recommend the use of 50% ethylene glycol to facilitate collection and identification. Using this trap, they obtained close agreement between the numbers and identity of aphids in the trap and on vegetation.

A'Brook (1968) described differences in aphid captures as affected by ground colour, and a possible limitation of the HCT is that catches may be biased when plants are small and separated by extensive areas of bare ground. Thus a correction is needed to allow for this effect early in the growing season.

Trapping live aphids for infectivity tests

The use of sticky traps does not allow live trapping and a different method is needed where the infectivity of captured aphids is to be determined. Live vectors can be collected directly from vegetation, or by using collecting devices such as nets, trays or mobile suction traps (D-Vac) (Howell, 1974; Nault *et al.*, 1979). However, insects collected in this way are mostly those spreading viruses within crops rather than those bringing virus into plantings. To assess incoming vectors other traps have been used including suction traps (Plumb, 1971), vertical nets (Halbert *et al.*, 1981) and water traps (Demski,

1981). Insects so caught were collected and placed on test plants, and those transmitting virus were identified.

This system has been used for barley yellow dwarf virus (BYDV) in England since 1969 (Plumb, 1976) and recently also for non-persistently transmitted viruses in France (Leclant, 1978; Labonne, 1981) and in Israel. To detect infectivity with non-persistently transmitted viruses, the trap should be operated for only short periods before the vectors are transferred, to avoid loss of infectivity. The disadvantage of the suction trap is that it traps all aphids, not just those that will land on the crop being studied, and the data must be corrected using data collected from an HCT-type trap. Halbert et al. (1981) tested the infectivity of live aphids blown on to stationary, vertical nets. This method has the advantage of being independent of electrical power but it is affected by wind speed.

Peanut mottle virus has been transmitted by aphids caught in water traps (Demski, 1981) but any effect of immersion in water on the proportion of vectors transmitting virus has not been determined.

THE USE OF BAIT PLANTS TO ASSESS VIRUS SPREAD

The seasonal spread of viruses can be assessed by recording changes in the number of plants with symptoms or by periodic indexing. However, more accurate information is desirable as once infection appears and begins to spread, multiple infection becomes increasingly important. Spread of virus at different periods in potato crops was determined by Broadbent et al. (1950), who exposed successive batches of potted glasshouse-grown potato plants for a limited time and then retained the plants to observe symptoms. The proportion of plants in each batch that become infected represents the "infection pressure" at the time and this method has been widely used (Schwartz, 1965; Madden et al., this volume). Marrou et al. (1979) studied the incidence of infection by CMV in trays of bait plants in relation to the capture of aphids in nearby yellow water traps. The bait plants were inserted in wet peat to keep them moist and the spread of CMV was followed in each of several years. The correlation between infection and the incidence of aphids in traps was good only in 1977.

O.W. Barnett (personal communication) found differences in the proportion of infected bait plants of different age exposed concurrently, and attributed this to differences in plant size. Thus it should be decided at the outset of an experiment whether the plants to be exposed should be the same size as those in the field, or the size that is most likely to be infected by virus. Bait plants provide data on the incidence and nature of infection that complement the information obtained from infectivity tests on live vectors at different periods throughout the season.

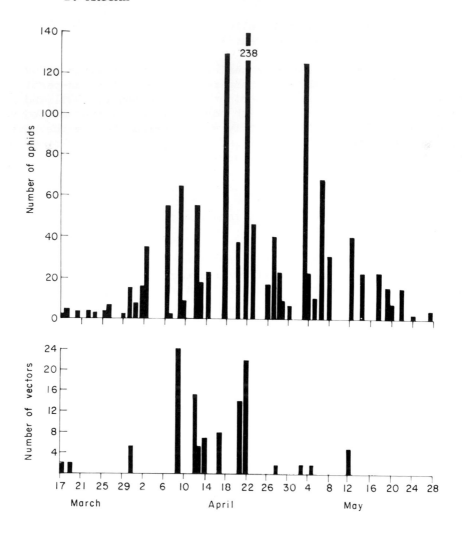

Figure 1. Total number of aphids and number of infective vectors caught in suction traps over a pepper field in Israel, March 17-June 5, 1981. (From Raccah & Loebenstein, 1981)

MONITORING THE APHID VECTORS OF VIRUSES OF BELL PEPPER IN ISRAEL

A monitoring system to assess the relative importance of the aphid vectors of cucumber mosaic virus and potato virus Y in bell pepper has been operated at Bet Dagan, Israel. The system included yellow plastic water traps, bait plants and suction traps. The yellow traps were exposed at canopy level, and the aphids caught were removed daily, counted and kept for identification. Successive batches of pepper bait plants of the cultivar being grown in the field were exposed for 3-6 days. Each batch was of 30-40 plants and after exposure they were sprayed with an aphicide before being returned to the glasshouse. When

symptoms appeared, the identity of the virus was determined by ELISA.
Two suction traps were operated in the field using a portable generator.
One sampled air at 1.7 m and the other at ground level. The traps were
operated for 15 min each morning, as in two previous years it had been
established that maximum numbers were caught at this time. The aphids
caught were transferred individually to pepper test plants. After an
inoculation access period of 18 h the aphids were removed and preserved
for identification. The viruses transmitted in this way were later
identified by ELISA.

Fig. 1 shows the total number of aphids caught in the suction trap
and the respective number of vectors at different dates. The proportion of vectors caught changed during the season, emphasising the
importance of assessing their infectivity for the entire period. Of
more than 44 aphid species trapped, 20 transmitted PVY and/or CMV on
at least one occasion (Raccah, Eastop & Loebenstein, unpublished
results). The numbers of the commonest species, known to transmit PVY
and CMV, caught in the yellow and suction traps are shown in Table 1.

Table 1. Catches of aphid vectors of PVY and CMV trapped in yellow
and suction traps in a pepper field at Bet Dagan, Israel.

Aphid species*	Yellow water pan		Suction trap†	
	17-20/4	28-30/4	17-20/4	28-30/4
Aphis citricola	2492	525	70	13
Macrosiphum euphorbiae	64	27	10	8
Myzus persicae	65	32	12	7
Uroleucon sonchi	26	14	2	3
Hyperomyzus lactucae	11	0	3	0
Aphis sp.	23	46	30	16
Other vector species	3	48	17	7

* Aphids caught in the suction trap were confined individually on
  test plants. These aphid species transmitted either CMV or PVY at
  least once (Raccah, Eastop & Loebenstein, unpublished results).
† Suction traps were operated daily for 15 min, 0700-0800 h.

A. citricola predominated in both traps and 75-92% of the aphids in the
yellow traps but only 24-50% of those in the suction traps were of this
species. Part of the difference in species composition between the
yellow water pan and suction trap catches may be because suction traps
were only operated in the morning, while the yellow traps were exposed
continuously. These catches and the results of the infectivity tests
using live aphids made it possible to determine the relative importance

Table 2. Abundance, transmission efficiency and relative importance as vectors of CMV and PVY of the aphids caught in a suction trap in a pepper field at Bet Dagan, Israel, in 1981.

Aphid species	Aphids tested No. (% of total)	Transmitters No. (%)		% of total transmission	
		CMV	PVY	CMV	PVY
A. citricola	373 (39.5)	42 (11.3)	23 (6.2)	52.5	48.9
A. craccivora	32 (3.4)	1 (2.9)	0 0	1.25	0
A. gossypii	25 (2.6)	2 (8.0)	3 (12.0)	2.5	6.4
Aphis sp.	43 (4.6)	7 (16.3)	3 (7.0)	8.75	6.4
Ma. euphorbiae	61 (6.5)	10 (16.4)	6 (9.8)	12.5	12.8
M. persicae	32 (3.4)	7 (22.0)	2 (6.3)	8.75	4.2
R. maidis	10 (1.1)	1 (10.0)	0 0	1.25	0
R. padi	34 (3.6)	1 (2.9)	0 0	1.25	0
Other vector species*	226 (23.9)	9 (4.0)	10 (4.4)	11.25	21.3
Total	836 (100.0)†				

* 12 additional species were vectors: 9 for CMV and 3 for PVY.
† 24 additional species totalling 109 aphids (11.4% of total) were not vectors of CMV or PVY.

of the different species as vectors (Table 2). The spirea aphid *A. citricola*, which transmitted less efficiently than *M. persicae* or *Macrosiphum euphorbiae* in the infectivity tests, was the most important vector because it occurs in very large numbers. A similar phenomenon was described by Halbert *et al.* (1981), who found that *Rhopalosiphum maidis* was the most important vector of soyabean mosaic virus because it occurred in large numbers, even though it transmitted less efficiently than other species.

The infection of bait plants with CMV and PVY in the 1981 Israeli experiments was related to the incidence of vectors, with maxima of aphid numbers and virus spread at the end of April (Fig. 2). The incidence of virus in the pepper planting was related to the total catches of vector species (Table 3). In 1980 more of the trapped aphids transmitted PVY than CMV, and PVY spread more than CMV; in 1981 the reverse occurred.

It is clearly advantageous to operate more than one type of trap to monitor vectors and virus spread. In our monitoring system an HCT-type trap was not used, but it would be helpful to have data on numbers alighting in assessing the relative importance of aphid species and the HCT trap will be used in any further work.

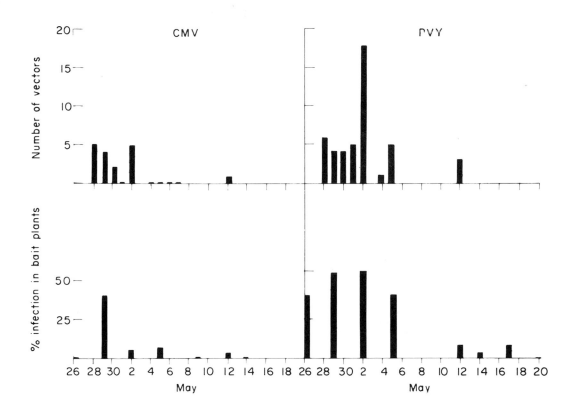

Figure 2. Incidence of aphid vectors caught in suction traps and percentage of pepper bait plants infected by CMV or PVY, April-May, 1980. (From Raccah & Loebenstein, 1981).

DETECTION OF VIRUSES IN VECTORS

Transmission tests provide the most reliable means of distinguishing between vector and non-vector species. However, the time required for test plants to develop symptoms limits the value of this method in monitoring vectors and forecasting epidemics. The use of ELISA or SSEM can provide information within a day or even less and these techniques are now being used. CMV (Gera *et al.*, 1978), citrus tristeza virus (Cambra *et al.*, 1979), PLRV (Clarke *et al.*, 1980; Tamada & Harrison, 1981), pea enation mosaic virus (Fargette *et al.*, 1981), PVY (Raccah *et al.*, 1981) and BYDV (R. Plumb, personal communication) have been detected in their aphid vectors, and maize rayado fino virus has been detected by ELISA in the leafhopper *Dalbulus maidis* (C. Rivera & R. Gámez, personal communication).

To obtain positive detection in *M. persicae* by ELISA, the extraction buffer was modified (Tamada & Harrison, 1981; Raccah *et al.*, 1981).

Table 3. Incidence of CMV and PVY in field-grown peppers and the corresponding percentage of infective vectors caught in a suction trap at Bet Dagan, Israel.

Year	Final incidence of infection in plantings*			Incidence of vectors in aphid traps†		
	Number of plants infected	% infection		Number of aphids tested	% of vectors	
		CMV	PVY		CMV	PVY
1980	92	4.3	53.2	456	3.5	10.7
1981	300	18.0	7.7	1459	11.3	5.7

From Raccah & Loebenstein (1981).
* Some plants were infected by both viruses.
† Some vectors carried both viruses.

For persistent viruses, attempts have been made to detect PLRV (Tamada & Harrison, 1981) and BYDV (R. Plumb, personal communication) in aphids caught in suction traps, using ELISA. The reliability of detection is not good enough to allow similar tests with non-persistent viruses. SSEM was used successfully to detect PLRV in *M. persicae* (Roberts & Harrison, 1979), however, throughput is limited and it is doubtful whether this technique will compete with ELISA when numerous aphids have to be tested.

CONCLUDING REMARKS

Work over the last decade has provided much information on the ecology and phenology of vectors. From comparisons of the various trapping systems, two complementary methods have emerged as being the most suitable:-

1. the HCT trap, which provides information on the number and species of aphid landing on the crop,
2. the suction trap, which records the aerial density of aphids, and allows their infectivity to be assessed.

Neither trap is subject to the limitations of the yellow water pan or sticky traps. However, further improvements are needed to permit periodic sampling. Using these traps in conjunction with bait plants provides information on the landing and relative importance of different vectors and allows "infection pressure" to be assessed. This information together with meteorological data will facilitate forecasting and the use of appropriate control measures. The recent developments in integrated control, and the use of oil sprays, white nets, mulches, etc.

(Loebenstein & Raccah, 1980) have considerable potential even though insecticides may be ineffective.

Acknowledgements

Work in this laboratory is supported by a grant from the United States-Israel Binational Agricultural Research and Development Fund (BARD). I am grateful to Drs. W.C. Adlerz, O.W. Barnett, M. Cambra, R.C. Clarke, R. Gámez, T.P. Pirone and A.H. Purcell for published and unpublished information. I am also obliged to Dr. M.E. Irwin for fruitful discussions on monitoring, and to Dr. S. Cohen for reading the manuscript. I appreciate the useful suggestions and advice given by Dr. G. Loebenstein, of this laboratory, and thank Drs. V.F. Eastop. A. Maoz, and R. Carlebach and A. Gera for their contributions to various parts of this study. The able technical assistance of Mrs. Sima Singer, Mr. S. levy, Mrs. M. Cohen and Mr. A. Gal-on is acknowledged.

This contribution is No. 268-E of the 1980 series of the Agricultural Research Organization, the Volcani Center, Bet Dagan, Israel.

REFERENCES

A'Brook J. (1968) The effect of plant spacing on the numbers of aphids trapped over the groundnut crop. *Annals of Applied Biology* 61, 209 94.

Adlerz W.C. (1976) Comparison of aphids trapped on vertical sticky boards and cylindrical aphid traps and correlation with watermelon mosaic virus 2 incidence. *Journal of Economic Entomology* 69, 495-8.

Broadbent L., Chaudhuri R.P. & Kapica L. (1950) The spread of virus diseases to single potato plants by winged aphids. *Annals of Applied Biology* 37, 355-72.

Cambra M., Moreno P. & Navarro L. (1979) Detección rápida del virus de la tristeza de los cítricos (CTV), mediante técnicas immunoenzymáticas (ELISA-sandwich). *Anales del Instituto Nacional de Investigaciones Agrarias, Protección Vegetal* 12, 115-25.

Clark M.F. & Adams A.N. (1977) Characteristics of the microplate method of enzyme-linked immunosorbent assay for the detection of plant viruses. *Journal of General Virology* 34, 475-83.

Clarke R.G., Converse R.H. & Kojima M. (1980) Enzyme-linked immunosorbent assay to detect potato leafroll virus in potato tubers and viruliferous aphids. *Plant Disease* 64, 43-5.

Derrick K.S. (1973) Detection and identification of plant viruses by serologically specific electron microscopy. *Phytopathology* 63, 441 (Abstract).

Demski J.M. (1981) Peanut mottle. *Abstracts of the International Meeting on Plant Virus Disease Epidemiology, Oxford, 28-30 July, 1981,* p. 14.

Eastop V.F. (1955) Selection of aphid species by different kinds of insect traps. *Nature* 176, 936.

Fargette D.J., Jenniskens M.J. & Peters D. (1981) Acquisition of PEMV by *Acyrthosiphon pisum* studied by ELISA and transmission tests. *Fifth International Congress of Virology, Strasbourg, 2-7 August, 1981*, p. 214 (Abstract).

Gera A., Loebenstein G. & Raccah B. (1978) Detection of cucumber mosaic virus in viruliferous aphids by enzyme-linked immunosorbent assay *Virology* 86, 542-5.

Halbert S.E., Irwin M.E. & Goodman R.M. (1981) Alate aphid (Homoptera : Aphididae) species and their relative importance as field vectors of soybean mosaic virus. *Annals of Applied Biology* 97, 1-9.

Howell P.J. (1974) Field studies of potato leafroll virus spread in south-eastern Scotland, 1962-1969, in relation to aphid population and other factors. *Annals of Applied Biology* 76, 187-97.

Irwin M.E. (1980) Sampling aphids in soybean field. Sampling methods. *Soybean Entomology* (Ed. by M. Kogan & D.C. Herzog), pp. 239-59. Springer-Verlag, New York.

Irwin M.E. & Goodman R.M. (1981) Ecology and control of soybean mosaic virus. *Plant Diseases and their Vectors: Ecology and Epidemiology* (Ed. by K. Maramorosch & K.F. Harris), pp. 181-220. Academic Press, New York.

Labonne G. (1981) Cucumber mosaic virus (CMV) spread by its vectors in muskmelon plots of southeastern France. *Abstracts of the International Meeting on Plant Virus Disease Epidemiology, Oxford, 28-30 July, 1981*, p. 29.

Leclant F. (1978) *Etude bioécologique des aphides de la région Méditerranéenne. Implications agronomiques.* PhD Thesis, Université des Sciences et Techniques du Languedoc.

Loebenstein G. & Raccah B. (1980) Control of non-persistently transmitted aphid-borne viruses. *Phytoparasitica* 8, 221-35.

Marrou J., Quiot J.B., Duteil M., Labonne G., Leclant F. & Renoust M. (1979) Ecologie et épidémiologie du virus de la mosaique du concombre dans le sud-est de la France. III. Intéret de l'exposition de plantes appats pour l'étude de la dissémination du virus de la mosaique du concombre. *Annales de Phytopathologie* 11, 291-306.

Moericke V. (1951) Eine Farbfalle zur Kontrolle des Fluges von Blattlausen insbesondere der Pfirsichblattlaus. *Nachrichtenblatt Deutsches Pflanzenschutzdienst* 3, 23-4.

Nault L.R., Gordon D.T., Gingery R.E., Bradfute O.I. & Castillo L. (1979) Identification of maize viruses and mollicutes and their potential vectors in Peru. *Phytopathology* 69, 824-8.

Plumb R.T. (1971) The control of insect-transmitted viruses of cereals. *Proceedings of the 6th British Insecticide and Fungicide Conference* pp. 307-13.

Plumb R.T. (1976) Barley yellow dwarf virus in aphids caught in suction traps, 1969-1973. *Annals of Applied Biology* 83, 53-9.

Raccah B. & Loebenstein G. (1981) The incidence of potato virus Y and cucumber mosaic virus in bell peppers and aphids trapped in suction traps. *Abstracts of the International Meeting on Plant Virus Disease Epidemiology, Oxford, 28-30 July, 1981*, p. 27.

Raccah B., Loebenstein G., Maoz A. & Gera A. (1981) Detection of non-persistent viruses in their aphid vectors: Problems and prospects. *Fifth International Congress of Virology, Strasbourg, 2-7 August, 1981*, p. 212 (Abstract).

Roberts I.M. & Harrison B.D. (1979) Detection of potato leafroll and
    potato mop top viruses by immunosorbent electron microscopy. *Annals
    of Applied Biology* 93, 289-97.
Schwartz R.E. (1965) Aphid-borne virus diseases of citrus and their
    vectors in South Africa. A. Investigations into the epidemiology
    of aphid-transmissible virus diseases by means of trap plants. *South
    African Journal of Agricultural Sciences* 8, 839-52.
Shaunak K.K. & Pitre H.N. (1971) Seasonal alate aphid collections in
    yellow traps in northern Mississippi: Possible relationships to
    maize dwarf mosaic disease. *Journal of Economic Entomology* 64, 1105.
Tamada T. & Harrison B.D. (1981) Quantitative studies on the uptake
    and retention of potato leafroll virus by aphids in laboratory and
    field conditions. *Annals of Applied Biology* 98, 261-76.
Taylor L.R. & Palmer J.M.P. (1972) Aerial sampling. *Aphid Technology*
    (Ed. by H.F. van Emden), pp. 189-254. Academic Press, London.

The statistical relationship between aphid trap catches and maize dwarf mosaic virus inoculation pressure

L.V. MADDEN, J.K. KNOKE & RAYMOND LOUIE

Ohio Agricultural Research and Development Center and
United States Department of Agriculture, Wooster, OH 44691, USA

INTRODUCTION

Maize dwarf mosaic virus (MDMV) and maize chlorotic dwarf virus are the most common viruses of maize (*Zea mays*) in the United States of America (Gordon *et al.*, 1981). Although chlorotic dwarf causes more yield loss in susceptible hybrids than does MDMV, the latter virus has a much wider geographical distribution (Gordon *et al.*, 1981). Twenty-three aphid species have been shown to transmit MDMV, although with great differences in transmission efficiency (Knoke *et al.*, 1977; Knoke & Louie, 1981).

Knoke *et al.* (1977) proposed the term *inoculation pressure* (IP) to describe the probability of susceptible maize plants becoming infected during a given time period. IP increases with aphid numbers, their movement among plants, and the frequency of probing by viruliferous aphids. These authors measured IP by the proportion of maize seedlings that became infected by MDMV during specified time intervals (Knoke *et al.*, 1977). This chapter reports:-

1. the association between MDMV IP and aphid water trap catches at two locations in Ohio,
2. the development of a statistical model relating this IP to numbers of particular species of aphids,
3. the determination of the most important aphid species in the epidemiology of maize dwarf mosaic.

MATERIALS AND METHODS

Data collection

Field data were collected in southern (Portsmouth, $38°44'N$, $82°58'W$) and northern (Wooster, $40°47'N$, $81°55'W$) Ohio. Galvanized metal pan traps, 30.5 x 30.5 x 10.2 cm, painted with "Canary Yellow" (Pratt & Lambert Inc., New York, NY), were located at Portsmouth from 1970 to 1979 (except for 1973 and 1975) and at Wooster from 1969 to 1979. The pans were three-quarters filled with water and positioned on a metal stand

Plumb R.T. & Thresh J.M. (1983) *Plant Virus Epidemiology.*
Blackwell Scientific Publications, Oxford.

with the upper edge of the trap 61 cm above ground. Aphids were removed, usually weekly, and counted. During some years aphids were removed and counted daily and these counts were later converted to weekly values. A random subsample of 50 aphids from the weekly collections was chosen, stored in 70% alcohol and later identified. The entire weekly sample was stored when less than 50 aphids were trapped. Some weekly sub-samples were later lost due to contamination or drying. The proportion of each species in the subsample was assumed to be the same as in the total weekly catch.

Individual maize seedlings ("trap plants") of the MDMV-susceptible hybrid WF9 x Oh51A were grown in 10.2 cm diameter plastic pots in a glasshouse and transported to the field stations 14 days after sowing. The seedlings were left in the field for 7 days before returning them to a glasshouse where they were observed for symptoms for 3-5 weeks (Knoke et al., 1977). During most years 50 trap plants/week/location were used. Partially buried plastic cups served as receptacles for the pots; a row of these cups was centrally located in a 3.05 x 32.98 m area that was weeded weekly.

Data analyses

Regression analysis was used to develop empirical models relating IP to total aphid catches and to catches of a particular aphid species. The simple regression model can be represented by:-

$$Y_i = B_0 + B_1 X_i + u_i \qquad\qquad (i)$$

where Y_i is the IP at the i-th time (week) for a given location and year; X_i is the natural log of the total number of aphids caught plus one for the i-th time; u_i is the error term (unexplained variability); and the Bs are parameters which are estimated from the data. The multiple regression model can be written:-

$$Y_i = B_0 + B_1 X_{i1} + \ldots + B_p X_{ip} + u_i \qquad\qquad (ii)$$

where X_{i1} is the natural log of the number of aphids plus one for the first aphid species at the i-th time; X_{ip} is the natural log of the number of aphids plus one for the p-th aphid species at the i-th time; the Bs are parameters representing the weight given to the Xs in relation to IP; the other terms are as in (i). In both equations, B_0 is the Y-intercept, i.e., the value of Y when all the Xs equal zero. An "all possible regressions" stepwise procedure (Neter & Wasserman, 1974) was used to eliminate non-significant variables (aphid species) from the multiple regression equation. The "Minitab" statistical program was used for all analyses (Ryan et al., 1980).

RESULTS

Six aphid species, known to be vectors of MDMV (Knoke & Louie, 1981), each totalled at least 5% of the aphids caught in at least 1 year of trapping. These species were:-

    Mp: *Myzus persicae*, green peach aphid,
    Ag: *Aphis gossypii*, cotton aphid,
    Ac: *Aphis craccivora*, cowpea aphid,
    Rm: *Rhopalosiphum maidis*, corn leaf aphid,
    Ha: *Hyalopterus atriplicis*, boat gall aphid,
    Da: *Dactynotus ambrosiae*, brown ambrosia aphid.

The total number of aphids, percentage of each species, as well as the
time of population maxima, varied greatly between locations and years.

   Tables 1 and 2 contain the results of the stepwise regression analysis
for the Portsmouth and Wooster data, respectively. To interpret these
data, 1973 at Wooster (Table 2) can be taken as an example. The prediction equation can be written as:-

$$\hat{Y} = 0.005 + 0.172X_3 + 0.121X_4$$

where $\hat{Y}$ is the predicted IP, and X_3 and X_4 are the natural logs of *A.
craccivora* and *R. maidis* numbers plus one, respectively. Regression
analysis showed that no aphid species was significant at either
Portsmouth or Wooster in 1976, at Portsmouth in 1979, or at Wooster in
1975 (Tables 1 and 2). At least one species was significant in each of
the other years.

   At Portsmouth, *M. persicae* was a significant predictor of IP in 4 of
the 8 years, or 4 of the 6 years when at least one aphid species was
significant (Table 1). This species was significant alone for 3 years.
R. maidis and *A. craccivora* were significant for 2 years, whereas *D.
ambrosiae*, *H. atriplicis* and *A. gossypii* were never significant.

   At Wooster, *M. persicae* and *A. craccivora* were significant in 3 of
11 years, or 3 of the 9 years when at least one aphid species was
significant (Table 2). *R. maidis* was significant in 5 of the 9 years
with significant aphid species; *A. gossypii* was significant for 2 years;
D. ambrosiae was significant once and *H. atriplicis* was never significant.

   Results of regressing the proportion of trap plants infected with
MDMV on the natural log of total aphids plus one are summarized in
Tables 3 and 4. At Portsmouth, IP was significantly related to total
aphids in 6 out of 8 years, as with the regressions using individual
aphid species. However, the years showing the relationship were not
all the same (Tables 1 & 3). At Wooster, IP was significantly associated with total aphids in 8 out of 11 years, 1 year less than with the
regressions of aphid species (Tables 2 & 4). More observations (weeks)
were used in the regressions with total aphids because more aphids were
trapped than were later identified.

Table 1. Estimated parameters for the stepwise regression analysis at *Portsmouth*, together with the standard deviations of the estimated parameters (in parentheses), standard deviation about the regression surface (s), F-statistic (F), degrees of freedom (D.F.), coefficient of determination (R^2), coefficient of determination adjusted for degrees of freedom (R^2_a), probability plot correlation test for normality of the residuals (Corr.), and subjective appraisal of the residual plot (Res.).

Year	Week	B_0 [†]	Estimated parameters for: [¶]			s	F	D.F.	R^2	R^2_a	Corr.	Res. [▽]
			Mp	Ac	Rm							
1970	6	-0.173 (.097)	0.130 (.027)	0.126 (.029)		0.081	15.60*	2,3	0.912	0.854	.96**	A
1971	13	0.400 (.075)			0.136 (.028)	0.178	24.16**	1,11	0.687	0.659	.95**	A
1972	16	0.399 (.100)		0.083 (.027)	0.111 (.028)	0.203	10.08**	2,13	0.608	0.548	.99**	A
1974	10	0.156 (.109)	0.119 (.039)			0.236	9.28*	1,8	0.537	0.479	.95**	A
1977	16	0.155 (.106)	0.066 (.023)			0.289	8.22*	1,14	0.370	0.325	.97**	A
1978	12	0.165 (.073)	0.129 (.016)			0.165	64.01**	1,10	0.865	0.851	.98**	A

† Y-intercept.   * Significant at P = 0.05 or ** P = 0.01.

¶ Coefficients for species in the regression model after stepwise elimination. See text for full names of the aphids indicated by abbreviations.

▽ Residual plot acceptable (A) or not acceptable (NA).

No aphid species was significant in 1976 or 1979.

Table 2. Estimated parameters for the stepwise regression analysis at *Wooster* (symbols as Table 1).

Year	Week	B_0[+]	Estimated parameters for:[¶]					s	F	D.F.	R^2	R_a^2	Corr.	Res.[▽]
			Mp	Ag	Ac	Rm	Da							
1969	19	-0.013 (.014)	0.021 (.004)					0.038	25.61**	1,17	0.601	0.578	.95*	A
1970	14	-0.004 (.019)	0.014 (.005)					0.045	9.74**	1,12	0.448	0.402	.94*	NA
1971	23	-.003 (.013)				0.012 (.004)		0.040	11.81**	1,21	0.360	0.329	.96**	A
1972	22	-0.004 (.005)	0.005 (.002)					0.013	12.66**	1,20	0.388	0.357	.93	NA
1973	23	0.005 (.057)			0.172 (.031)	0.121 (.017)		0.181	37.33**	2,20	0.789	0.768	.96*	A
1974	22	-0.001 (.004)		0.009 (.002)	-0.011 (.004)		0.015 (.003)	0.011	17.55**	3,18	0.745	0.703	.95*	A
1977	21	-0.029 (0.41)		-0.045 (.022)		0.067 (.016)		0.101	11.35**	2,18	0.558	0.508	.93	NA
1978	16	-0.016 (.034)				0.038 (.014)		0.096	7.91**	1,14	0.361	0.315	.93	A
1979	18	-0.028 (.045)			0.077 (.024)	0.139 (.019)		0.122	31.99**	2,15	0.810	0.785	.96**	A

No aphid species was significant in 1975 or 1976.

Table 3. Estimated parameters for the regression analysis at *Portsmouth* using total numbers of aphids (symbols as Table 1).

Year	Week	Estimated Parameters† B_0	B_1	s	F	D.F.	R^2	R_a^2	Corr.	Res.$^\nabla$
1970	15	-0.434 (.234)	0.129 (.046)	0.168	7.98*	1,13	0.380	0.333	.96**	A
1971	14	1.552 (.530)	-0.178 (.101)	0.314	3.11	1,12	0.206	0.140	.99**	A
1972	21	0.071 (.196)	0.137 (.038)	0.283	13.03**	1,19	0.407	0.376	.97**	NA
1974	20	-0.403 (.166)	0.192 (.031)	0.220	38.42**	1,18	0.681	0.663	.99**	A
1976	12	-0.045 (.254)	0.120 (.037)	0.184	10.36**	1,10	0.509	0.460	.96**	A
1977	19	0.082 (.147)	0.071 (.029)	0.316	6.14*	1,17	0.265	0.222	.98**	A
1978	17	0.023 (.067)	0.145 (.017)	0.184	72.98**	1,15	0.830	0.818	.99**	A
1979	18	0.271 (.152)	0.051 (.032)	0.244	2.56	1,16	0.138	0.084	.97**	A

† B_0: Y-intercept; B_1: natural log of total aphids plus one.

DISCUSSION

Previous studies (Blair, 1970; Shaunak & Pitre, 1971) have shown the possible role of certain aphid species in the spread of MDMV. The studies consisted essentially of collecting and identifying aphids near maize fields and, in one case (Shaunak & Pitre, 1971), determining which species transmitted the virus to test plants. These results were based on 1 year of observations and were essentially descriptive and inconclusive. Louie *et al.* (1974) and Knoke *et al.* (1974) showed that the infection of trap plants and total numbers of trapped aphids followed parallel trends. Other workers (Hagel & Hempton, 1970; Louie, 1980; Irwin & Goodman, 1981) have shown similar relationships between aphid numbers and disease incidence for bean yellow mosaic, sugarcane mosaic, and soybean mosaic, respectively. Their studies were also descriptive or qualitative.

Table 4. Estimated parameters for the regression analysis at *Wooster* using total numbers of aphids (symbols as Table 1).

Year	Week	Estimated Parameters†		s	F	D.F.	R^2	R_a^2	Corr.	Res. ∇
		B_0	B_1							
1969	20	-0.052 (.023)	0.022 (.005)	0.042	18.19**	1,18	0.503	0.475	.95*	A
1970	15	-0.066 (.048)	0.021 (.009)	0.049	5.18*	1,13	0.285	0.230	.93	A
1971	24	-0.021 (.020)	0.010 (.004)	0.043	7.63*	1,22	0.258	0.224	.92	A
1972	23	-0.013 (.009)	0.005 (.002)	0.015	6.72*	1,21	0.242	0.206	.92	NA
1973	25	-0.284 (.113)	0.140 (.023)	0.237	36.92**	1,23	0.616	0.600	.96*	A
1974	23	-0.021 (.008)	0.008 (.002)	0.015	15.45**	1,21	0.424	0.396	.93	NA
1975	19	-0.248 (.134)	0.099 (.028)	0.180	12.61**	1,17	0.426	0.392	.94	A
1976	18	-0.115 (.156)	0.067 (.033)	0.160	4.10	1,16	0.204	0.154	.97**	A
1977	23	-0.008 (.071)	0.020 (.021)	0.138	0.95	1,21	0.043	-0.002	.78	NA
1978	16	-0.108 (.133)	0.036 (.029)	0.114	1.50	1,14	0.097	0.032	.80	NA
1979	19	-0.078 (.099)	0.094 (.027)	0.204	12.28**	1,17	0.419	0.385	.98**	A

† B_0: Y-intercept; B_1: natural log of total aphids plus one.

In our study we have attempted to quantify the relationship between MDMV-infected trap plants and aphid trap catches. Inspection of the residual plots indicated that the natural log transformation of aphid species numbers resulted in the most appropriate empirical model for inoculation pressure. Although the residual plot for some years at Wooster exhibited an undesirable pattern, no other single transformation of the aphid variables produced more acceptable residuals for the majority of years and locations. A common transformation of variables was necessary to compare years and locations readily. Gregory's multiple infection transformation of the dependent variable IP (Gregory, 1948) was also attempted to find a more appropriate model. This transformation constrains predicted values of IP to <1, as well as potentially giving the model a more biological interpretation. The transformation has the disadvantage that observed values of IP = 1 must be eliminated from the data since the transformation is undefined when IP = 1. This constraint reduced the number of observations to unacceptable levels for some years at Portsmouth. In addition, the multiple infection transformation did not result in a more appropriate model as judged by the residual plots.

Normality of the residuals, which is an assumption of regression (Neter & Wasserman, 1974), was evaluated with a probability plot correlation test (Ryan & Joiner, 1976). Normality of the residuals in the aphid species equations was rejected in only 3 years. Residuals for the total aphid equations were less acceptable than those for aphid species. However as with the aphid species, no other transformation was satisfactory for as many of the years. The Y-intercept (B_0) was kept in the regression equations, even when not significantly different from zero, because regression models with a "forced" zero intercept have less desirable properties than those for ordinary regression models with a Y-intercept (Neter & Wasserman, 1974).

The regression models significantly fitted the data, as indicated by significant F-tests (Tables 1-4), but were not sufficient for prediction. The proportion of "explained variability" (R^2, coefficient of determination) for the multiple regression model was seldom above 0.80 and often below 0.60. Knowledge of aphid behaviour, the percentage of aphids viruliferous at each time during the season, as well as environmental conditions would be needed to explain a larger proportion of the data variability. Nevertheless, this work has shown that a significant amount of information on IP can be obtained by using trap catches of aphid species.

The coefficient of determination adjusted for degrees of freedom (R_a^2), although not an exact measure of explained variability, is very useful for comparing models (Neter & Wasserman, 1974). This is because R^2 will increase, or at least remain the same, as additional variables are included in a regression model, even when a variable is not significant. This allows a model to be developed with an artificially high R^2 by merely including many nonsignificant variables. R_a^2 will, however, decrease as nonsignificant variables are added to a regression model; R_a^2 can thus be used to compare models with different numbers of variables as exhibited by our data. The value of R_a^2 for each significant multiple regression model (Tables 1 & 2) was larger than R_a^2 for the simple regression model (Tables 3 & 4), except in 1 year. A better description of IP was thus obtained by using numbers of each aphid species rather than simply using total aphid numbers. This improvement was not unexpected since total aphids comprise a range of species with different transmission efficiencies (Knoke & Louie, 1981).

Aphid flights differed greatly among years and locations, causing numbers, species composition, time of maximum number, and significant species to differ from year to year and between locations. Environmental conditions and host crop phenologies are likely to have a large effect on these aphid flights. Despite this variability, *M. persicae* was repeatedly significant with MDMV IP at Portsmouth. At Wooster, *R. maidis*, *M. persicae*, and *A. craccivora* were less consistently significant. For example, *M. persicae* was significant at Portsmouth for 4 of the 6 years when at least one species was significant, whereas at Wooster *R. maidis* was significant in 5 out of the 9 years when at least one aphid was significant.

Portsmouth has a warmer climate (mean annual air temperature (T) = 13.2°C) than Wooster (T = 9.3°C). Also, the overwintering host, *Sorghum halepense*, of five of the six known strains of MDMV occurs at Portsmouth but not at Wooster (Knoke & Louie, 1981). Because of these characteristics, inoculation pressure is higher at Portsmouth than at Wooster (Knoke *et al.*, 1974; Louie *et al.*, 1974). To illustrate this, more than 50% of the susceptible trap plants were infected throughout the majority of each growing season at Portsmouth, whereas at Wooster, disease incidence on trap plants seldom reached 50%. We are now determining which environmental conditions favour IP at Portsmouth and Wooster. The aim is to predict aphid numbers, IP, and the resulting disease incidence.

Acknowledgements

We thank R.J. Anderson for counting and identifying aphids, and J.J. Abt and S.S. Mendiola for technical assistance.

This paper describes cooperative investigations of the Ohio Agricultural Research and Development Center (OARDC), and United States Department of Agriculture (USDA), Agricultural Research Service and has been approved for publication by OARDC as journal article No. 167-81. Mention of a commercial or proprietary product does not constitute an endorsement by the USDA.

REFERENCES

Blair B.D. (1970) Aphids collected from a Scioto County, Ohio, corn field and areas bordering the field. *Journal of Economic Entomology* 63, 1099-101.
Gordon D.T., Bradfute O.E., Gingery R.E., Knoke J.K., Louie R., Nault L.R. & Scott G.E. (1981) Introduction: History, geographic distribution, pathogen characteristics, and economic importance. *Virus and Viruslike Diseases of Maize in the United States* (Ed. by D.T. Gordon, J.K. Knoke & G.E. Scott), pp. 1-12. Southern Cooperative Series Bulletin 247.
Gregory P.H. (1948) The multiple infection transformation. *Annals of Applied Biology* 35, 412-7.

Hagel G.T. & Hampton R.O. (1970) Dispersal of aphids and leafhoppers
    from red clover to red mexican beans, and the spread of bean yellow
    mosaic by aphids. *Journal of Economic Entomology* 63, 1057-60.
Irwin M.E. & Goodman R.M. (1981) Ecology and control of soybean mosaic
    virus. *Plant Diseases and Their Vectors: Ecology and Epidemiology.*
    (Ed. by K. Maramorosch & K.F. Harris), pp. 181-220. Academic Press,
    New York.
Knoke J.K., Anderson R.J. & Louie R. (1977) Virus disease epiphytology:
    Developing field tests for disease resistance in maize. *Proceedings
    International Maize Virus Disease Colloquium & Workshop*, 16-19
    August 1976. (Ed. by L.E. Williams, D.T. Gordon & L.R. Nault),
    pp. 116-21. Ohio Agricultural Research & Development Center, Wooster.
Knoke J.K. & Louie R. (1981) Epiphytology of maize virus diseases.
    Virus and Viruslike Diseases of Maize in the United States. (Ed.
    by D.T. Gordon, J.K. Knoke & G.E. Scott), pp. 92-102. Southern
    Cooperative Series Bulletin 247.
Knoke J.K., Louie R., Anderson R.J. & Gordon D.T. (1974) Distribution
    of maize dwarf mosaic and aphid vectors in Ohio. *Phytopathology*
    64, 639-45.
Louie R. (1980) Sugarcane mosaic virus in Kenya. *Plant Disease* 64,
    944-7.
Louie R., Knoke J.K. & Gordon D.T. (1974) Epiphytotics of maize dwarf
    mosaic and maize chlorotic dwarf diseases in Ohio. *Phytopathology*
    64, 1455-9.
Neter J. & Wasserman W. (1974) *Applied Linear Statistical Models.*
    Richard D. Irwin Inc., Homewood, Il.
Ryan T.A. & Joiner B.L. (1976) *Normal probability plots and tests for
    normality.* Technical Report, Statistics Department, Pennsylvania
    State University, 12 pp.
Ryan T.A. & Joiner B.L. (1980) *Minitab Reference Manual.* Pennsylvania
    State University.
Shaunak K.K. & Pitre H.N. (1971) Seasonal alate aphid collections in
    yellow pan traps in northeastern Mississippi: possible relation-
    ship to maize dwarf mosaic disease. *Journal of Economic Entomology*
    64, 1105-9.

Field experiments on the integrated control of aphid-borne viruses in muskmelon

H. LECOQ* & M. PITRAT†
*Station de Pathologie Végétale,
†Station d'Amélioration des Plantes Maraîchères,
INRA, Domaine Saint Maurice, 84140 Montfavet, France

INTRODUCTION

Aphid-borne viruses regularly cause severe losses in muskmelon (*Cucumis melo*) plantings in southern France. Cucumber mosaic virus (CMV) is prevalent in this region, and every year in the Lower Rhône Valley most of the muskmelon plants grown in the open show typical mosaic symptoms by mid-June (Messiaen *et al.*, 1963; Quiot *et al.*, 1979a). When infection occurs before or at fruit setting (i.e. within 50 days of planting), fruit yield is severely reduced and it is therefore very important to restrict virus spread at this early stage. Epidemics caused by watermelon mosaic virus 2 (WMV2) and muskmelon yellow stunt virus (MYSV) also occasionally occur, usually later in the growing season. These two potyviruses can drastically decrease the yield of marketable fruit but are less important than CMV because their attacks are less frequent (Quiot *et al.*, 1979a; Lecoq *et al.*, 1981).

Several measures have been used to limit virus spread within muskmelon fields including the use of white polyethylene as a mulch which acts as an aphid repellent (Messiaen & Maison, 1965; Marrou & Messiaen, 1968). Weed control along field boundaries has also been proposed as a means of limiting sources of virus and/or aphid vectors near plantings (Doolittle & Walker, 1926; Wellman, 1937; Quiot *et al.*, 1979b). However, these practices alone seldom give adequate control.

Another approach is the development of resistant cultivars. A programme was started in 1972 to introduce two types of resistance to CMV into "Charentais"-type muskmelon, the main type cultivated in France.

The first type of resistance prevents infection by CMV "Common" strains and is oligogenic and recessive (Risser *et al.*, 1977). It is ineffective against CMV strains of the "Song" pathotype, which induce mosaic symptoms in all cultivars. Nearly one third, of over a thousand, CMV isolates collected from naturally infected cultivated plants or weeds in south-east France were "Song" strains (Leroux *et al.*, 1979).

Plumb R.T. & Thresh J.M. (1983) *Plant Virus Epidemiology.*
Blackwell Scientific Publications, Oxford.

The second type of resistance prevents the transmission of either
"Common" or "Song" strains by one of their main natural vectors, *Aphis
gossypii* (Lecoq *et al.*, 1979). The resistance is governed by a single
dominant gene which also confers resistance to infestation by *A.
gossypii* and prevents aphids colonizing plants (Pitrat & Lecoq, 1980).
This resistance is specific to *A. gossypii* and is ineffective against
other aphid species, including *Myzus persicae*, *A. fabae* and *A. craccivora*, which are efficient CMV vectors (Lecoq *et al.*, 1980).

A cultivar possessing both resistance mechanisms remains susceptible
to CMV "Song" strains transmitted by vectors other than *A. gossypii*,
and epidemics caused by this pathotype are observed in such cultivars.

No resistance to WMV2 or MYSV is known at present except that
resistance to virus transmission by *A. gossypii* is also effective
against these viruses (Lecoq *et al.*, 1980).

The aim of this study, started in 1979, was to follow the development
of epidemics caused by aphid-borne viruses in CMV-susceptible cultivars
and in cultivars with the types of resistance described above, and to
assess the effects on these epidemics of plastic mulches and of
removing weeds before planting.

METHODS

The plantings were in an important vegetable-growing area near Avignon;
the experimental plots were south of a 6 m high windbreak of cypress
(*Cupressus sempervirens*) and separated from each other by 3 m high
artificial windbreaks of dried reeds (*Arundo donax*). Such windbreaks
are used commonly in the Lower Rhône valley, and at their base the
ground is often heavily infested with weeds.

Each plot of approximately 600 m² was planted with 260 plants of
either the CMV-susceptible cultivar Védrantais or a resistant breeding
line with both types of resistance. The resistant cultivar in 1979
was Songwhan charmi (PI 161375) and in 1980 Pamise, the product of the
second backcross from PI 161375 to a "Charentais" cultivar. Seeds were
sown in peat blocks and plantlets were left in an insect-proof greenhouse until they reached the 2/3-leaf stage. All plots were planted on
the same day. Two successive plantings, in spring (early May) and
summer (early August), were made annually. Each crop was exposed to
different conditions. Large aphid infestations occur in spring when
virus sources are scarce, whereas there tend to be fewer aphids in
summer when virus sources are numerous.

In 1979, all weeds growing along the windbreaks between some plots
were carefully destroyed by herbicides before planting and thereafter
mechanically. In 1980, a white polyethylene mulch (1.60 m wide, 60 μm
thick) was used in plots where the borders were carefully weeded. The
mulch was laid on bare ground and circular holes, *c.* 15 cm in diameter,
were punched through it so as to transplant muskmelon plantlets. Tn

both years control plots were neither weeded nor mulched. Otherwise
standard cultural practices were adopted for all plots (including
mechanical or manual weeding within the planting).

All plants were observed individually for mosaic symptoms every 2 or
3 days. There is a good correlation between the occurrence of mosaic
symptoms and the presence of CMV either alone or in double infections
with WMV2 as determined by indexing on test plants and electron microscopy (Quiot et al., 1979c). WMV2 rarely occurred alone.

The delay in occurrence of epidemics due to cultural practices or
host-plant resistance was estimated by a "protective effect index" (PEI).
PEI represents the delay in days in the time taken for 50% of the plants
to become infected in treated plots compared with equivalent untreated
controls. This is estimated from the regression lines obtained by
plotting against time the observed incidences of infection from 5 to 95%.

RESULTS

The development of aphid-borne virus disease epidemics in four successive muskmelon plantings is shown in Fig. 1(a-d). Each season the
development of the epidemic was delayed in the plots of resistant
cultivars compared with the susceptible plots grown similarly. The PEI
was 12.3-23.5 days (Table 1). In only one comparison did the epidemic
develop earlier in a resistant plot than in a susceptible one, although
the latter was also weeded and mulched (Fig. 1c).

Table 1. The PEIs for various treatment comparisons in plots of
CMV-resistant (R) or susceptible (S) muskmelon cultivars which were
weeded (w) or weeded and mulched (w-m).

| Planting | Treatment comparisons | | | | | | | |
	R/S	Rw/Sw	Rw-m/Sw-m	Sw/S	Rw/R	Sw-m/S	Rw-m/R	Rw-m/S
Spring 1979	21.2	18.9	-	0.9	-1.3	-	-	-
Summer 1979	14.0	23.5	-	2.8	12.3	-	-	-
Spring 1980	14.0	-	16.8	-	-	18.7	21.5	35.5
Summer 1980	12.3	-	-	-	2.0	-	17.3	29.6

In 1979, weeding the plot edges before planting reduced virus spread
but only in some conditions. Weeding had no effect on infection of the
spring planting (Fig. 1a) possibly because few weeds were infected with
CMV at this time. By contrast many weeds were infected by the summer
when weed control around the plantings did restrict infection and gave
PEIs of 2.0-12.3 (Fig. 1b).

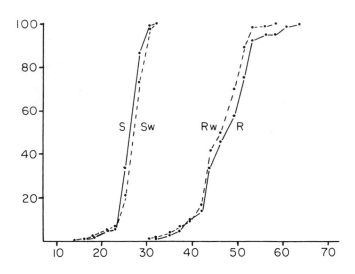

(a) Spring planting 1979.

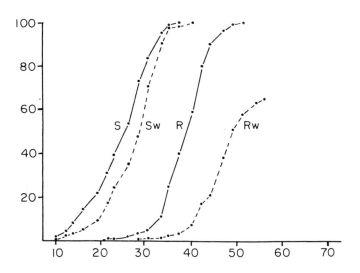

(b) Summer planting 1979.

Figure 1. Development of CMV-epidemics in partially CMV-resistant (R)

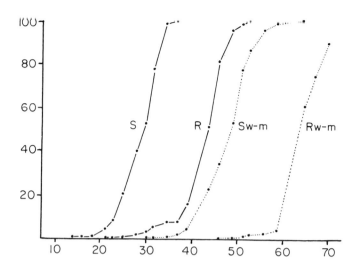

(c) Spring planting 1980.

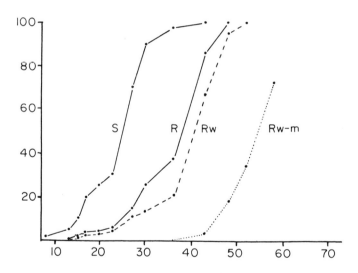

(d) Summer planting 1980.

or susceptible (S) plots, weeded (w) or weeded and mulched (w-m).

In 1980, a combination of weeding and mulching protected both resistant and susceptible cultivars very effectively (PEIs of 17.3-21.5). The longest delay in the epidemic occurred when plots of the CMV-resistant cultivar were weeded and mulched compared with the susceptible cultivar not weeded or mulched (PEIs of 35.5 and 29.6 for spring and summer planting, respectively).

DISCUSSION

Aphid-borne viruses cause some of the most damaging diseases of musk-melon in southern France and in many other parts of the world. Many attempts have been made to develop effective control measures, but no one method has proved satisfactory.

Weeding to remove virus sources from near plantings had only a limited effect on subsequent virus spread in our experiments, possibly because not all plots were weeded. Better protection might be expected when weeding is carried out over whole areas and not only around some borders. Wellman (1937) showed that weeding to a distance of at least 25-40 m around the field margins efficiently protected celery against CMV in Florida.

Mulching was more efficient than weeding in delaying virus spread but the protective effect seems to be lost as soon as muskmelon vines reach the stage when they cover its surface. This happens rapidly because white plastic mulch warms the soil and favours early plant growth.

Partial resistance delayed virus spread significantly although its effect would be expected to be greater in the absence of adjacent susceptible control plots which are likely to act as virus sources. However, the use of partial resistance may increase the frequency of adapted virus strains. Studies of CMV infection of weeds growing around plots that have grown five successive CMV-susceptible or resistant crops have not shown a significant increase in the frequency of the "Song" pathotype associated with the use of resistant cultivars (Lecoq & Pitrat, unpublished). This suggests that the resistance is durable.

The observations reported here indicate that an integrated control scheme using cultural practices and partially resistant cultivars delayed virus spread and adequately protected the crop. Indeed, mulched and weeded plantings of a resistant muskmelon cultivar remained free of virus during most of the fruit setting and early fruit growth periods when infection drastically affects fruit development. Later infection has much less effect on yield.

These results also indicate the potential importance of partial resistance, which is often neglected or overlooked because it does not provide complete protection. Moreover, the benefits of partial resistance are increased when used with certain cultural practices and their value should not be underestimated.

Other measures could be included in the muskmelon protection scheme, particularly the use of oil sprays (Zitter & Simons, 1980) which could be applied when plant growth has obscured the mulch and made it inefficient in repelling aphids. Trials to test this are planned and the search will be continued for additional types of resistance.

REFERENCES

Doolittle S.P. & Walker M.N. (1926) Control of cucumber mosaic by eradication of wild host plants. *United States Department of Agriculture Bulletin* 1461, 14 pp.

Lecoq H., Cohen S., Pitrat M. & Labonne G. (1979) Resistance to cucumber mosaic virus transmission by aphids in *Cucumis melo*. *Phytopathology* 69, 1223-5.

Lecoq H., Labonne G. & Pitrat M. (1980) Specificity of resistance to virus transmission by aphids in *Cucumis melo*. *Annales de Phytopathologie* 12, 139-44.

Lecoq H., Pitrat M. & Clément M. (1981) Identification et caractéri-sation d'un potyvirus provoquant la maladie du rabougrissement jaune du melon. *Agronomie* 1, 827-34.

Leroux J.P., Quiot J.B., Lecoq H. & Pitrat M. (1979) Mise en évidence et répartition dans le Sud-Est de la France d'un pathotype particu-lier du virus de la mosaique du concombre. *Annales de Phytopathologie* 11, 431-8.

Marrou J. & Messiaen C.M. (1968) Essai de protection des cultures de melons et de courgettes contre le virus de la mosaique du concombre. *Etudes de Virologie, Annales des Epiphyties* 19, 147-57.

Messiaen C.M. & Maison P. (1965) Essai de prévention des attaques de virus I du concombre réalisés en 1965. *Comptes Rendus de Journées Phytiatrie et de Phytopharmacie Circum Méditerranéennes, Marseille*, 195-7.

Messiaen C.M., Maison P. & Migliori A. (1963) Le virus 1 du concombre dans le Sud-Est de la France. *Phytopathologia Mediterranea* 2, 251-60.

Pitrat M. & Lecoq H. (1980) Inheritance of resistance to cucumber mosaic virus transmission by *Aphis gossypii* in *Cucumis melo*. *Phyto-pathology* 70, 958-61.

Quiot J.B., Douine L. & Gebré-Selassié K. (1979a) Fréquence des principales viroses identifiées dans une exploitation maraîchère du Sud-Est de la France. *Annales de Phytopathologie* 11, 283-90.

Quiot J.B., Marchoux G., Douine L. & Vigouroux A. (1979b) Ecologie et épidémiologie du virus de la mosaique du concombre dans le Sud-Est de la France. V. Rôle des espèces spontanées dans la conservation du virus. *Annales de Phytopathologie* 11, 325-48.

Quiot J.B., Verbrugghe M., Labonne G., Leclant F. & Marrou J. (1979c) Ecologie et épidémiologie du virus de la mosaique du concombre dans le Sud-Est de la France. IV. Influence des brise-vent sur la répartition des contaminations virales dans une culture protégée. *Annales de Phytopathologie* 11, 307-24.

Risser G., Pitrat M. & Rode J.C. (1977) Etude de la résistance du melon
    (*Cucumis melo* L.) au virus de la mosaique du concombre. *Annales de
    l'Amélioration des Plantes* 27, 509-22.
Wellman F.L. (1937) Control of southern celery mosaic in Florida by
    removing weeds that serve as sources of mosaic infection. *United
    States Department of Agriculture Technical Bulletin* 548, 16 pp.
Zitter T.A. & Simons J.N. (1980) Management of viruses by alteration
    of vector efficiency and by cultural practices. *Annual Review of
    Phytopathology* 18, 289-310.

The comparative ecology of cucumber mosaic virus in Mediterranean and tropical regions

J.B. QUIOT, G. LABONNE &
LAURENCE QUIOT-DOUINE
INRA, CRAAG, Domaine Duclos 97170 Petit-Bourg,
Guadeloupe, French West Indies

INTRODUCTION

Various methods are used to control epidemics of diseases caused by non-persistent viruses. However, experience has shown that they are not always equally effective in all regions. This emphasizes the need for a better understanding of the effects of environment on the spread of virus diseases in order to develop the most effective control measures.

With this objective an ecological study has been undertaken on cucumber mosaic virus (CMV) in two completely different regions: the Avignon area of southern France, which has a Mediterranean climate, and the West Indian island of Guadeloupe in the tropics. From such studies, it is hoped to determine the main factors influencing the spread and control of CMV and to develop methods that can be applied to solve similar problems due to other virus diseases.

MATERIALS AND METHODS

CMV is an RNA-containing virus with a divided genome and is transmitted by many aphids in a non-persistent manner; it has a worldwide distribution with a very wide host range. It is prevalent in many vegetable crops and herbaceous ornamentals and also occurs occasionally in some fruit trees.

Avignon-Montfavet has mild winters, few frosts and a hot dry summer from June to September. Strong, northerly "Mistral" winds blow frequently throughout the year. CMV is regarded as a plague by vegetable growers because every year it seriously affects plantings of muskmelon, tomato and pepper (Messiaen et al., 1963; Quiot et al., 1979a).

Guadeloupe is a tropical island (latitude $16°N$) in the Caribbean sea. Experiments were done in the calcareous lowlands where the climate is tropical, with no cold period but with marked dry and wet seasons. North-eastern "Alizee" winds prevail throughout the year.

Plumb R.T. & Thresh J.M. (1983) *Plant Virus Epidemiology.*
Blackwell Scientific Publications, Oxford.

Sugarcane is the main crop in the area, but the area of pastures and vegetable plantings has increased over the last decade. Tomato, muskmelon, watermelon, okra, yam and sweet potato are cultivated in home gardens and in small scattered fields. CMV was first recorded in banana plantings on the hillsides 40 km south of our study area (Yot-Dauthy & Bové, 1966). It has since been found occasionally in some vegetable fields and in numerous weeds (Migliori et al., 1978), but not causing serious crop losses.

EXPERIMENTAL

In the two study areas, small fields were planted with crops susceptible to CMV using local methods of cultivation. The progress of the virus disease epidemics was followed, and data were collected on vector populations, meteorological conditions and the incidence of infection in nearby wild plants.

The experimental conditions and methods used in the Avignon area between 1974 and 1978 have been described previously (Devergne et al., 1978; Quiot et al., 1979c; Quiot, 1980).

In Guadeloupe, experiments were done during 1979-81 in a field surrounded by various crops including sugarcane, cotton, vegetables and pasture. The experimental plots were planted between January and August, which is the usual vegetable-growing period. CMV was identified by inoculating differential hosts or by enzyme-linked immunosorbent assay (ELISA). In 1980 and 1981, vector populations were monitored during the periods of growth. Insects were caught with traps made of sticky fishing thread stretched at intervals of 2.5 mm on wooden frames (30 x 30 cm). These traps were placed at different heights in the plots and renewed twice weekly.

RESULTS

Progress of epidemics in the two areas

There were striking differences in the progress of CMV epidemics in the two areas. In Avignon, CMV soon appeared in muskmelon crops. In each of the four years of the experiments, all the muskmelon crops were totally infected within 40 days of planting (Quiot et al., 1979d). By contrast, CMV appears far less important in Guadeloupe. During the 1979 and 1981 experiments and the first one of 1980, CMV was not detected in any of the plots, even though CMV-infected sources were planted in the March-April experiment of 1981. Some late infections of CMV occurred in the July-September experiment of 1980.

Different results were also obtained using as bait plants 25 young muskmelon seedlings renewed weekly (Marrou et al., 1979). In Avignon, this method showed that CMV spreads from late May until late September and that the number of bait plants infected was largely

influenced by the amount of infection in the surrounding fields.   In
Guadeloupe, no bait plants were infected although CMV was present in
nearby weeds.

Watermelon mosaic virus 2 (WMV2) was prevalent in cucurbits in
Avignon (Quiot et al., 1979a), whereas WMV1 was identified in Guadeloupe
and shown to spread rapidly in cucurbit crops.   To understand the big
differences in the behaviour of CMV in Mediterranean and tropical
climates, a comparison was made of factors that influence the course of
the epidemics.

Virus populations

CMV isolates can be separated into two main groups, "B" and "C"
according to the symptoms they induce in *Nicotiana tabacum* Xanthi nc
(Marrou et al., 1975), their serotype (Devergne & Cardin, 1973) or
their sensitivity to heat (Marchoux et al., 1976; Douine et al., 1979).
Other differential hosts can be used to separate CMV subgroups (Fig. 1).

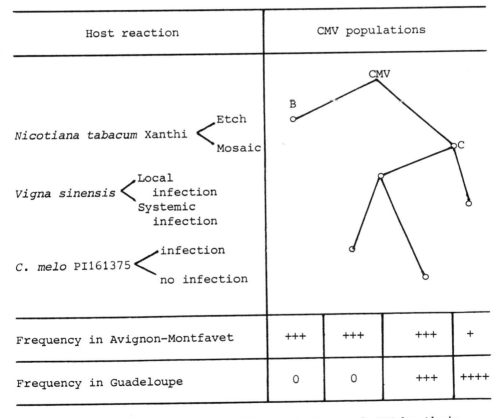

Host reaction	CMV populations			
Frequency in Avignon-Montfavet	+++	+++	+++	+
Frequency in Guadeloupe	0	0	+++	++++

Figure 1. Differentiation of the populations of CMV by their
reaction on differential hosts and their relative frequencies in
Avignon-Montfavet and Guadeloupe.

Though only 30 isolates from Guadeloupe have been characterized compared with 862 from Avignon, there appear to be important differences in the CMV populations (Fig. 1):-

1. Heat-sensitive "B" strains were not detected in Guadeloupe.

2. The frequency of the isolates causing systemic infection of cowpea is much lower in Avignon than in Guadeloupe where there are many leguminous weed hosts of CMV (Migliori et al., 1978).

3. In Avignon, mixed infection by "B" and "C" strains is rare (Quiot et al., 1979e). In Guadeloupe, mixed infection by CMV strains causing local infection and others causing systemic infection in cowpea is encountered frequently. Such mixtures are very infrequent in Avignon. Nevertheless, these two types of CMV can be inoculated to cucurbits under artificial conditions, and it is assumed that the CMV populations in Avignon and Guadeloupe are equally able to infect cucurbits.

Wild hosts

In Avignon, CMV was detected in 40 (29%) of the 137 wild species growing near the experimental field (Quiot et al., 1979b). Annual and perennial species, including some in which CMV is seed-borne, maintain the virus in the area throughout the year, and spread between wild plants occurs even in the absence of infected crops. Determining the number of infected weeds in a given area is difficult because the likelihood of plants becoming infected depends on the surrounding conditions that influence their growth and appearance and the microclimatic conditions, particularly temperature, which can influence virus content (Quiot et al., 1979b).

In Guadeloupe, no detailed study has been made but a general survey has shown that CMV is frequently found in wild plants throughout the island (Migliori et al., 1978). Other surveys using ELISA have detected CMV in weeds growing near experimental fields. Survival of the virus is assured in vegetatively propagated species, including Commelina spp., and in perennials, including leguminous trees. The distribution of CMV throughout the island is also favoured by its occurrence in vegetatively propagated crops such as yam and banana.

Aphid vectors of CMV

In Avignon, alate aphids of more than 140 species have been caught in yellow water pans (diameter 30 cm) or suction traps situated in vegetable fields. Maximum aphid populations (2000 aphids/week/pan) occur between May and July. Few aphids are caught during the dry summer yet spread of CMV in cantaloupe fields occurs in both seasons (Quiot et al., 1979d). During the cold season, few aphids fly.

In Guadeloupe, few aphids are trapped in yellow pans (Migliori et al., 1977). However, aphid catches were increased by using sticky thread traps in 1980 and 1981 and about 20 species have been caught in this way. Limited infestations of some species have been observed at various seasons but they seem to be controlled quickly by predators and parasites.

Several species of aphids have been found to be carrying CMV under natural conditions in Avignon. The most efficient species, such as *Aphis gossypii* and *A. craccivora*, occur in both Avignon and Guadeloupe. Moreover, limited laboratory experiments have shown that *A. gossypii* collected in Avignon, or in Guadeloupe, are equally effective in transmitting CMV and that, in Guadeloupe, the two main strains of CMV can be transmitted to curcurbits by this vector. These results suggest that the limited spread of CMV in Guadeloupe cannot be attributed to the absence of vector species or to the inability of the tropical aphids to transmit.

DISCUSSION AND CONCLUSIONS

This comparative study of the ecology of CMV in southern France and Guadeloupe shows that the virus is endemic in the wild vegetation of the two areas and causes damaging epidemics in the vegetable crops of Avignon but not in similar crops in Guadeloupe.

Comparative investigations of virus strains, wild hosts and vectors of CMV in the two areas have not provided a clear explanation for the observed differences. Furthermore, the prevalence of other non-persistent viruses (e.g. WMV) in both areas suggests that, in Guadeloupe, some conditions act specifically on the spread of CMV. Tropical conditions, especially temperature, could decrease virus multiplication in plant hosts and so reduce the number of effective sources.

That agricultural conditions are important in influencing spread of CMV is suggested by experience in Florida over the last 50 years. Wellman (1937) first recognized that CMV caused a destructive disease of celery and other vegetables in Florida and, as a control measure, suggested eliminating the *Commelina* spp., considered to be the main source of CMV. CMV became progressively less important in crops although it was detected later in curcurbits in central Florida (Anderson, 1952). More recently it was not considered to be important in south Florida (Zitter, 1979), whereas WMV1 and WMV2 continue to cause epidemics (Adlerz, 1969).

Further experiments are planned in Guadeloupe to determine the influence of climate and cultural practices on the ecology of both CMV and WMV.

Acknowledgements

Antisera to CMV and WMV1 were kindly supplied by Drs. Devergne and Purcifull, respectively.

REFERENCES

Adlerz W.C. (1969) Distribution of Watermelon Mosaic Viruses 1 and 2 in Florida. *Proceedings Florida State Horticultural Society* 81, 161-5.

Anderson C.W. (1952) The distribution of cucurbit viruses in central Florida. *Plant Disease Reporter* 36, 377-9.

Devergne J.C. & Cardin L. (1973) Contribution à l'étude du virus de la mosaique du concombre (CMV). IV. Essai de classification de plusieurs isolats sur la base de leur structure antigénique. *Annales de Phytopathologie* 5, 409-30.

Devergne J.C., Cardin L. & Quiot J.B. (1978) Détection et identification sérologique des infections naturelles par le virus de la mosaique du concombre. *Annales de Phytopathologie* 10, 233-46.

Douine L., Marchoux G., Quiot J.B. & Clément M. (1979) Phénomènes d' interférences entre souches du virus de la mosaique du concombre. II. Effet de la température d'incubation sur la multiplication de deux souches de sensibilité thermique différentes, inoculées simultanément ou successivement à un hôte sensible : *Nicotiana tabacum* var. Xanthi n.c. *Annales de Phytopathologie* 11, 421-30.

Marchoux G., Douine L. & Quiot J.B. (1976) Comportement thermique différentiel de certaines souches du virus de la mosaique du concombre. Hypothèse d'un mécanique pléiotropique reliant plusieurs propriétés. *Compte Rendu de l'Académie des Sciences de Paris, D* 283, 1601-4.

Marrou J., Quiot J.B., Marchoux G. & Duteil M. (1975) Caractérisation par la symptomatologie de quatorze souches du virus de la mosaique du concombre et de deux autres cucumovirus : tentative de classification. *Mededelingen van de Fakulteit Landbouwetenschappen Rijksuniversiteit Gent* 40, 107-21.

Marrou J., Quiot J.B., Duteil M., Labonne G., Leclant F. & Renoust M. (1979) Ecologie et épidémiologie du virus de la mosaique du concombre dans le Sud-Est de la France. III. Intéret de l'exposition de plantes-appâts pour l'étude de la dissémination du virus de la mosaique du concombre. *Annales de Phytopathologie* 11, 291-306.

Messiaen C.M., Maison P. & Migliori A. (1963) Le virus 1 du concombre dans le Sud-Est de la France. *Phytopathologia Mediterranea* 2, 251-60.

Migliori A., Quiot J.B., Leclant F., Marchoux G. & Coleno A. (1977) Premières observations sur l'épidémiologie du virus de la mosaique du concombre et du virus de la mosaique de la pastèque en Guadeloupe. *Annales de Phytopathologie* 9, 123-40.

Migliori A., Marchoux G. & Quiot J.B. (1978) Dynamique des populations du virus de la mosaique du concombre en Guadeloupe. *Annales de Phytopathologie* 10, 455-66.

Quiot J.B. (1980) Ecology of cucumber mosaic virus in the Rhône valley of France. *Acta Horticulturae* 88, 9-21.

Quiot J.B., Douine L. & Gebré-Selassié K. (1979) Fréquence des principales viroses identifiées dans une exploitation maraîchère du Sud-Est de la France. *Annales de Phytopathologie* 11, 283-90.

Quiot J.B., Marchoux G., Douine L. & Vigouroux A. (1979) Ecologie et épidémiologie du virus de la mosaique du concombre dans le Sud-Est de la France. V. Rôle des espèces spontanées dans la conservation du virus. *Annales de Phytopathologie* <u>11</u>, 325-48.

Quiot J.B., Marrou J., Labonne G. & Verbrugghe M. (1979) Ecologie et épidémiologie du virus de la mosaïque du concombre dans le Sud-Est de la France : Description du dispositif expérimental. *Annales de Phytopathologie* <u>11</u>, 265-82.

Quiot J.B., Verbrugghe M., Labonne G., Leclant F. & Marrou J. (1979) Ecologie et épidémiologie du virus de la mosaïque du concombre dans le Sud-Est de la France. IV. Influence des brise-vent sur la répartition des contaminations virales dans une culture protégée. *Annales de Phytopathologie* <u>11</u>, 307-24.

Quiot J.B., Devergne J.C., Cardin L., Verbrugghe M., Marchoux G. & Labonne G. (1979) Ecologie et épidémiologie du virus de la mosaïque du concombre dans le Sud-Est de la France. VII. Répartition de deux types de populations virales dans des cultures sensibles. *Annales de Phytopathologie* <u>11</u>, 359-74.

Wellman F.L. (1937) Control of southern celery mosaic in Florida by removing weeds that serve as sources of mosaic infection. *United States Department of Agriculture Technical Bulletin* <u>548</u>, 16 pp.

Yot-Dauthy D. & Bové J.M. (1966) Mosaïque du Bananier : Identification et purification de diverses souches du virus. *Fruits* <u>21</u>, 449-65.

Zitter T.A. (1979) Methods for controlling the most common vegetable viruses in South Florida. *Belle Glade AREC Research Report EV 1978-9*.

Barley yellow dwarf virus—a global problem

R.T. PLUMB

Plant Pathology Department, Rothamsted Experimental Station,
Harpenden, Herts AL5 2JQ, UK

INTRODUCTION

The disease of Gramineae now known as barley yellow dwarf is probably
ancient but it was not until 1951 (Oswald & Houston, 1951) that the
cause was identified as an aphid-transmitted virus, barley yellow dwarf
virus (BYDV). BYD is the most widespread virus disease of small grain
cereals and a detailed consideration of its epidemiology would require
a monograph rather than a brief chapter, even though in an earlier
review Bruehl (1961) wrote prophetically "The complexity of this
disease ..... will prevent detailed treatment of epidemiology for some
time to come". In this chapter factors affecting BYDV epidemiology are
considered with the emphasis on the progress made and where effort is
still required.

THE DISEASE AND THE CAUSAL VIRUSES

Barley yellow dwarf is probably best considered as a convenient, all-
embracing name for diseases with similar symptoms and effects that are
caused by persistently aphid-transmitted viruses only some of which are
serologically related. BYDV - MAV is the type member of the luteovirus
group (Rochow & Israel, 1977; Matthews, 1979), which also contains other
virus isolates that cause BYD. Recent evidence of relationships among
BYDV, beet western yellows virus and potato leaf roll virus (Duffus &
Rochow, 1978; Rochow & Duffus, 1978; Roberts et al., 1980) suggests that
BYD is caused by only some of a continuous, overlapping range of
viruses. Nevertheless, for practical purposes, the concept of a disease
called BYD is useful provided that the epidemiological consequences of
the existence of several strains, or distinct viruses, are realized.

Diagnosis

Until recently the only practical method of diagnosing BYDV was by transmission to indicator plants using aphids; the development of typical
symptoms was considered diagnostic. Serological tests were possible
(Rochow & Ball, 1967) but were only convenient for experimental use.

Plumb R.T. & Thresh J.M. (1983) *Plant Virus Epidemiology.*
Blackwell Scientific Publications, Oxford.

The recent development of enzyme-linked immunosorbent assay (ELISA) and serologically specific electron microscopy (SSEM) for BYDV (Paliwal, 1977, 1979; Lister & Rochow, 1979; Plumb & Lennon, 1981) has greatly increased the speed of diagnosis and the number of samples that can be handled.

Virus isolates

Isolates from BYD-infected plants have been grouped by their vector specificity and their effects on the host. The groups are designated by the initial letters of their principal vector(s) (Rochow, 1970, 1979):-

1. RPV - transmitted specifically by *Rhopalosiphum padi*.
2. RMV - transmitted specifically by *R. maidis*.
3. MAV - transmitted specifically by *Macrosiphum (Sitobion) avenae*.
4. SGV - transmitted specifically by *Schizaphis graminum*.
5. PAV - transmitted non-specifically by *R. padi* and *M. avenae*.

Relationships between isolates from different parts of the world are unknown but British PAV-type isolates reacted as well in SSEM tests with a broad-spectrum antiserum (PAV + RPV) from Kentucky, United States of America (USA), as with the homologous antiserum. The British antiserum also detected a virus isolate from Chile.

Cross protection

There are reports of cross protection between some isolates (Smith, 1963a; Jedlinski & Brown, 1965) and lack of protection between others (Bruehl, 1961). Most results suggest that cross protection is uncommon and is unlikely to influence BYD epidemiology greatly.

THE VECTORS

The number of known aphid vectors of isolates of BYDV is 23 (A'Brook, 1981a). All of these aphids do not occur together in any one country and no virus isolate has been tested using all aphids. Consequently the identification of isolates by their vector specificity should be seen as locally and epidemiologically useful, and not universally applicable. Epidemiologically this method of identifying isolates is complicated by the phenomenon of dependent transmission (Rochow, 1977). It appears that in mixed infections of MAV and RPV the protein of RPV occasionally encapsidates the nucleic acid of MAV which can then be acquired and transmitted by *R. padi* (Rochow, 1977). This process requires the two viruses to multiply simultaneously and this occurs in the host but not the vector (Paliwal & Sinha, 1970).

BYDV is not transmitted to the progeny of infective vectors but all instars of *R. padi* are as efficient as adults at acquiring and transmitting BYDV (Toko & Bruehl, 1959), although Watson & Mulligan

(1960) suggested that nymphs near ecdysis may acquire virus inefficiently. Of considerable epidemiological importance is the ability of
other morphological, especially winged sexual, forms to transmit BYDV.
Gynoparae and males of *R. padi* will transmit BYDV to oats (Smith *et
al.*, 1977; A'Brook & Dewar, 1980), so may cause primary infection but
will not produce progeny to disperse virus in the crop. For *R. fitchii
(insertum)* only oviparae in the autumn and alatoid fundatrigeniae in
the spring transmitted BYDV (Orlob & Arny, 1960). It seems reasonable
to conclude that sexual forms are not very important in the spread of
BYD because they rarely feed and do not breed on susceptible hosts.

VECTOR BIOLOGY AND MIGRATION

Bruehl (1961) described the biology of the important vectors in North
America and Carter *et al.* (1980) consider the main cereal aphid species
in Britain; Eastop (this volume) gives the biology of the most important
vectors of BYDV. Little seems to be known about several of the vectors.

All aspects of vector biology affect their ability to transmit BYDV.
Of most epidemiological significance is whether a species alternates
between graminaceous or other hosts and the timing and size of its
migratory flights. *Sitobion avenae* (the preferred name for *M. avenae*),
R. maidis, Macrosiphum miscanthi, Schizaphis graminum and *Metopolophium
festucae* always live on Gramineae, and reproduce parthenogenetically or
sexually, depending on day length and temperature. *R. padi, R. insertum,
Metopolophium dirhodum* and *Sitobion fragariae* respond to the same
stimuli but migrate to woody primary hosts, that are not susceptible to
BYDV, on which they lay eggs. The production of sexual forms of *R. padi*
is determined by temperature and day length (Dixon & Glen, 1971) and in
Britain varying proportions of host-alternating species overwinter
viviparously on Gramineae (Carter *et al.*, 1980).

Fundatrices produced from eggs give rise to one or two apterous
generations before alate, migrant forms are produced. Aphids overwintering viviparously can produce alates as soon as conditions are
favourable. Consequently BYDV is much more likely to be introduced
early into a spring-sown cereal by the progeny of viviparous aphids
than from aphids that survive as eggs, either on alternate hosts or
Gramineae. In eastern England *R. padi* is the first infective species
recorded in most years (Plumb, 1977a), presumably derived from the few
that survive viviparously on Gramineae, whereas *S. avenae*, possibly
from eggs on Gramineae, and *M. dirhodum*, from eggs on roses, are not
infective until approximately two and four weeks later, respectively.
In 1979, after an unusually severe winter, no infective *R. padi* were
caught (Plumb, 1981a), presumably because all viviparously overwintering
aphids had been killed and alates migrated exclusively from the primary
host, *Prunus padus*.

The monitoring of migrant aphids is discussed by Taylor (this volume)
and the use of aerial monitoring by suction traps is of great value in
Europe. The timings of flights of most migrant species are now well

known and there are three distinct migrations of cereal aphids (Taylor, 1977). The aphids that migrate in May and June colonize and introduce virus to the spring- and autumn-sown crops. However, at least in eastern England, virus introduced at this time to autumn-sown crops seems of little epidemiological significance. The summer migration, from June to August, spreads virus within the growing crop and later to perennial hosts as the cereal crop ripens. The autumn migration, from September to November, is mainly a return migration, almost exclusively of *Rhopalosiphum* spp., to their primary hosts, but virus is also introduced to autumn-sown cereals.

In North America, long-distance migration of *S. graminum* and *S. avenae* and subsequent infection by BYDV is associated with low-level jet winds (Wallin & Loonan, 1971). In Europe such long-distance dispersal, although possible, has yet to be demonstrated (Carter *et al.*, 1980; Cochrane, 1980; Dewar *et al.*, 1980). Knowledge of cereal aphid movements is sparse in areas that have a summer drought. In south-east Australia, where *R. padi, M. miscanthi* and *R. maidis* are not known to have a sexual stage (Eastop, this volume), suction traps caught aphids for most of the winter in mild, coastal regions. In the inland areas where most cereals are grown, few aphids flew until spring and flights ceased as temperatures increased and foliage died in the summer. Aphids carried virus at all times, which is to be expected if the alates were derived solely from viviparous colonies (Smith & Plumb, 1981).

CROP INFECTION

Virus sources

For epidemiological purposes it is convenient to assume that all species of Gramineae are susceptible to one or more strains of BYDV, even though there are species which appear to be immune. However, what is a source of BYDV depends on the assessor: one man's source may be another man's crop!

In northern and north-western Europe and New Zealand perennial grasses are the principal virus source (Lindsten, 1964; Smith & Wright, 1964; Doodson, 1967) and, at least in Britain, appear to provide a reservoir of infection of several isolates (Plumb, 1977b). Interactions between weather and vector biology probably determine which virus isolate is introduced to cereal crops in the spring and this may account for the regional differences in the frequency of occurrence of different virus isolates in cereals (Plumb, 1974). Grasses also appear to be the predominant, often the only, source of virus for infection of autumn-sown cereals in Britain. Aphids, principally *R. padi*, develop on grasses and acquire virus which they carry into the newly emerged cereal crops. In some areas of southern and western England maize may act as a source and in France maize is the principal source of virus carried by *R. padi* migrating during September to November (Dedryver & Robert, 1981). As maize is an annual crop its importance as a source of BYDV depends on how much is infected by late summer. Where maize is the predominant source there may, therefore, be a relationship between disease incidence in maize and infection in autumn-sown cereals. In Britain, and probably elsewhere where grasses are the principal source, there is no obvious relationship between incidence in the summer and infection in autumn-sown cereals. In areas of summer drought, irrigated grass, or roadside verges that stay green because of water run-off from the surface, may provide a local source of BYDV (Price, 1970) and aphids may also spread into the area from grass or cereal sources in less arid regions. *R. padi*, while not generally recognized as a root-feeding aphid, can acquire virus from barley roots (Orlob, 1966) and in winter is often found feeding on cereals at or below soil level. Aphids may also oversummer in this way on the roots of perennial grasses. *R. rufiabdominalis*, also a vector of BYDV, overwinters on the underground parts of wheat and barley and this has been suggested as being of potential epidemiological importance in North America (Jedlinski, 1981).

   While most grass species are potential sources of BYDV it does not follow that they are all equally good sources of aphids or virus. Coon (1959) found large differences between grasses in "relative host efficiency" for *S. avenae* and *R. fitchii (insertum)* and Clark *et al.* (1979) showed that the concentration of BYDV in maize was less than in some other susceptible Gramineae. Acquisition and transmission tests also suggest that ryegrass may be a poorer source of virus than cereals. Volunteer cereals may, for this reason and because of their proximity to new cereal crops, be a more important source of virus than nearby grass. Nevertheless, one of the most dangerous cropping practices is to follow a grass crop too quickly by a cereal. If the grass is not dead before the cereal is sown aphids may move directly from the dying grass to the emerging cereal crop which can then be severely affected.

Primary infection

The pattern of primary infection depends on the aphid vector but con- flicting results have been obtained on the distribution within a cereal crop of the initial migrants of *R. padi, S. avenae* and *M. dirhodum* (Carter *et al.*, 1980). The proportion of infective migrant aphids is usually small (Plumb, 1981a) and differs between regions and seasons in Britain (Plumb, 1976; Smith *et al.*, 1977; A'Brook & Dewar, 1980) and between countries (Smith & Plumb, 1981). Consequently, even when large migrations occur, primary infection is usually scattered and sparse. This is in contrast to the pattern of infection that can occur when cereals follow grass. However, where migration into the crop is pro- longed, widespread infection can result from primary infection alone, even when few of the immigrants are infective.

   The extent of primary infection is influenced by crop growth stage at the time of aphid migration. In experiments on spring barley from 1976 to 1979 at Rothamsted, aphid populations and virus incidence were always greatest on the latest sown crops. This was apparently caused by immigrant aphids alighting preferentially on the latest-sown crops, possibly because the early-sown crop covered the ground completely and was less attractive to alate aphids than the mosaic of crop and ground (A'Brook, 1968) presented by the late-sown crop.

Secondary spread

The pattern of secondary virus spread is also influenced by the vector species. One of the main features of spread by several species, but especially by *R. padi*, is the development of distinct patches of infection (Smith, 1963b; Hooper, 1978). These patches are probably caused by the local spread of virus, probably by wingless aphids, from initial foci. Saucer-shaped, discoloured depressions develop in the crop as the first-infected plants in the centre are much more severely stunted than the later-infected plants round the periphery. In Britain such a pattern of infection occurs in autumn-sown cereals and is usually associated with the survival of aphids on the crop for most or all of the winter. Spring infection seldom produces discernable patches because spring-sown crops are usually more mature at the time of aphid imigration in the spring than are autumn-sown crops during the autumn migration so that infection is less likely, spread is restricted and damage is limited. The most common vector in the spring and summer is usually *S. avenae* which is more restless than other vector species (Dean, 1973), a factor that also contributes to the scattered infection pattern usually seen in spring-sown crops.

For a particular BYDV isolate to spread extensively the aphid(s) that transmit it most efficiently must be present. In Britain, autumn-sown crops are often infected by a virus isolate transmitted efficiently by *R. padi* but inefficiently, or not at all, by *S. avenae* and *M. dirhodum*. These last two species are most numerous in the summer, thus providing another reason why virus infection of autumn-sown crops seems to be influenced relatively little by summer aphid populations.

Virus-infected plants differ from healthy ones as hosts of aphids and this influences virus spread. Aphids were more common on infected hosts, partly because more alighted on diseased than on healthy plants, and partly because those on diseased plants multiplied more quickly (Ajayi, 1981). More alate *R. padi* and *S. avenae* were produced on BYDV-infected than on healthy oats and this may increase virus spread (Gildow, 1980). Miller & Coon (1964) reported that *S. avenae* bred on BYDV-infected oats lived longer than those on healthy oats, although Elamin (1975) found the reverse and Markkula & Laurema (1964) detected no effect of the disease status of the host on the longevity of *S. avenae* and *M. dirhodum; R. maidis* produced smaller colonies on infected than on healthy barley plants (Gill & Metcalfe, 1977). Clearly, infection of the host may affect aphid populations but these effects seem inconsistent.

EFFECTS ON YIELD

The effect of BYDV on yield declines as crop age at infection increases (Doodson & Saunders, 1970) and oats and barley are more severely damaged

than wheat. Experimental infection of oat and barley seedlings by a
severe (PAV) isolate can decrease yield by 90% (Doodson & Saunders,
1970). Milder isolates of BYDV also decrease yield but their effects
are less well described. Yield loss caused by natural infection is more
difficult to measure because of interactions between growth stage and
infection, damage caused directly by aphids or indirectly by other
cereal diseases that may be affected by virus-infection of the host
(Comeau & Pelletier, 1976; Potter, 1980) and infection by more than one
virus isolate. Consequently, losses attributed to BYDV can be only
approximate. Early infection of cereals with a severe strain, such as
can occur after grass, may result in crop failure, but national losses
in UK were estimated at up to 10% in 1967-9 (Doodson & Saunders, 1969)
and similar losses have been reported from New Zealand (Smith, 1963b)
and Canada (Gill, 1980).

There have been few attempts to determine yield lost to BYDV in
grasses. It can be large (Catherall, 1966), although in New Zealand,
BYDV was considered of little economic importance in well managed grass/
white clover pastures (Latch, 1980).

CONTROL

Breeding

Tolerance of BYDV is conditioned in barley by a single, major gene of
Ethiopian origin designated *Yd2* (Rasmusson & Schaller, 1959). Tolerance
has also been selected for in oats and selected lines were equally
tolerant of two virus isolates (Jedlinski, 1972). There is reason to
hope that the *Yd2* gene from barley can be incorporated and expressed in
wheat (C.O. Qualset, personal communication). There are obvious dangers
in relying for tolerance on a narrow genetic base, even when it appears
to be effective and durable against most virus isolates, and the
International Maize and Wheat Improvement Center (CIMMYT) plans to
improve tolerance of BYDV in cereal cultivars.

Carter *et al.* (1980) conclude that breeding for host resistance to
cereal aphids is a more promising way of minimizing aphid damage than
is the use of natural enemies, but the currently available resistance
seems unlikely to provide effective control of BYDV.

Husbandry

Avoiding infection seems the most attractive method of preventing
damage from BYDV and in many regions is achieved by sowing late in
autumn after the autumn aphid migration and sowing early in spring so
that plants have passed their most vulnerable growth stages when the
spring migration occurs (Plumb, 1981a). In New Zealand, spring infection is "unavoidable" but in most regions virus infection in the autumn
can be avoided by delaying sowing to the "earliest safe sowing date".
This date is based on a knowledge of aphid migration flights, and crops
sown at, or after, this date emerge after aphid flights have ceased

(Lowe, 1967).

In Britain, it is best to sow spring cereals as early as possible
(Plumb, 1977a). Field experiments (Plumb, 1977a) and aerial surveys
(Hooper, 1978), have shown that winter cereals sown after the middle of
October largely avoid infection, as they emerge after most aphid flight
stops. In the Pacific North-West region of the USA, 15 October is
considered the earliest safe sowing date (A. Cholic, personal communication). In the absence of virus infection, autumn crops sown early
(September) in Britain, usually yield better than crops sown later and
few farmers consider the risk of BYDV infection justifies delaying
sowing and jeopardizing the potential yield of the crop. Avoidance is
acceptable as a control strategy only when it is compatible with other
husbandry requirements or when expected yield losses from BYDV are
greater than the potential gain from early sowing, and when these losses
cannot be economically prevented by alternative control methods.

Chemicals

Pesticides applied either as granules or sprays have been used to
control the vectors of BYDV (Smith & Wright, 1964; Plumb, 1981a). If
experience suggests that infection seems certain then the use of
insecticide granules as a prophylactic at sowing may be justified
(Mulholland & Jessep, 1967; Plumb, 1977a). An alternative strategy in
the UK is to apply a single aphicide spray early in November to crops
at risk to prevent secondary spread of BYDV. At present chemicals
seem unlikely to prevent primary infection and their effect on secondary spread could be swamped by a continuous influx of infective vectors.
Applications of synthetic pyrethroid insecticides appear to give the
best control of BYDV in autumn-sown crops (Barrett et al., 1981). For
spring-sown crops in Britain aphicidal sprays seem justified in some
seasons, but then only on crops sown after mid-April.

PREDICTION

Knowledge of aphid populations and flights has been useful in predicting the likelihood of BYDV infection and much effort has been
devoted to producing predictive models for *S. avenae* populations
(Rabbinge et al., 1979; Carter et al., 1980). In Britain, the
Rothamsted Insect Survey (RIS) provides detailed information on aphid
flight and migration (Taylor, 1977; Taylor, this volume). However,
BYDV infection cannot be predicted from aphid numbers alone and they
must be supplemented by knowledge of aphid infectivity, i.e. the
aphid's ability to infect a host with BYDV. In Britain infectivity
of cereal aphids is determined at two sites in England (Plumb, 1976;
Smith et al., 1977), and one in Wales (A'Brook & Dewar, 1980). This
technique is now being used increasingly to aid epidemiological studies
of other virus/vector combinations (Raccah, this volume; Lecoq & Pitrat,
this volume).

The method currently used is a direct transmission testing method, i.e. the aphid is allowed to feed on a cereal (usually oats) test plant (Plumb, 1976). This has the advantage of simulating natural conditions so that the aphid's biological behaviour and its likelihood of transmitting virus should closely resemble what happens naturally. Results from indirect methods using ELISA or SSEM (Plumb, 1981b; Raccah, this volume) must be interpreted with care because virus can be detected in aphids that fail to transmit in tests and vice versa. This is an especial problem for BYDV as there is no evidence that vector specificity is determined by ability to acquire virus. For many BYDV vectors the sexual forms produced in the autumn may acquire virus from the grass host on which they develop but seem to be of little epidemiological consequence because they are unlikely to reach or feed on cereals.

The principal disadvantage of transmission testing is the time taken for test plants to develop symptoms which, for a severe virus isolate, is 2-4 weeks. This could make the information of little use, except in retrospect, but in Britain the infectivity of migrant aphids in the autumn can be determined in time to advise whether spraying is justified at the optimum time (Plumb & Lennon, 1982). Obtaining results more rapidly would permit earlier planning, and tests have shown that infection of test plants can be detected by SSEM one week after the beginning of the acquisition feed, well before symptoms appear.

Trap catches in the autumn from the RIS have been combined with the proportion of infective aphids to give an Infectivity Index (II) (Plumb & Lennon, 1981, 1982) that can be used to guide decisions on the need for aphicidal sprays. The II, which when exceeded indicates that sprays are required, needs to be more accurately assessed by experiment and will probably vary from region to region (Plumb & Lennon, 1982). The II should be interpreted with care and if possible coupled with some measure of secondary virus spread (Kendall & Smith, 1981).

If the II is calculated weekly and a cumulative index produced it can also allow for crop sowing date (Plumb & Lennon, 1982). When used with local knowledge the II seems a promising basis for forecasting the need to control BYDV in the autumn. For spring-sown crops no useful Index is yet available. The date of introduction of virus in the spring is quite predictable but the course of epidemics in spring-sown crops is not and the most useful indicators of the likelihood of virus infection are sowing date and location (Plumb, 1977a). Greater refinement of aphid population models may help virus prediction and A'Brook (1981b) has shown an association between the numbers of *M. dirhodum*, *R. insertum*, *R. padi* and *S. avenae* caught in suction traps and weather data. This may help to avoid unnecessary spray treatments, especially in autumn.

DISCUSSION

Bruehl's rather pessimistic prophecy, made more than 20 years ago, has been amply justified. However, considerable progress has been made and prospects for minimizing losses due to BYDV are good, although the virus still causes serious losses in many regions.

At present, of the many factors affecting BYD epidemiology, the development of vector populations in the cereal crop is best, albeit incompletely, understood. While this is essential knowledge, its value in predicting the incidence of BYD is limited unless accompanied by data on the numbers and origins of the initial migrants into the crop, and here our knowledge is relatively poor. Monitoring the infectivity of aphids gives some of this information indirectly, but to obtain direct information requires a closer examination of BYDV sources and in this area our knowledge of BYDV epidemiology is rudimentary.

It is possible to avoid infection and in some areas this can be done in ways that are agriculturally acceptable. Nevertheless, for the foreseeable future control of BYDV seems likely to depend on chemical control of the aphid vectors. To ensure that such chemicals are used only when needed and to minimize yield lost to BYDV, the epidemiology of the virus and the phenology and infectivity of the aphid vectors require continued study.

Acknowledgement

I thank Dr T.P. Pirone for the BYDV antiserum from Kentucky, USA.

REFERENCES

A'Brook J. (1968) The effect of plant spacing on the numbers of aphids trapped over the groundnut crop. *Annals of Applied Biology* 61, 289-94.

A'Brook J. (1981a) Vectors of barley yellow dwarf virus. *Euraphid: Rothamsted 1980* (Ed. by L.R. Taylor), p. 21. Rothamsted Experimental Station, Harpenden.

A'Brook J. (1981b) Some observations in West Wales on the relationships between numbers of alate aphids and weather. *Annals of Applied Biology* 97, 11-5.

A'Brook J. & Dewar A.M. (1980) Barley yellow dwarf virus infectivity of alate aphid vectors in West Wales. *Annals of Applied Biology* 96, 51-8.

Ajayi O. (1981) Interactions between barley yellow dwarf virus, aphids and *Cladosporium* on cereals. *Ph.D. Thesis, London University.*

Barrett D.W.A., Northwood P.J. & Horellou A. (1981) The influence of rate and timing of autumn applied pyrethroid and carbamate insecticide sprays on the control of barley yellow dwarf virus in English and French winter cereals. *Proceedings 1981 British Crop Protection Conference - Pests and Diseases* 2, 405-12.

Bruehl G.W. (1961) Barley yellow dwarf. *Monograph, American Phytopathological Society* 1, 52 pp.

Carter N., McLean I.F.G., Watt A.D. & Dixon A.F.G. (1980) Cereal Aphids: a case study and review. *Applied Biology* 5, 271-348.

Catherall P.L. (1966) Effects of barley yellow dwarf virus on the growth and yield of single plants and simulated swards of perennial rye-grass. *Annals of Applied Biology* 57, 155-62.

Clark M.F., Bates D., Plumb R.T. & Lennon E. (1979) *Report of East Malling Research Station for 1978*, p. 101.

Cochrane J. (1980) Meteorological aspects of the numbers and distribution of the rose-grain aphid, *Metopolophium dirhodum* (Wlk.) over South-East England in July 1979. *Plant Pathology* 29, 1-8.

Comeau A. & Pelletier G.J. (1976) Predisposition to septoria leaf blotch in oats affected by barley yellow dwarf virus. *Canadian Journal of Plant Science* 56, 13-9.

Coon B.F. (1959) Grass hosts of cereal aphids. *Journal of Economic Entomology* 52, 994-6.

Dean G.J.W. (1973) Distribution of aphids in spring cereals. *Journal of Applied Ecology* 10, 447-62.

Dedryver C.A. & Robert Y. (1981) Ecological role of maize and cereal volunteers as reservoirs for gramineae virus transmitting aphids. *Proceedings 3rd Conference on Virus Diseases of Gramineae*, pp. 61-6. Rothamsted Experimental Station, Harpenden.

Dewar A.M., Woiwod I. & Choppin de Janvry E. (1980) Aerial migrations of the rose-grain aphid, *Metopolophium dirhodum* (Wlk.) over Europe in 1979. *Plant Pathology* 29, 101-9.

Dixon A.F.G. & Glen D.M. (1971) Morph determination in the bird cherry-oat aphid, *Rhopalosiphum padi* L. *Annals of Applied Biology* 68, 11-21

Doodson J.K. (1967) A survey of barley yellow dwarf virus in S.24 perennial ryegrass in England and Wales, 1966. *Plant Pathology* 16, 42-5.

Doodson J.K. & Saunders P.J.W. (1969) Observations on the effects of some systemic chemicals applied to cereals in trials at the NIAB. *Proceedings 5th British Insecticide and Fungicide Conference* 1, 1-7.

Doodson J.K. & Saunders P.J.W. (1970) Some effects of barley yellow dwarf virus on spring and winter cereals in field trials. *Annals of Applied Biology* 66, 361-74.

Duffus J.E. & Rochow W.F. (1978) Neutralization of beet western yellows virus by antisera against barley yellow dwarf virus. *Phytopathology* 68, 45-9.

Elamin E.M. (1975) Studies on the ecology of aphids on spring cereals and maize. *Ph.D. Thesis, London University.*

Gildow F.E. (1980) Increased production of alatae by aphids reared on oats infected with barley yellow dwarf virus. *Annals of the Entomological Society of America* 73, 343-7.

Gill C.C. (1980) Assessment of losses on spring wheat naturally infected with barley yellow dwarf virus. *Plant Disease* 64, 197-203.

Gill C.C. & Metcalfe D.R. (1977) Resistance in barley to the corn leaf aphid, *Rhopalosiphum maidis*. *Canadian Journal of Plant Science* 57, 1063-70.

Hooper A.J. (1978) Aerial photography. *Journal of the Royal Agricultural Society of England* 139, 115-23.

Jedlinski H. (1972) Tolerance to two strains of barley yellow dwarf virus in oats. *Plant Disease Reporter* 56, 230-4.

Jedlinski H. (1981) Rice root aphid, *Rhopalosiphum rufiabdominalis*, a vector of barley yellow dwarf virus in Illinois, and the disease complex. *Plant Disease* 65, 975-8.

Jedlinski H. & Brown C. (1965) Cross protection and mutual exclusion by three strains of barley yellow dwarf virus in *Avena sativa* L. *Virology* 26, 613-21.

Kendall D.A. & Smith B.D. (1981) The significance of aphid monitoring in improving barley yellow dwarf virus control. *Proceedings 1981 British Crop Protection Conference - Pests and Diseases* 2, 399-403.

Latch G.C.M. (1980) Effects of barley yellow dwarf virus on simulated swards of Nui perennial ryegrass. *New Zealand Journal of Agricultural Research* 23, 373-8.

Lindsten K. (1964) Investigations on the occurrence and heterogeneity of barley yellow dwarf virus in Sweden. *Lantbrukshögskolans Annaler* 30, 581-600.

Lister R.M. & Rochow W.F. (1979) Detection of barley yellow dwarf virus by enzyme-linked immunosorbent assay. *Phytopathology* 69, 649-54.

Lowe A.D. (1967) Avoid yellow dwarf virus by late sowing. *New Zealand Wheat Review* 10, 59-63.

Markkula M. & Laurema S. (1964) Changes in the concentration of free amino acids in plants induced by virus diseases and the reproduction of aphids. *Annales Agriculturae Fenniae* 3, 265-71.

Matthews R.E.F. (1979) Classification and nomenclature of viruses: Third report of the International Committee on Taxonomy of Viruses. *Intervirology* 12, 129-296.

Miller J.W. & Coon B.F. (1964) The effect of barley yellow dwarf virus on the biology of the aphid vector, the English grain aphid, *Macrosiphum granarium*. *Journal of Economic Entomology* 57, 970-4.

Mulholland R.I. & Jessep C.T. (1967) Insecticide granules and virus in autumn-sown wheat. *New Zealand Wheat Review* 10, 64-9.

Orlob G.B. (1966) The role of subterranean aphids in the epidemiology of barley yellow dwarf virus. *Entomologia Experimentalis et Applicata* 9, 85-94.

Orlob G.B. & Arny D.C. (1960) Transmission of barley yellow dwarf virus by different forms of the apple-grain aphid, *Rhopalosiphum fitchii* (Sand.). *Virology* 11, 273-4.

Oswald J.W. & Houston B.R. (1951) A new virus disease of cereals, transmissible by aphids. *Plant Disease Reporter* 35, 471-5.

Paliwal Y.C. (1977) Rapid diagnosis of barley yellow dwarf virus in plants using serologically specific electron microscopy. *Phytopathologische Zeitschrift* 89, 25-36.

Paliwal Y.C. (1979) Serological relationships of barley yellow dwarf virus isolates. *Phytopathologische Zeitschrift* 94, 8-15.

Paliwal Y.C. & Sinha R.C. (1970) On the mechanism of persistence and distribution of barley yellow dwarf virus in an aphid vector. *Virology* 42, 668-80.

Plumb R.T. (1974) Properties and isolates of barley yellow dwarf virus. *Annals of Applied Biology* 77, 87-91.

Plumb R.T. (1976) Barley yellow dwarf virus in aphids caught in suction traps, 1969-73. *Annals of Applied Biology* 83, 53-9.

Plumb R.T. (1977a) Aphids and virus control on cereals. *Proceedings 1977 British Crop Protection Conference - Pests and Diseases* 3, 903-13

Plumb R.T. (1977b) Grass as a reservoir of cereal viruses. *Annales de Phytopathologie* 9, 361-4.

Plumb R.T. (1981a) Chemicals in the control of cereal virus diseases. *Strategies for the control of cereal disease* (Ed. by J.F. Jenkyn & R.T. Plumb), pp. 135-45. Blackwell Scientific Publications, Oxford.

Plumb R.T. (1981b) Problems in the use of sensitive serological methods for detecting viruses in vectors. *Proceedings 3rd Conference on Virus Diseases of Gramineae in Europe*, pp. 123-6. Rothamsted Experimental Station, Harpenden.

Plumb R.T. & Lennon E. (1981) Serological diagnosis of barley yellow dwarf virus (BYDV). *Report of Rothamsted Experimental Station for 1980*, Part 1, p. 181.

Plumb R.T. & Lennon E. (1982) Aphid infectivity and the Infectivity Index. *Report of Rothamsted Experimental Station for 1981*, Part 1, pp. 195-7.

Potter L.R. (1980) The effects of barley yellow dwarf virus and powdery mildew in oats and barley with single and dual infections. *Annals of Applied Biology* 94, 11-7.

Price R.D. (1970) Stunted patches and deadheads in Victorian cereal crops. *Technical Publication, Department of Agriculture, Victoria* 23.

Rabbinge R., Ankersmit G.W. & Pak G.A. (1979) Epidemiology and simulation of population development of *Sitobion avenae* in winter wheat. *Netherlands Journal of Plant Pathology* 85, 197-220.

Rasmusson D.C. & Schaller C.W. (1959) The inheritance of resistance in barley to the yellow-dwarf virus. *Agronomy Journal* 51, 661-4.

Roberts I.M., Tamada T. & Harrison B.D. (1980) Relationship of potato leafroll virus to luteoviruses: evidence from electron microscope serological tests. *Journal of General Virology* 47, 209-13.

Rochow W.F. (1970) Barley yellow dwarf virus. *CMI/AAB Descriptions of Plant Viruses* 32, 4 pp.

Rochow W.F. (1977) Dependent virus transmission from mixed infections. *Aphids as Virus Vectors* (Ed. by K.F. Harris & K. Maramorosch), pp. 253-73. Academic Press, New York.

Rochow W.F. (1979) Field variants of barley yellow dwarf virus: detection and fluctuation during twenty years. *Phytopathology* 69, 655-60.

Rochow W.F. & Ball E.M. (1967) Serological blocking of aphid transmission of barley yellow dwarf virus. *Virology* 33, 359-62.

Rochow W.F. & Duffus J.E. (1978) Relationships between barley yellow dwarf and beet western yellows viruses. *Phytopathology* 68, 51-8.

Rochow W.F. & Israel H.W. (1977) Luteovirus (Barley yellow dwarf virus) Group. *The Atlas of Insect and Plant Viruses* (Ed. by K. Maramorosch), pp. 363-9. Academic Press, New York.

Smith B.D., Kendall D.A., Singer M.C., Halfacree S. & Mathias K. (1977) Cereal aphids and the spread of barley yellow dwarf virus. *Report of Long Ashton Research Station for 1976*, pp. 96-7.

Smith H.C. (1963a) Interaction between isolates of barley yellow dwarf virus. *New Zealand Journal of Agricultural Research* 6, 343-53.

Smith H.C. (1963b) Control of barley yellow dwarf in cereals. *New Zealand Journal of Agricultural Research* 6, 229-44.

Smith H.C. & Wright G.M. (1964) Barley yellow dwarf virus on wheat in New Zealand. *New Zealand Wheat Review* 9, 60-79.

Smith P.R. & Plumb R.T. (1981) Barley yellow dwarf virus infectivity of cereal aphids trapped at two sites in Victoria. *Australian Journal of Agricultural Research* <u>32</u>, 249-55.

Taylor L.R. (1977) Aphid forecasting and the Rothamsted Insect Survey. *Journal of the Royal Agricultural Society of England* <u>138</u>, 75-97.

Toko H.V. & Bruehl G.W. (1959) Some host and vector relationships of strains of the barley yellow dwarf virus. *Phytopathology* <u>49</u>, 343-7.

Wallin J.R. & Loonan D.V. (1971) Low level jet winds, aphid vectors, local weather and barley yellow dwarf virus outbreaks. *Phytopathology* <u>61</u>, 1068-70.

Watson M.A. & Mulligan T.E. (1960) The manner of transmission of some barley yellow-dwarf viruses by different aphid species. *Annals of Applied Biology* <u>48</u>, 711-20.

The epidemiology of some aphid-borne viruses in Australia

R.G. GARRETT* & G.D. McLEAN†
*Plant Research Institute, Department of Agriculture,
Burnley, Victoria, Australia 3121
†Department of Agriculture, Jarrah Road, South Perth,
Western Australia, WA 6151

INTRODUCTION

Aphids have been caught in traps at high altitudes (Berry & Taylor, 1969) and hundreds of kilometres from land (Hardy & Milne, 1937). It is hardly surprising, therefore, that aphids are generally considered to be capable of migrating long distances and of carrying viruses with them. For example, in North America, barley yellow dwarf virus (BYDV) is believed to be spread northwards each spring by cereal aphids migrating from Oklahoma and neighbouring states to southern Canada (Orlob & Medler, 1961). In south-eastern Australia, outbreaks of the persistently trans-mitted subterranean clover stunt virus (SCSV) often occur in legume crops in the spring and these outbreaks have been attributed to the migration of the vector, *Aphis craccivora*, from overwintering regions along the New South Wales coast (Gutierrez *et al.*, 1974).

However, there are reasons to doubt whether long-distance migrations of aphids are the main reason for epidemics in crops or pastures. Indeed, schemes for producing pathogen-tested stock are generally effective, provided that the pathogen is not prevalent locally. Thus nations, where possible, operate schemes for producing elite stocks of vegetatively propagated crops, such as potato, ornamentals and small fruits, and for producing seed of crops in which seed-borne viruses are important. Such schemes provide one of the few reliable means of con-trolling insect-borne viruses, showing that long-distance spread is unimportant, at least for these virus/host combinations.

In southern Australia, aphid populations decline during the hot, dry summers and cool, wet winters. However, neither condition is sufficient to eliminate aphid populations provided that suitable host plants survive. Therefore, there is scope for local aphid populations to play a more significant role there than in the colder regions of northern Europe or North America.

As a basis for a discussion of the sources from which epidemics develop, this paper considers specific examples of virus spread in Australia, mainly from experience in Victoria. Some of the viruses have

Plumb R.T. & Thresh J.M. (1983) *Plant Virus Epidemiology.*
Blackwell Scientific Publications, Oxford.

many wild hosts and are widespread. Others do not infect many common
plants, but vegetative propagation, or continuous cropping within a
region, provide adequate reservoirs. In some examples, the reservoirs
are known; in others they are inferred. However, it will become
apparent that in all the examples discussed, the epidemics could have
originated in or immediately around the crop.

CAULIFLOWER MOSAIC IN CAULIFLOWER

In Victoria, cauliflower mosaic virus (ClMV) is more of a problem in the
intensive market garden areas near Melbourne than in scattered brassica
crops. However, even where cauliflower is widely grown reliable virus
control can be achieved by the regular use of insecticides, combined
with the isolation of young from maturing crops. Epidemics are usually
limited in extent and occur where appropriate control measures are not
used. For example, at Keilor, where cauliflowers are grown continuously
by several independent farmers on river flats, infection by ClMV
invariably occurs, but seldom exceeds 0.1%. However, in 1976 and 1977,
infection reached 100% in field A (Fig. 1). Infection spread to crops
on adjacent properties with marked gradients of infection from field A
(Fig. 2a). The development of the epidemics was only curtailed when the
hot, dry summer greatly decreased the aphid populations in the valley.
In both years, incidences of ClMV were < 1% (mostly < 0.1%) until the
spring, when rain interrupted planting. When the weather improved,
priority was given to new plantings, and one grower did not clear old
crops. Each year, one such field (B in Fig. 1) provided the source for
virus spread into a succession of crops to give total infection in field
A and the gradients in field C. Virus and aphid vectors were common on
Sinapis and *Sisymbrium* spp. along the river banks, and especially in
overgrazed paddocks on the hill at D (Fig. 1), yet the disease was well
controlled in crops near these sources. Thus, epidemic development was
restricted to plantings immediately around the severely diseased crop.
None of the outbreaks observed could be attributed to any other sources.

TULIP BREAKING VIRUS IN TULIP

Tulip breaking virus (TuBV) is a non-persistently aphid-transmitted
virus with a narrow host range and few wild hosts. TuBV spreads into
tulip crops slowly even on properties where infected stocks are grown.
Spread within crops is also slow provided that there is little initial
infection. For example, a grower with imported healthy tulip stocks
from New Zealand had only 6% infection after 6 years (J. Sutton,
personal communication), despite being close to large growers with
infected stocks.

    Sutton and Garrett (1978) found that most spread was over very short
distances, of less than 45 cm, representing only 4 plant spaces (Fig.
2b). Such steep gradients in a crop lacking resident aphid populations
indicate that immigrant alate aphids were moving within the crop, but
mostly over very short distances.

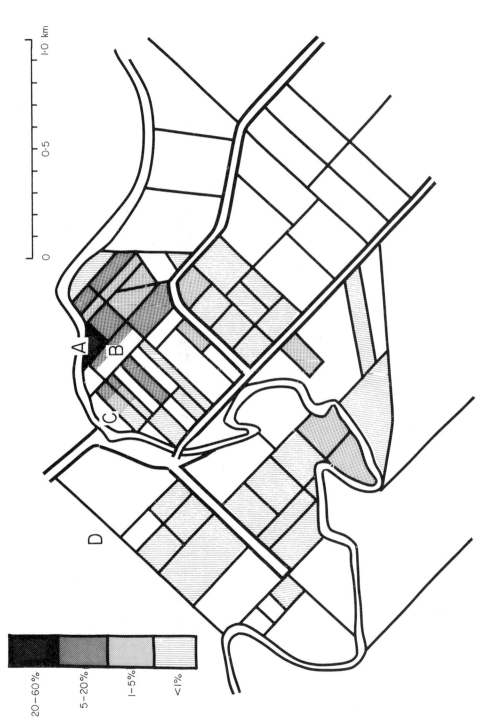

Figure 1. The distribution of ClMV in cauliflower crops at Keilor, Victoria, July 1976. A the crop with the most virus infection; B crop residues, believed to have initiated the epidemic; C a crop with steep gradients of infection; D pastures containing brassica weeds infected with ClMV.

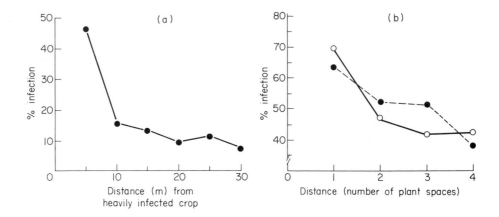

Figure 2.   (a) Gradient of ClMV infection in cauliflower at Keilor,
             Victoria, 1976.
          (b) Gradients of infection by tulip breaking virus along
             beds (●) from infected plants and across beds (o)
             from furrows. Distances are measured as plant spaces
             and plants were *c*. 10 cm apart.

MAIZE DWARF MOSAIC IN SWEETCORN

In the last 11 years in Victoria, only three outbreaks of maize dwarf
mosaic virus (MDMV) were recorded in sweetcorn. Two of these originated
from seed infections of 0.25% where the growers had failed to control
aphids during the spring. Infected plants occurred in very obvious
clusters, of up to 7 plants along the row. Runs of 1, 2, 3, 4, 5, 6
and 7 infected plants were recorded on 16, 6, 9, 3, 2, 2 and 1
occasion(s) respectively on examining 642 plants in 7 rows. The plants
at the end of each run are likely to have been late infections as they
tended to be bigger and less severely affected than those in the centre.

   There is no evidence that MDMV has become endemic in Victoria, even
though there are many hosts of the virus. The nearest large areas of
sweetcorn and maize are several hundred kilometres to the north and
east in mid New South Wales and it is unlikely that they led to
epidemics on only two or three of the many crops in Victoria.

LETTUCE NECROTIC YELLOWS VIRUS IN LETTUCE

Lettuce necrotic yellows virus (LNYV) is a plant rhabdovirus infecting
lettuce, *Sonchus oleraceus* and few other hosts (Martin, this volume).
The main vector is *Hyperomyzus lactucae* in which the virus replicates
(O'Loughlin & Chambers, 1967). *H. lactucae* breeds on *S. oleraceus*,
and can spread LNYV to lettuce on which it fails to breed. Indeed, the
aphid cannot survive on lettuce long enough to complete the access and
incubation periods required to transmit the virus, and so lettuce is

not an important virus source. *S. oleraceus* is a widespread and common
weed of cultivated ground, providing an efficient source of viruliferous
aphids. Nevertheless, Stubbs *et al.* (1963) found infection gradients
from large reservoirs of infection (Fig. 3a) and disease incidences were
substantially reduced when nearby *S. oleraceus* was removed. Thus, even
with a persistently transmitted virus, most virus spread occurred close
to reservoirs, although spread was further and greater than for ClMV
or TuBV.

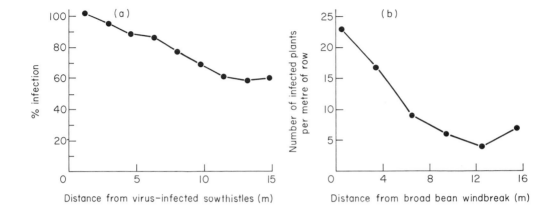

Figure 3.   (a) Gradient of LNYV infection in lettuce growing at
              Braeside near a headland containing infected sow-
              thistles (*S. oleraceus*). (Data courtesy of L.L.
              Stubbs).
          (b) Gradient of infection of SCSV in French beans near a
              windbreak of broad beans infected with the virus, and
              bearing colonies of the vector, *Aphis craccivora*.
              (Approximately 21 plants/m.)

SUBTERRANEAN CLOVER STUNT VIRUS IN FRENCH BEANS

Annually in late spring, French bean crops in south-eastern Victoria are
infected by subterranean clover stunt virus (SCSV), which is transmitted
persistently by aphids. *Aphis craccivora* is the main vector in Victoria
(Smith, 1966), where outbreaks of SCSV have caused yield losses of up to
60%. However, incidences are more usually < 10%, causing perhaps 6%
yield loss. Virus spread is generally most rapid during the early
spring; plants become infected when very young and usually develop
symptoms on the first trifoliate leaves.

    Although severe epidemics have been too rare for detailed study, some
useful information has emerged from observing minor outbreaks.
Gradients and edge effects are common and Fig. 3b shows a typical
gradient in a crop with a moderate disease incidence. At the crop edge,
where the incidence of SCSV was greatest, there was a windbreak of
SCSV-infected broad beans infested with *A. craccivora* and this was the

probable virus source. An alternative explanation, that the gradients were largely the result of virus spread by large aphid populations at the crop edges, is unlikely because when the observations were made the crop was too young for secondary spread to have occurred.

There were other indications that the aphids infesting French bean crops originated locally. Thus aphids were found infesting seedlings as they emerged from the ground and aphid numbers were highest at the crop edge, with a marked gradient into the crop (Table 1). The aphids had presumably settled onto the bare earth in large numbers within the

Table 1. Numbers of alate *A. craccivora* found at the margin of a young French bean crop.

Distance from field edge (m)	Row 1	6	11	16	21
0 - 1	43*	17	25**	21	31
1 - 2	23	19	17	14	16
2 - 3	8	6	11	12	14
3 - 4	9	6	4	4	3

*   One plant had 27 alate *A. craccivora*
** One plant had 9 alate *A. craccivora*

previous few hours, and had then walked onto the germinating seedlings. If the aphids had been part of a large migrating aerial population, they are likely to have settled more widely and in a more random manner. Secondly, an attempt was made to detect large aphid migrations by using suction traps at three sites in the same river valley. Throughout the season the traps caught different mixtures of aphid species but no dense aerial populations of *A. craccivora*.

BARLEY YELLOW DWARF VIRUS IN CEREALS AND GRASSES

The incidence of barley yellow dwarf virus (BYDV) in cereals often exceeds 10% in Victoria (Price, 1970). However, infection is probably much greater by harvest, for symptoms are difficult to recognize in wheat, and later infections induce only slight symptoms which can be confused with ripening. Infections are widespread in ryegrass pastures, causing significant losses (W. Bowker, personal communication). *Rhopalosiphum padi* is the main vector of BYDV in Victoria and in one study up to 61% of trapped aphids were infective (Smith & Plumb, 1981). The same virus strains occur in both cereals and grasses. This, and the high proportion of *R. padi* carrying virus, suggest that the known, local, pasture reservoirs are the annual sources of the virus.

BEAN YELLOW MOSAIC AND CLOVER YELLOW VEIN VIRUSES IN LUPIN

The spread of bean yellow mosaic virus (BYMV) and clover yellow vein virus (CYVV) into field lupin crops was studied for 2 years at Ruther-glen in northern Victoria. Both viruses caused apical necrosis and pod abortion in lupin. Plants produced no seed when infected more than 6 weeks before harvest. The small amount of seed produced by plants infected 4-6 weeks before harvest was of such poor quality that seed transmission could at most have been rare. Furthermore, no infections were found during the first 4 months of growth in 100 000 plants inspected annually.

Both viruses are transmitted in the non-persistent manner by *Myzus persicae, A. craccivora* and *Aulacorthum solani*, and they spread into lupin crops when aphids colonize the plants. In both years, spread to lupin was very slow (0.01% of the stand infected per week) until mid-October when the rate exceeded 1% per week. At its maximum, the weekly increment rose to 3%, and then decreased during late November and December. During the period of most rapid spread, clusters of infected plants developed. All virus spread was attributed to immigrant aphids, since the aphid populations on the crops declined rapidly (mostly due to predation) before the first alatiform nymphs were produced.

The sources of virus and aphids are not known with certainty. How-ever the crops are grown where improved pastures are common, and the viruses are likely to be widely established in pasture legumes. BYMV has been detected in samples of white clover near lupin crops. Maximum spread coincided with hay-making, which suggests that aphids may have moved from pastures at this time.

WATERMELON MOSAIC VIRUS IN WATERMELON

In the last 8 years, severe outbreaks of watermelon mosaic virus (WMMV) have occurred in watermelon (*Citrullus lanatus*) and rockmelon (*Cucumis melo*) crops at Carnarvon in Western Australia. The crops are grown in a small area *c*. 900 km north of the nearest other cucurbit crops at Perth. Maximum virus spread has usually occurred during the early spring and declined as aphid numbers decreased to very low numbers during the late summer. Infected plants were at first distributed throughout the crops with no obvious gradients of infection. Marked clusters developed later and the remaining healthy plants were in groups at the final observation, when much spread had occurred (Fig. 4). This pattern of infection shows that most spread was within the crop, as the initial foci expanded.

There are three possible local sources of the initial inoculum. It may be that the low incidence of virus seen during the late winter is seed-borne (Rader *et al.*, 1941; Lindberg *et al.*, 1956) or has spread from common weed hosts or, more likely, has spread from pumpkin (*Cucurbita pepo*) crops grown throughout the year in the same river flats, and known to be infected with WMMV.

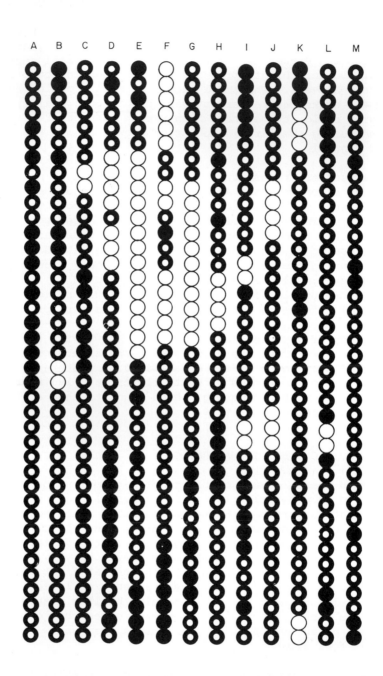

Figure 4. The distribution of WMMV-infected watermelons in a crop near the end of the season, showing clusters of uninfected plants ◯ , plants infected early ● and late ◉

DISCUSSION

In these examples the virus sources could have been far from the crops, but in no case is it necessary to postulate that this occurred to account for the observed patterns of distribution. In no instance is long-distance spread of virus likely to have contributed substantially to the active growth of epidemics. In each instance there was sufficient local aphid activity to account for the virus spread. This is particularly so for ClMV in cauliflower and MDMV in sweetcorn. The marked patches of diseased plants and the spread of virus from the margins of the patches would have been less obvious if there had been large influxes of viruliferous aphids throughout the year. The evidence for spread from local sources is less good for the movement of BYMV and CYVV into lupins and of WMMV into watermelon. Although potential reservoirs of each virus exist near crops, no comparative studies have been made of their incidence or strains in crops and other hosts.

In general, spread within but not into crops can be effectively controlled by insecticides. However, even if aphid spread into crops is not controlled, problems do not arise unless sources are common in the immediate vicinity. Hence the success of several schemes for pro- ducing virus-free stocks of potato, lettuce seed, ornamentals, and top and soft fruits. Viruses such as cucumber mosaic virus and bean yellow mosaic virus, which are often common in plants growing near crops, are far more difficult to control. Indeed, when susceptible perennial hosts of the virus and vector abound, there seems little reason to expect virus spread to stop at the crop margins. Local vector activity should be sufficient for regular outbreaks of disease, irrespective of virus spread from afar.

The sources from which viruses spread into crops are clearly import- ant, for distant sources are more difficult to locate and eliminate than those nearby. However, proof of the source of aphid-borne viruses is difficult to obtain, especially when immigrant aphids may be responsible for virus spread within the crops, as with TuBV, and BYMV in lupin. There is therefore a need to show:-

1. Whether or not aphids which spread virus into or within crops
    originate locally. Aphids can be caught far from a susceptible
    crop, but is this epidemiologically important? The density of
    aphids at high altitudes (the aphids which could, perhaps, move
    far) is very low, of the order of 1 per 3000 m³ (Berry & Taylor,
    1969). Aerial populations of monophagous aphids, such as
    the damson-hop aphid *Phorodon humuli*, decrease to very low
    densities (1 trapped per month) within *c*. 160 km of the main
    source (Taylor *et al.*, 1979). Aerial populations of polyphagous
    aphids, such as *M. persicae*, also tend to be determined by the
    local abundance of its hosts (Taylor, 1977), indicating that only
    a small proportion of the aphids migrate far.

2. Whether or not viruses brought into crops originate from nearby
    sources. Gradients may provide circumstantial evidence for this,

but they can be due to variations in aphid activity, unless it is
known that secondary spread does not occur. Sensitive survey
methods, using the enzyme-linked immunosorbent assay (Clark &
Adams, 1977) or radio-immuno assays (Ball, 1973), combined with
batch sampling (Gibbs & Gower, 1960), may prove useful for
detecting low incidences of viruses in large potential reservoirs.

3.   The aphid species which have acquired virus *en route* to the crop.
Serological methods may again prove useful, although trapping of
live aphids and feeding them on test plants is more definitive
(Raccah, this volume).

4.   The factors which most affect the rate of growth of epidemics.
Mathematical modelling of epidemics is moving away from the
classical approach of the "simple" and "compound interest" models,
for it is doubtful whether the exponential rate of disease spread
remains constant for long. By including in the model those factors
thought to affect the rate of spread, and by examining how changes
to them affect the model, some insight may be gained into the ways
epidemics develop, and perhaps how they can be controlled.

5.   The effects of low levels of seed transmission. Viruses may be
seed-borne in few of their hosts, such as CMV in wild cucumber
(Lindberg *et al.*, 1956) and in *Stellaria media* (Tomlinson & Walker,
1973). Some are seed-borne at very low levels, such as MDMV in
sweetcorn (Shepherd & Holdeman, 1965). Although aphid-borne
viruses such as lettuce mosaic virus can be controlled by the use
of seed with less than 0.1% carrying virus, levels of seed transmission too low to be reliably measured could under field conditions introduce viruses into new areas.

Acknowledgements

We thank Dr. P.R. Smith, Dr. G.T. O'Loughlin, Dr. R.J. Sward, Ms. J.R.
Moran, Mr. P. Ridland and Mr. G. Berg for helpful discussion. Mr. P.
Slotergraph and his colleagues in the Victoria Department of Agricul-
ture, Bairnsdale, kindly examined many sweetcorn crops for MDMV, and
Mr. P. Burt of the West Australian Department of Agriculture examined
watermelon crops at Carnarvon.

Authors' note: since this chapter was drafted it has been shown that
bean yellow mosaic virus is a strain of clover yellow vein virus.

REFERENCES

Ball E.M. (1973) Solid phase radioimmunoassay for plant viruses.
    Virology 55, 516-20.

Berry R.E. & Taylor L.R. (1969) High altitude migration of aphids in
    maritime and continental climates. *Journal of Animal Ecology* 37,
    713-22.
Clark M.F. & Adams A.N. (1977) Characteristics of the microplate
    method of enzyme-linked immunosorbent assay for detection of plant
    viruses. *Journal of General Virology* 34, 475-83.
Gibbs A.J. & Gower J.C. (1960) The use of a multiple-transfer method in
    plant virus transmission studies - some statistical points arising
    in the analysis of results. *Annals of Applied Biology* 48, 75-83.
Gutierrez A.P., Nix H.A., Havenstein D.E. & Moore P.A. (1974) The
    ecology of *Aphis craccivora* Koch and subterranean clover stunt virus.
    III. A regional perspective of the phenology and migration of the
    cowpea aphid. *Journal of Applied Ecology* 11, 21-35.
Hardy A.C. & Milne P.S. (1937) Insect drift over the North Seas, 1936.
    Nature (London) 139, 510.
Lindberg G.D., Hall D.H. & Walker J.C. (1956) A study of melon and
    squash mosaic viruses. *Phytopathology* 46, 489-95.
O'Loughlin G.T. & Chambers S.C. (1967) The systemic infection of an
    aphid by a plant virus. *Virology* 33, 262-71.
Orlob G.B. & Medler J.T. (1961) Biology of cereal and grass aphids in
    Wisconsin (Homoptera). *Canadian Entomologist* 93, 703-14.
Price R.D. (1970) Stunted patches and deadheads in Victorian cereal
    crops. *Technical Publication, Department of Agriculture, Victoria*
    No. 23.
Rader W.E., Fitzpatrick H.F. & Hildebrandt E.M. (1941) A seed-borne
    virus of muskmelon. *Phytopathology* 37, 809-16.
Shepherd R.J. & Holdeman Q.L. (1965) Seed transmission of the Johnson
    grass strain of sugarcane mosaic virus in corn. *Plant Disease
    Reporter* 49, 468-9.
Smith P.R. (1966) A disease of French bean (*Phaseolus vulgaris* L.)
    caused by sub-clover stunt virus. *Australian Journal of Agricultural
    Research* 17, 875-83.
Smith P.R. & Plumb R.T. (1981) Barley yellow dwarf virus infectivity
    of cereal aphids trapped at two sites in Victoria. *Australian
    Journal of Agricultural Research* 32, 249-55.
Stubbs L.L., Guy J.A.D. & Stubbs K.J. (1963) Control of lettuce
    necrotic yellows virus disease by the destruction of sow-thistle
    (*Sonchus cleraceus* L.). *Australian Journal of Experimental Agri-
    culture and Animal Husbandry* 3, 215-8.
Sutton J. & Garrett R.G. (1978) The epidemiology and control of tulip
    breaking virus in Victoria. *Australian Journal of Agricultural
    Research* 29, 555-63.
Taylor L.R. (1977) Migration and the spatial dynamics of an aphid
    Myzus persicae. *Journal of Animal Ecology* 46, 411-23.
Taylor L.R., Woiwod I.P. & Taylor R.A.J. (1979) The migratory ambit
    of the hop aphid and its significance in aphid population dynamics.
    Journal of Animal Ecology 48, 955-72.
Tomlinson J.A. & Walker V.M. (1973) Further studies on seed trans-
    mission in the ecology of some aphid-transmitted viruses. *Annals
    of Applied Biology* 73, 293-8.

Studies on the main weed host and aphid vector of lettuce necrotic yellows virus in South Australia

D.K. MARTIN
South Australian College of Advanced Education, Magill Campus,
Lorne Avenue, Magill, South Australia 5072

INTRODUCTION

Lettuce necrotic yellows virus (LNYV) can cause severe losses in lettuce crops in both Australia (Stubbs & Grogan, 1963; Randles & Crowley, 1970) and New Zealand (Fry et al., 1973), the only countries where it has been recognized.

The main vector of the virus is the aphid, Hyperomyzus lactucae, and the principal source of both virus and vector is the common sowthistle, Sonchus oleraceus. However, H. lactucae does not colonize or breed on lettuce, which is infected, almost incidentally, by migrant aphids from sowthistle searching for suitable host plants. Thus virus spreads into but not within lettuce crops and infected plants are scattered throughout plantings with no evidence of secondary spread.

The relative simplicity of the conceptual model of disease outbreaks of lettuce necrotic yellows makes this disease a useful one for studying host/virus/vector interactions. Consequently, field observations have been in progress for some years in major lettuce-growing areas in the states of Victoria and South Australia (Stubbs & Grogan, 1963; Randles & Crowley, 1970; Randles & Carver, 1971; Boakye & Randles, 1974). The work described here involved detailed observations from June 1975 to November 1977 on H. lactucae and sowthistle in a lettuce-growing area of the foothills of the Mount Lofty Range near Adelaide, South Australia. There were also ancillary studies of the anholocyclic reproduction of the aphid and of the growth of sowthistles in various environmental conditions in laboratory growth chambers. The results of these studies are used to describe and interpret the build-up and dispersal of the vector from sowthistles and the spread of LNYV to lettuce.

THE VIRUS

Lettuce necrotic yellows is a rhabdovirus with bacilliform particles c. 227 nm long and c. 62 nm wide (Francki & Randles, 1980). It is

Plumb R.T. & Thresh J.M. (1983) Plant Virus Epidemiology.
Blackwell Scientific Publications, Oxford.

transmitted circulatively (Stubbs & Grogan, 1963), transovarially (Boakye & Randles, 1974) and probably propagatively (Francki & Randles, 1980) in *H. lactucae*. The only other recorded vector is *H. carduellinus* (Randles & Carver, 1971) which has a limited distribution in Australia and seems to be relatively unimportant. There is a temperature-dependent latent period in the vector of 5-8 days after which transmission occurs consistently, usually until death (Boakye & Randles, 1974).

HOST PLANTS

Because of the latent period in the vector, the main sources of viruliferous aphids are infected plants of species that support breeding colonies. Aphids developing on such plants complete their latent period within 24 h of reaching the fourth instar and are therefore viruliferous when they migrate (Boakye & Randles, 1974).

Infected sowthistle seedlings within a lettuce crop are unimportant as sources of virus infection to that crop because as young plants they seldom support aphid populations that produce winged migrants while the crop is still susceptible.

The natural host range of LNYV is small and only lettuce (*Lactuca sativa*), *L. serriola*, *S. oleraceus*, *S. hydrophilus*, *Reichardia tingitana* and *Embergeria megalocarpa* have been found infected (Stubbs & Grogan, 1963; Randles & Carver, 1971). There is no evidence of transmission through the seed of lettuce cultivars, sowthistles or manually inoculated *Nicotiana glutinosa* (Francki & Randles, 1970).

LNYV seems to be unrelated to other rhabdoviruses (Francki & Randles, 1980) and Randles & Carver (1971) suggested that it may have originated in Australia in a native perennating host such as *E. megalocarpa* or *S. hydrophilus*. It may have become established elsewhere after the introduction and naturalization of other *Sonchus* spp. and *H. lactucae*.

S. oleraceus is a common weed in Australia in and around cultivated areas and along roadsides within the 375 mm isohyet. Plants grow on a wide variety of soils with a pH range of 6.5-9.0 (Buckli, 1936) and, as they can tolerate salt, are often found in the first line of sandhills along the coast. Reproduction is only by seed, with 100% germination in all but the summer months (Gill, 1938). Germination requires high light intensity and shaded plants seldom survive beyond the seedling stage (Lewin, 1948).

The maximum numbers of sowthistles occur in South Australia in the hot summer months of November and December (Fig. 1a) following spring germination after winter rains. Flowering occurs throughout the year and in the sampling period the only time when < 40% of the sowthistles were flowering was in November 1976 following extensive weed control measures that included mowing and ploughing (Fig. 1a).

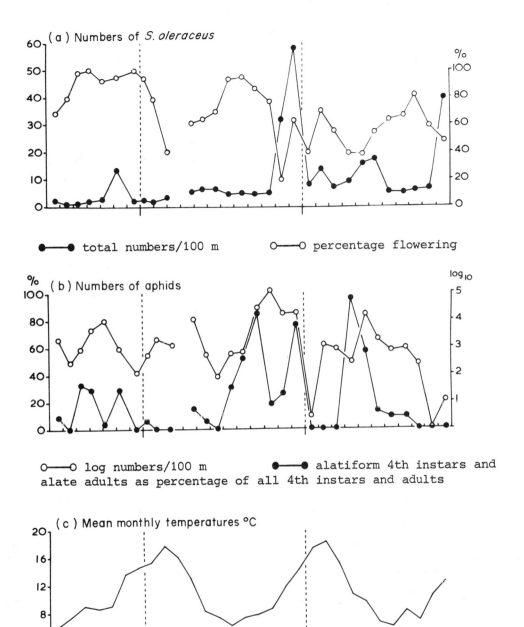

Figure 1. Field data collected over 30 months in a lettuce-growing area of the foothills of the Mount Lofty Range, Adelaide.

In winter, some plants have elongate stems bearing flowers, whereas many are non-flowering, frost-resistant rosettes (Lewin, 1948). The immature rosette form persists longest during the cool winter months with short day-lengths. The rosettes produce many new flower-bearing stems as day-length and temperature increase during spring (Martin, 1979). Furthermore, germination and rapid growth of new plants in spring results in large numbers of flowering plants by early summer.

THE MAIN VECTOR

H. lactucae is considered a palaearctic species (Eastop, this volume) that could have been introduced to Australia as eggs on the primary winter host *Ribes nigrum*. Its introduction seems to have been relatively recent because none of its attendant parasites occurs in Australia (Stary & Schlinger, 1967; Stary, 1970).

In cold European and North American climates *H. lactucae* is holocyclic, alternating between the summer host (sowthistle) and *R. nigrum* (Hille Ris Lambers, 1949). In Australia host alternation has not been recorded, although male alatae have been observed in south eastern Australia (Eastop, 1966) and South Australia (M. Carver, unpublished data). In southern Australia, *H. lactucae* is normally anholocyclic on sowthistle, reproducing parthenogenetically and viviparously throughout the year, albeit in small numbers in the cool winter and hot summer months.

Distribution on host plant

H. lactucae is generally found only on mature, flowering plants of sowthistle, clustered on or around the flower buds. However, solitary individuals or small, young colonies sometimes occur on leaves of immature, non-flowering sowthistle especially after rain.

On three representative sampling occasions in September and December 1976 and May 1977, 10 173 aphids were found on 67 flowering plants, whereas only 31 aphids were found on 65 non-flowering plants. This emphasises the crucial importance of flowering plants as hosts. The stage of flower or bud development is critical and five clearly defined phases have been distinguished:-

1. flower bud first evident, tightly closed, length = diameter,
2. bud elongates, length > diameter,
3. open flower; sepals folded back and corollas apparent,
4. post-flower; ligules disappear, base starts to swell,
5. seed head; pappus and achene appear.

On over 3000 flowering plants examined, aphids occurred on and/or immediately below flowers at stages 1-4 but never at stage 5. Colonies of aphids extended a short distance down the pedicels and stems of buds at stages 1-4 when populations became crowded.

Movement on and between hosts

Laboratory studies showed that at the onset of stage 5, aphids migrate to more suitable sites elsewhere on the plant or to other plants. Some of the apterae and nymphs that disperse must reach adjacent sowthistles or lettuce but they can only be important in local virus spread. Moreover, such aphids are unlikely to feed long enough to infect lettuce unless they have not previously fed for at least 6 h (Boakye & Randles, 1974). This suggests that the main spread to lettuce is by alatae, especially those that have flown for so long that they readily settle and feed.

To determine the form of *H. lactucae* responsible for the spread of LNYV in lettuce, Boakye (1973) manipulated populations of *H. lactucae* on infected sowthistle plants placed within lettuce plots. A high background incidence of *c.* 40% infection in control plots was apparently related to the large number of alate *H. lactucae* trapped. Plots in which an additional source of alate, viruliferous *H. lactucae* was placed showed no significantly greater incidence of LNYV. By contrast, in plots with additional apterous viruliferous aphids the disease incidence was *c.* 80% and a steep gradient was shown from the aphid source plants. This suggests that apterae can contribute to local spread from sowthistles in at least some circumstances. However, marginal patterns of spread and clumping of diseased plants have not been observed in South Australian lettuce crops and the main spread of LNYV has been attributed to alatae. This spread correlates well with aphid flights 4-5 weeks earlier (Stubbs & Grogan, 1963; Randles & Crowley, 1970).

Alate production

For *H. lactucae*, as for other aphid species (Van Emden, 1972), the induction of alatae is the result of complex interactions of photoperiod, temperature, food quality and aphid density.

In the survey areas, most alate *H. lactucae* were produced at high population densities and cool mean temperatures (< 15°C) and least at low densities and high mean temperatures (> 15°C) (Fig. 1b, 1c) (Martin, 1979; Maelzer, 1981). At least 50% of all adult and 4th instar *H. lactucae* were alate or alatiform when mean monthly temperatures were 8-14°C (Fig. 1b, 1c). By contrast, < 10% of all adults and 4th instars were alate or alatiform when it was warmer.

These observations are consistent with results from southern Australia that showed that *H. lactucae* is consistently trapped in the greatest numbers during autumn and spring with relatively few being caught in the hot summer, and cold winter, months (O'Loughlin, 1963; Hughes *et al.*, 1964, 1965; Randles & Crowley, 1970). However, in the laboratory the rate of increase (r_m) of both apterae and alatae increases with temperature between 15 and 20°C if other environmental variables are optimum (Table 1). Clearly some process must be restricting population development in the field when temperatures are otherwise favourable.

The populations of aphid species are limited at high temperatures by increases in the populations and/or voracity of predators and/or parasites (Dunn, 1952; Frazer & Gilbert, 1976; Maelzer, 1978). This is unlikely to be the explanation for *H. lactucae* because parasites are not found and the three predators present in South Australia are rare.

Table 1. The effect of temperature on the maximum intrinsic rate of increase (r_m) of apterous and alate *H. lactucae*.

Type of aphid	Temperature $^{\circ}$C	Number of aphids tested	Intrinsic rate of increase (r_m)*
Apterous	15	37	0.17
	20	25	0.23
	25	73	0.26
Alate	15	50	0.13
	20	48	0.19
	25	93	0.23

* Estimated using Watson's (1964) method

The factors influencing the number of flower buds produced and the period they remain favourable for aphid production were then investigated. The approach was similar to that used for the rose aphid, *Macrosiphum rosae*, which also feeds primarily on buds. Maelzer (1977) suggested that the relative time rose buds remain favourable for *M. rosae* is shorter at high than at low temperatures so that the time for reproduction is curtailed and the largest populations occur at temperatures that are not the optimum for aphid reproduction. Similar factors have been shown to influence the reproduction of *H. lactucae* on sowthistle with the additional complication that flowering is influenced by photoperiod.

The time buds of sowthistle remain favourable for *H. lactucae* was determined at three temperatures and for two photoperiods. The intrinsic rates of increase determined earlier (Table 1) were used to calculate the potential size of aphid colonies from the time (*t*) buds remained favourable at each combination of temperature and photoperiod (Table 2). These calculations assumed, for each treatment, that on the first day a bud became favourable for aphids it was colonized by one adult female and that the colony grew exponentially for time *t*.

Most buds produced at 25°C remained favourable for relatively short periods, consequently larger populations of aphids developed at

Table 2. Potential size of colonies of *H. lactucae* on favourable sowthistle flower buds at three temperatures and two day lengths.

	8 h light			16 h light		
	15°C	20°C	25°C	15°C	20°C	25°C
Number of buds observed	473	314	464	371	312	402
Time bud remains favourable to *H. lactucae* (days)	35.2	22.1	14.1	20.5	15.4	13.2
Calculated number of individuals in aphid colony	397	161	39	33	35	31

lower temperatures. This interaction between aphid and host plant helps explain why the numbers of *H. lactucae* in South Australia are greatest in September and October, when mean temperatures are < 15°C and photoperiod is 12-13 h/day, rather than in November-January, when temperatures are much higher but photoperiod longer (14-16 h). Thus at moderate temperatures when days are relatively short, aphids become crowded as populations increase. This crowding initiates alate production and consequent spread of virus.

CONCLUSIONS

The spread of lettuce necrotic yellows virus to lettuce is determined by the ecology of the main virus reservoir plant, sowthistle, and the aphid *H. lactucae*. Because studies on lettuce necrotic yellows are not complicated by the effects of secondary spread within the crop or by other vectors, it is an ideal system for studying the complex interactions between virus, vector and host plants. This work has highlighted the advantages of an ecological approach to epidemiology of the type long advocated by Carter (1939) but seldom adopted by subsequent workers, who have tended to concentrate exclusively on virus, vector or host plant and not on multi-disciplinary studies.

REFERENCES

Boakye D.B. (1973) Transmission of LNYV by *H. lactucae* (L.) (Homoptera Aphididae) with special reference to aphid behaviour. *PhD Thesis,*

218   D.K. MARTIN

Adelaide University.

Boakye D.B. & Randles J.W. (1974) Epidemiology of LNYV in South Australia. III. Virus transmission parameters and vector feeding behaviour on host and non-host plants. *Australian Journal of Agricultural Research* 25, 791-803.

Buckli M. (1936) Oekologie der Acherunkräuter der Nordost-Schweiz. *Beiträge geobotanischer Landesaufnahme* No. 19.

Carter W. (1939) Populations of *Thrips tabaci*, with special reference to virus transmission. *Journal of Animal Ecology* 8, 261-76.

Dunn J.A. (1952) The effect of temperature on the pea aphid-ladybird relationship. *Annual Report 1951, National Vegetable Research Station, Wellesbourne.*

Eastop V.F. (1966) A taxonomic study of Australian Aphoideae (Homoptera). *Australian Journal of Zoology* 14, 399-592.

Francki R.I.B. & Randles J.W. (1970) Lettuce necrotic yellows virus. *CMI/AAB Descriptions of Plant Viruses* 26, 4 pp.

Francki R.I.B. & Randles J.W. (1980) Rhabdoviruses infecting plants. *Rhabdoviruses, Vol. III* (Ed. by D.H.L. Bishop), pp. 135-65. CRC Press, Florida.

Frazer B.D. & Gilbert N. (1976) Coccinellids and aphids: a quantitative study of the impact of adult ladybirds (Coleoptera : Coccinellidae) preying on field populations of pea aphids (Homoptera : Aphididae). *Journal of the Entomological Society of British Colombia* 73, 33-56.

Fry P.R., Close R.C., Procter C.H. & Sunde R. (1973) Lettuce necrotic yellows virus in New Zealand. *New Zealand Journal of Agricultural Research* 16, 143-6.

Gill N.T. (1938) The viability of weed seeds at various stages of maturity. *Annals of Applied Biology* 25, 447-56.

Hille Ris Lambers D. (1949) Contributions to a monograph of the aphididae of Europe. IV. *Temminckia* 8, 182-322.

Hughes R.D., Carver M., Casimir M., O'Loughlin G.T. & Martyn E.J. (1965) A comparison of the numbers and distribution of aphid species flying over eastern Australia in two successive years. *Australian Journal of Zoology* 13, 823-39.

Hughes R.D., Casimir M., O'Loughlin G.T. & Martyn E.J. (1964) A survey of aphids flying over eastern Australia in 1961. *Australian Journal of Zoology* 12, 174-200.

Lewin R.A. (1948) Biological flora of British Isles. *Journal of Ecology* 36, 203-23.

Maelzer D.A. (1977) The biology and main causes of changes in numbers of the rose aphid *Macrosiphum rosae* (L.) on cultivated roses in South Australia. *Australian Journal of Zoology* 25, 269-84.

Maelzer D.A. (1978) The growth and voracity of *Leis conformis* (Boisd.) (Coleoptera : Coccinellidae) fed on the rose aphid *Macrosiphum rosae* (L.) (Homoptera : Aphididae) in the laboratory. *Australian Journal of Zoology* 26, 293-304.

Maelzer D.A. (1981) The ecology of pest species in Australia. *The ecology of pests in Australia* (Ed. by R.L. Kitching & R.E. Jones), pp. 89-106. CSIRO, Melbourne.

Martin D.K. (1979) The ecology of *Hyperomyzus lactucae* (L.) and the epidemiology of lettuce necrotic yellows virus. *PhD Thesis,*

Adelaide University.

O'Loughlin G.T. (1963) Aphid trapping in Victoria. I. The seasonal occurrence of aphids in three localities and a comparison of two trapping methods. *Australian Journal of Agricultural Research* 14, 61-9.

Randles J.W. & Carver M. (1971) Epidemiology of LNYV in S.A. II. Distribution of virus, host plants and vectors. *Australian Journal of Agricultural Research* 22, 231-7.

Randles J.W. & Crowley N.C. (1970) Epidemiology of LNYV in S.A. I. Relationship between disease incidence and activity of *H. lactucae* (L.) *Australian Journal of Agricultural Research* 21, 447-53.

Stary P. (1970) *Biology of aphid parasites (Hymenoptera : Aphididae) with respect to integrated control.* W. Junk, The Hague.

Stary D. & Schlinger E.I. (1967) *A revision of the Far East Asian Aphididae (Hymenoptera).* W. Junk, The Hague.

Stubbs L.L. & Grogan R.G. (1963) Necrotic yellows: a newly recognized virus disease of lettuce. *Australian Journal of Agricultural Research* 14, 439-59.

Van Emden H.F. (1972) *Aphid Technology.* Academic Press, London.

Watson T.F. (1964) Influences of host plant condition on population increase of *Tetranychus telarius* (L.) (Acarina : Tetranychidae). *Hilgardia* 35, 273-322.

Epidemiology and control of aphid-borne virus diseases in California

JAMES E. DUFFUS

Agricultural Research Service, US Department of Agriculture,
Salinas, California, USA

INTRODUCTION

> On the wide level acres of the valley the topsoil lay deep
> and fertile - - the whole valley floor, and the foothills
> too, were carpeted with lupins and poppies - -
> (Steinbeck - "East of Eden" 1952)

Salinas and the Salinas Valley of California have been termed "East of
Eden" by Steinbeck, but aphid-transmitted viruses have long "considered"
this area of mild climate nestled in the coastal ranges of central
California as Eden. Virtually all the numerous vegetable and field
crops grown in the area have experienced severe and destructive attacks
by aphid-transmitted virus diseases.

The intensive agriculture of the area involves a complex overlapping
of growing seasons and crops with a diversity of weed and crop species
and a temperate year-round climate that provide favourable ecological
conditions for disease spread.

Studies in California's important agricultural valleys have indicated
that weeds play a significant role in the build-up of the green peach
aphid (*Myzus persicae*) in these areas. In the Salinas Valley, 27
species of plants in 13 families serve as important hosts and a succession of favourable hosts allows the aphid to reproduce parthenogenetically throughout the year. In spring, aphids build up on weed hosts
(wild mustards (*Brassica* spp.), fiddleneck (*Amsinckia* spp.), cheeseweed
(*Malva* spp.), dock (*Rumex* spp.), and pigweed (*Chenopodium* spp.)),
whereas in autumn, they build up on cultivated crops such as lettuce,
spinach, potato and crucifers, as well as on cover crops and weeds
(Duffus, 1971).

The moderate summer temperatures and mild winters allow a continuous
movement of aphids and viruses so that vast reservoirs of both can
develop.

Plumb R.T. & Thresh J.M. (1983) *Plant Virus Epidemiology*.
Blackwell Scientific Publications, Oxford.

EPIDEMIOLOGY AND CONTROL

Three aphid-transmitted diseases or disease complexes have received particular attention in California. They represent each of the main groups of aphid-transmitted virus diseases and the research on the nature, vector transmission and ecological aspects of the causal viruses has had marked economic and scientific effects on Californian and world agriculture. Interestingly, the main emphasis in the control of these viruses has been biological or ecological rather than chemical, although the vastly increased use of pesticides and changes in cropping practices may also have affected the incidence of virus diseases in recent years.

Celery mosaic

Celery mosaic virus (CeMV) (Severin & Freitag, 1938) is a potyvirus with flexuous rod-shaped particles *c*. 760 nm long. The virus is transmitted in a non-persistent manner by several species of aphids. It has a limited host range restricted to the Umbelliferae and occurs naturally only in celery, celeriac and carrot (Severin & Freitag, 1938; Brandes & Luisoni, 1966; Shepard & Grogan, 1967). Strains of the virus occur in parsley and poison hemlock but these isolates do not seem to be of importance in celery (Sutabutra & Campbell, 1971).

Celery mosaic became increasingly prevalent during the early 1930s in celery-growing districts of California, causing severe yield losses and losses of quality in transit. The virus is not seed-borne in celery and the only apparent sources of infection detected in the 1930s were actively growing celery, celeriac, and carrot. Celery was being grown as a year-round crop that provided a continuous source of inoculum.

Following these observations, growers were asked to agree on a definite period each year when celery would not be grown. This voluntary programme was superseded when a section added to the Agricultural Code granted the Director of Agriculture authority to proclaim a host-free period and/or district.

Following the initiation of the programme, mosaic decreased dramatically, the quality of celery improved and yields increased 2-3 fold compared with earlier, epidemic years (Milbrath, 1948).

Lettuce mosaic/"June yellows"

From the early 1930s to the 1960s lettuce mosaic was an extremely important factor in the production of lettuce in the Salinas Valley. This disease, either alone or in combination with others due to the aster yellows pathogen or to the viruses causing beet western yellows, tomato spotted wilt, lettuce speckles, beet stunt and sowthistle yellow vein, completely devastated numerous plantings of the crop.

As early as 1923, Newhall (1923) showed that lettuce mosaic virus (LMV) was seed-transmitted and that seed was probably the most important source of primary inoculum. Epidemiological studies by Kassanis (1947) and Broadbent (1951) confirmed that the virus was seed borne but also stressed the need for sanitation and isolation in the production of healthy lettuce crops.

In California in the early 1950s, Grogan *et al.* (1952) tested experimentally the control of lettuce mosaic through the use of virus-free seed. Further work (Zink *et al.*, 1956) showed that the percentage of plants that develop mosaic in the field largely depends upon the amount of seed-borne virus and the numbers and mobility of the aphid vectors. If > 0.1% of seed is infected, control is likely to be unsatisfactory and the attempt at control by using only seed with 0.1% or less infection was unsatisfactory (Grogan, 1980). Only after utilizing all the epidemiological information known were completely satisfactory results obtained. The methods used included:-

1. the use of seed that had given no evidence of seed-borne LMV in tests on a representative sample of 30 000 seeds,
2. removal or destruction of nearby overwintering weeds,
3. prompt destruction of all residual lettuce plants after harvest,
4. planting new crops away from established crops.

The adoption of these measures has greatly reduced the incidence and spread of LMV as well as losses due to it, and has also reduced the damage caused by other diseases.

Yellows complex of sugarbeet

The yellowing viruses of sugarbeet are serious hazards to stable production of sugarbeet and numerous other crops throughout the world.

The yellowing disease of sugarbeet was first attributed to an infectious entity in 1936 in Europe (Roland, 1936; Van Schreven, 1936) and was recorded in California and elsewhere in the United States in 1951 (Coons & Kotila, 1951). However, photographic evidence indicates that it occurred in the Salinas Valley as early as 1945 and perhaps in 1921 (Bennett, 1960). Following the discovery of yellows, epidemics occurred annually in California until 1968.

Two important yellowing diseases are involved in the complex, beet yellows and beet western yellows. They are caused by two distinct viruses both of which are spread largely by *M. persicae*. Excluding vector populations, two of the main factors influencing the epidemiology of the yellowing viruses are virus/vector relationships and virus sources.

Beet yellows virus (BYV) is transmitted in a semi-persistent manner and is retained by the vector for 24-72 h. Such transmission characteristics cause virus spread to be local, i.e. the disease incidence is high in areas adjacent to the virus source but quickly decreases with distance. A few km is apparently an effective barrier to the distribution of BYV. The principal source of the virus is beet itself, including overwintering beet fields and volunteer beets regenerating in fields used previously or growing in waste places.

Beet western yellows virus (BWYV) is much more widely distributed and is the most common beet virus. BWYV is transmitted in a persistent manner by aphids which may retain the virus for life. Virus distribution is much more general and widespread than the local distribution of the beet viruses that are transmitted non-persistently or semi-persistently. Beet western yellows spreads from beets, common weeds and other crop plants (Duffus, 1973).

The sugarbeet yields in California from 1910 to 1974 show interesting trends and provide new insights into the economic impact of beet-free periods and yellowing viruses on sugarbeet yields (Fig. 1) (Duffus, 1978).

Yields of sugar/acre were c. 2.0 tons in 1910, but declined sharply when the much publicized ravages of beet curly top virus occurred following World War I (Duffus, this volume). Production markedly

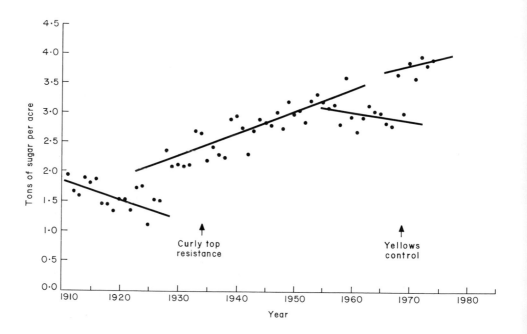

Figure 1. Production of sugar per acre in California 1911-1974.

improved in the late 1920s and early 1930s when beet production was abandoned in areas that were often devastated by curly top, and curly top-resistant cultivars were introduced in 1934. Production steadily increased to 3.0 tons of sugar/acre about 1950.

This was a period of progressively better curly top resistance and cultural practices with improvements in cultivars, soil and crop management techniques, fertilization methods and materials, stand establishment, irrigation methods, deep tillage, and insect and nematode control.

Despite the introduction of cultivars combining resistance to bolting, curly top and downy mildew, the introduction of hybrid cultivars and monogerm seed, the increased use of nematicides, herbicides and mechanical harvesting systems, the implementation of a beet leafhopper control programme (Duffus, this volume) and better fertilizer management, yields significantly declined from about 1950 until the late 1960s.

The decreasing yields during this period of increasingly sophisticated sugarbeet technology were directly related to losses caused by the yellows virus complex. About 1950, farmers began to grow crops in overlapping sequence and this resulted in increased incidence of the yellows virus complex and increasingly severe yield losses.

Epidemiological studies in California in the late 1950s (Duffus, 1963) confirmed European reports that there was a close correlation between virus yellows incidence and the proximity of overwintered beet fields. During this period of extremely serious yellows losses, sugarbeet growers and processors reached agreements to maintain a beet-free period between harvesting and sowing the new crop. These "beet-free periods" differed for different areas within the beet-growing districts in the state so that beet-free and overwintering areas were separated. These programmes were introduced over several years but they were first put into general use for the 1968 crop. During the beet-free periods, there was an effort to prevent sugarbeets being left in the ground within the beet-free area and a cooperative effort was made to destroy weed beets.

After the implementation of beet-free periods, yields significantly increased. However, the large 1968 crop and rain prevented completion of this harvest, so that the beet-free period was impossible to enforce and there was much yellows and poor yields in 1969. Also from 1968, two hybrid sugarbeet cultivars with moderate resistance to virus yellows were introduced.

For the state as a whole, mean sugar production was 0.86 tons/acre greater for the growing seasons following the introduction of yellows control than for the period 1950-1967 (Fig. 1).

Although several factors have contributed to improved yields in the state since 1969, the general absence of destructive attacks by the yellowing viruses brought about by the implementation of beet-free periods has been the factor that is most obvious to growers, processors and sugarbeet researchers.

CONCLUSIONS

The general decrease in virus incidence caused by measures introduced to combat virus diseases and disease complexes has led to the discovery of previously unknown disease syndromes such as those due to sowthistle yellow vein virus (Duffus et al., 1970), lettuce speckles mottle virus (Falk et al., 1979) and turnip mosaic virus (Zink & Duffus, 1969) in lettuce, and beet yellow stunt virus in lettuce and sugarbeet (Duffus, 1972).

The control procedures introduced for some of these diseases have changed the appearance, agronomy and productivity of the area and no longer do lupins and poppies carpet the valley floor. The "market garden" approach to production involving overlapping crops and prolonged harvesting procedures has given way to modern rotation and planting schedules and the result has been vastly increased yields for most of the valley's important crops.

The interactions between crops and disease control systems are complex and little understood, but no longer are sugarbeet fields completely yellow by early July, and spring lettuce and spinach crops do not have to be abandoned and disced under without harvesting a plant. The control principle of avoiding sources of inoculum and a general clean-up of the hosts on which aphids build up, together with cropping changes related to economic factors, have markedly changed and reduced disease problems in the area.

REFERENCES

Bennett C.W. (1960) Sugar beet yellows disease in the United States. *United States Department of Agriculture Technical Bulletin* 1218, 1-63.
Brandes J. & Luisoni E. (1966) Untersuchungen über einige Eigenschaften von zwei gestreckten Sellerieviren. *Phytopathologische Zeitschrift* 57, 277-88.
Broadbent L. (1951) Lettuce mosaic in the field. *Agriculture: Journal of Ministry of Agriculture* 57, 578-82.
Coons G.H. & Kotila J.E. (1951) Virus yellows of sugar beets and tests for its occurrence in the United States. *Phytopathology* 41, 559.
Duffus J.E. (1963) Incidence of beet virus diseases in relation to overwintering beet fields. *Plant Disease Reporter* 47, 428-31.
Duffus J.E. (1971) Role of weeds in the incidence of virus diseases. *Annual Review of Phytopathology* 9, 161-79.
Duffus J.E. (1972) Beet yellow stunt, a potentially destructive virus disease of sugarbeet and lettuce. *Phytopathology* 62, 161-5.
Duffus J.E. (1973) The yellowing virus diseases of beet. *Advances in Virus Research* 18, 347-86.

Duffus J.E. (1978) The impact of yellows control on California sugar-
    beets. *Journal of the American Society of Sugar Beet Technologists*
    20, 1-5.
Duffus J.E., Zink F.W. & Bardin R. (1970) Natural occurrence of sow-
    thistle yellow vein virus on lettuce. *Phytopathology* 60, 1383-4.
Falk B.W., Duffus J.E. & Morris T.J. (1979) Transmission, host range,
    and serological properties of the viruses causing lettuce speckles
    disease. *Phytopathology* 69, 612-7.
Grogan R.G. (1980) Control of lettuce mosaic with virus-free seed.
    Plant Disease 64, 446-9.
Grogan R.G., Welch J.E. & Bardin R. (1952) Common lettuce mosaic and
    its control by the use of mosaic-free seed. *Phytopathology* 42,
    573-8.
Kassanis B. (1947) Studies on dandelion yellow mosaic and other virus
    diseases of lettuce. *Annals of Applied Biology* 34, 412-21.
Milbrath D.G. (1948) Control of western celery mosaic. *Bulletin of
    California Department of Agriculture* 37, 3-7.
Newhall A.G. (1923) Seed transmission of lettuce mosaic.
    Phytopathology 13, 104-6.
Roland G. (1936) Recherches sur la jaunisse de la betterave et
    quelques observations sur la mosaique de cette plante. *Sucrerie
    Belge* 55, 213-7.
Severin H.H.P. & Freitag J.H. (1938) Western celery mosaic. *Hilgardia*
    11, 495-558.
Shepard J.F. & Grogan R.G. (1967) Partial purification, properties, and
    serology of western celery mosaic virus. *Phytopathology* 57, 1104-10.
Sutabutra T. & Campbell R.N. (1971) Strains of celery mosaic virus from
    parsley and poison hemlock in California. *Plant Disease Reporter* 55,
    328-32.
Van Schreven D.A. (1936) De vorgelingsziekte bij de biet en haar
    oorzaak. *Mededelingen van het Instituut voor Rationele Suikerprod-
    uctie* 6, 1-36.
Zink F.W. & Duffus J.E. (1969) Relationship of turnip mosaic virus
    susceptibility and downy mildew (*Bremia lactucae*) resistance in
    lettuce. *Journal of the American Society for Horticultural Science*
    94, 403-7.
Zink F.W., Grogan R.G. & Welch J.E. (1956) The effect of the percentage
    of seed transmission upon subsequent spread of lettuce mosaic.
    Phytopathology 46, 662-4.

A technique for examining the long-distance spread of plant virus diseases transmitted by the brown planthopper, *Nilaparvata lugens* (Homoptera: Delphacidae), and other wind-borne insect vectors

L.J. ROSENBERG & J.I. MAGOR

Centre for Overseas Pest Research, London W8 5SJ, UK

INTRODUCTION

The spread of vector-borne plant diseases is related to the mobility of their vectors and the dispersal of several plant viruses transmitted by Homoptera has been associated with the wind-assisted migration of vectors (Johnson, 1969). This chapter shows how weather data can be used to study displacements of wind-borne vectors. The insect chosen to illustrate this technique is the brown planthopper, *Nilaparvata lugens*, the only known vector of two viruses of rice, grassy stunt and ragged stunt.

Vector and diseases

The brown planthopper only feeds on rice and is found from Pakistan eastwards to Fiji and from Japan and Korea southwards to Australia. It is known to make wind-assisted migratory flights each year to colonize the summer rice-growing areas of China, Japan and Korea (Kisimoto, 1976; Lee & Park, 1977; Cheng *et al.*, 1979). Whether it migrates elsewhere is less certain (Kisimoto & Dyck, 1976; Dyck *et al.*, 1979), but the capture of specimens over the sea between the Philippine Islands (Saxena & Justo, 1980) supports the hypothesis that it is a migrant throughout its range.

Infection of rice by grassy stunt virus has been confirmed, or symptoms have been recorded, in eleven countries: Bangladesh, Brunei, India, Indonesia, Japan, Malaysia, the Philippines, Sri Lanka, Taiwan, Thailand and Vietnam. Ragged stunt, which is a more recently identified virus disease (Hibino, 1979), has so far been recorded in only eight countries: China, India, Indonesia, Japan, Malaysia, the Philippines, Sri Lanka and Thailand (Fig. 1). Both diseases cause severe stunting and infected plants produce few or no panicles.

The viruses which cause grassy stunt and ragged stunt are transmitted in a persistent manner by their vector, but neither is transmitted transovarially. Details of the transmission of these diseases are given by Ling (1972) and Hibino (1979).

Plumb R.T. & Thresh J.M. (1983) *Plant Virus Epidemiology.*
Blackwell Scientific Publications, Oxford.

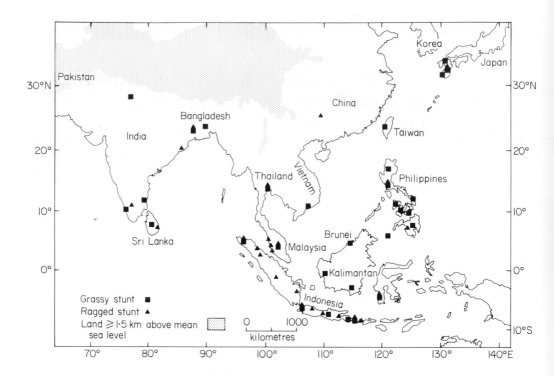

Figure 1. Known distribution of rice grassy stunt and ragged stunt diseases.

Both macropterous (long-winged) and brachypterous (short-winged) brown planthoppers are important in the spread of the diseases but in this chapter only wind-borne displacement of macropters is considered. The movements of these insects provide opportunities for disease spread over hundreds of kilometres in a few days. Their annual migration between China and Japan crosses at least 750 km of sea and the capture of marked brown planthoppers in China 184-720 km from their point of release (Anon., 1981) shows that similar distances are flown over land.

Brown planthopper migratory behaviour

The air speed of brown planthoppers is unknown but from their size a speed of about 1 m/s (4 km/h) can be expected (Lewis & Taylor, 1967). Wind speeds commonly exceed this and then the movements of the insects will be approximately down-wind. Trajectory analysis, in which wind speed and direction are used to examine the movement of air-borne particles, can therefore be used to simulate migratory flights. Wind direction and speed change with altitude and time, and to apply trajectory analysis effectively, the time of take-off and the height and duration of insect flight must be known or estimated. Probable values for these parameters and the factors which limit them were

Table 1. Laboratory and field observations on the migratory behaviour of brown planthoppers.

Flight parameter	Method of observation	Values	Reference
Take-off time	(a) Field studies	Dawn and dusk, temperature ≥17°C, wind speed < 3.1 m/s	Ohkubo & Kisimoto (1971)
	(b) Emergence traps	Dawn and dusk	D.E. Padgham(*)
Temperature threshold for flight	Tethered flights	50% flew at c. 16.5°C, few at 10°C	Ohkubo (1973)
Flight duration	(a) Transoceanic flights	c. 30 h	Y. Nagai(*)
	(b) Tethered flights	10 h (max. 23 h) at R.H.85% (times of flights unknown)	Ohkubo (1973)
	(c) Tethered flights of insects caught in emergence traps	Dawn flights: 1-3 (max. 5) h Dusk flights: 10-12 (max.17) h	D.E. Padgham(*)
	(d) Potential duration based on lipid contents of field populations	17 h (♀♀) 22 h (♂♂) (bred on IR20) 14 h (♀♀) 18 h (♂♂) (bred on IR36)	D.E. Padgham(*)
Height of flight	Field studies	Many caught at 1.5 km, few at 2 km. Height of flight increases with temperature	Tu (1980)
Physiological status	(a) Field studies	♀♀ unmated (ovaries immature)	Kisimoto (1976) Cheng et al. (1979)
	(b) Laboratory studies	3-4 days after emergence	Kusukabe & Hirao (1976)

*Personal Communication

obtained from field and laboratory observations (Table 1). These
suggest that brown planthoppers migrate when they are sexually
immature, that flights at dusk result in long-distance displacement and
that some individuals are capable of sustaining flights for up to 30 h
if the temperature is ≥ 17°C.

Migratory flights of brown planthoppers can, therefore, be simulated
by constructing 30 h trajectories in the low level wind-fields from
areas where immature macropters have recently emerged. Many macropters are known to be produced when rice is mature or ready to be
harvested (Nasu, 1969; MacQuillan, 1975; Dyck et al., 1979). By
simulating movements from areas where grassy stunt and ragged stunt are
present, the potential spread of viruliferous insects can be studied.

To illustrate this technique, weather data for 16-23 September 1979
were used to model displacements from areas where the diseases have
been confirmed or suspected and where macropters were present in
September 1979 or where rice is mature or being harvested in September.
These areas were Bangladesh, Brunei, south India, Indonesia, Japan, the
Philippines, Sri Lanka and Taiwan (Figs. 1 & 2).

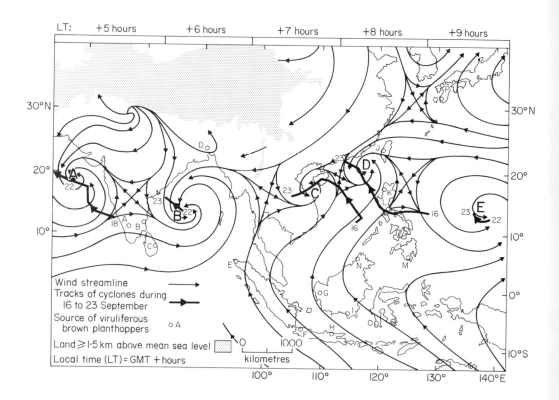

Figure 2. Surface streamlines at 0000 h GMT on 22 September with
tracks of cyclones, 16-23 September 1979.

Weather data and analysis

Wind speed and direction, recorded at standard heights from 10 m above ground level (surface) to 16.6 km above mean sea level, were obtained from the United Kingdom Meteorological Office and from daily weather reports published by national meteorological services. At altitudes above 2 km, temperatures were < 17°C and all migration was assumed to occur below this height. Wind-fields from 10 m to 1.5 km were analysed by the streamline-isotach method described by Palmer *et al.* (1955). Although wind-fields change progressively from hour to hour, in this study wind directions and speeds at 0000 h and 1200 h GMT were used to represent the 12 h periods starting at 1800 h and 0600 h GMT, respectively.

As an example of the analysis, the surface streamline pattern for 0000 h GMT on 22 September 1979 is shown (Fig. 2). The streamlines, drawn parallel to the observed winds, show that the wind-field between 10° and 20°N was dominated by five cyclonic circulations (A-E). These low pressure systems moved generally west or north-west during the study period and sudden changes in wind direction and speed occurred in their vicinity. To the north and south of the cyclone zone relatively undisturbed North-Easterly and South-Easterly Trade winds prevailed, the latter becoming south-westerlies as they crossed the equator.

SIMULATING BROWN PLANTHOPPER MIGRATIONS

Daily trajectories were drawn in the 10 m and 1.5 km wind-fields from the source areas (Fig. 3). Sunset times taken from the Nautical Almanac (Anon., 1977) were used as the starting time (dusk) for each trajectory. The temperature threshold for flight (17°C) was the only variable used in the model to limit flight activity. Take-off was assumed to occur only when dusk surface temperatures were ≥ 17°C at nearby synoptic stations. The trajectory was stopped if an area was reached where the temperature was < 17°C, although a 1.5 km trajectory was continued at the surface if temperatures there exceeded the flight threshold.

Two variables which may affect the insect's flight duration were not used in the model. Although brown planthoppers do not take off when wind speed exceeds 3.1 m/s (11 km/h) (Ohkubo & Kisimoto, 1971), wind speeds measured at 10 m were not considered sufficiently representative of those at crop height to be used to inhibit take-off in the model. Kisimoto & Dyck (1976) and Tu (1980) have also noted that showers and descending streams of air at weather fronts favour the descent of planthoppers, but in this study trajectories were continued for 30 h regardless of the weather encountered. Areas where rain or showers occurred were noted on the working charts but the model assumed that only some brown planthoppers landed in these areas.

Examples of the potential spread of viruliferous brown planthoppers

Trajectories from India, Taiwan and the Philippines were chosen to, show how vertical, horizontal and temporal changes in wind speed and direction may affect the dispersal of brown planthoppers (Fig. 3). These examples illustrate two important aspects of wind-borne migration.

1. *Wind-borne migrants leaving the same site on different days very often arrive in different places, even though flying at the same height and for the same time.*

2. *The track of wind-borne migrants can seldom be predicted by assuming that it occurs in the direction and at the speed of the winds measured at either the point of departure or the point of arrival.*

The most marked difference in wind direction, and therefore in destination reached, between the surface and 1.5 km wind-fields occurred in south India on 16 and 20 September and the greatest similarity was in the northern Philippines on 22 September. Generally, wind direction differed during each 30 h period. Direction was most consistent in Taiwan (at the surface, on 18 September; at 1.5 km, on 16, 18 and 20 September) and in the northern Philippines (at the surface and at 1.5 km on 16 and 22 September). These were periods when the seasonal Trade winds were relatively undisturbed. The most complex changes of track were from the Philippines on 18 and 20 September when the trajectories reflect the cyclonic circulation associated with the westward movement across the area of a tropical storm (Figs. 2 & 3).

Variations in track length were caused by differences in wind speed. An increase of speed with height was noted on all occasions and was most marked in south India on 18, 20 and 22 September and Taiwan on 16, 20 and 22 September. Differences in wind speed between days are best illustrated by the distances covered by the surface trajectories starting in west Taiwan.

A measure of the potential spread of brown planthoppers from each source and the mixing of populations from different sources within an area are given in Fig. 4. This highlights a third aspect of wind-borne migration.

3. *Wind-borne migrants caught at a site over a period of time are likely to have come from different sources.*

The trajectories indicate that many migrants probably perish at sea, but some reach distant rice-growing areas. Rice-growing areas were invaded both on the same day (southern China reached from J and K on 22 September) and on different days (south-west Kalimantan received immigrants from I on 19 September and from G on 21 September). Five source areas (B, I, K, L, P) were themselves invaded from elsewhere during this period.

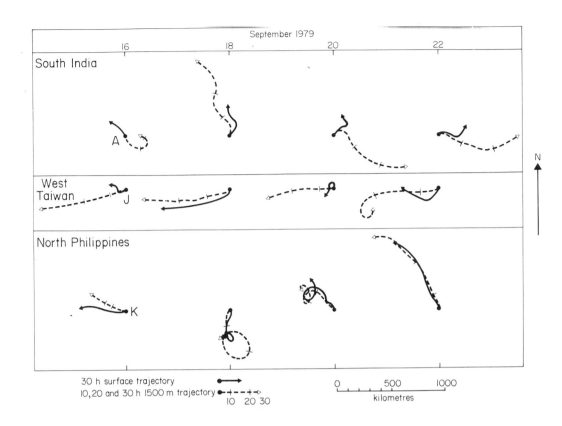

Figure 3. Tracks of surface and 1.5 km winds illustrating probable routes of brown planthoppers emigrating from south India, west Taiwan and north Philippines (Points A, J, K in Fig. 2) at dusk on 16, 18, 20 and 22 September 1979.

Too few trajectories were prepared for this study to postulate the spread of the diseases from currently known sources. If, however, trajectories do represent migratory flights of viruliferous brown plant-hoppers, then those shown in Fig. 4 suggest that grassy stunt and ragged stunt are more widespread than so far reported. For example, it seems probable that the diseases may have a wider distribution in southern India and Sri Lanka than Fig. 1 suggests. If, as the trajectories indicate, insects reach China from the Philippines and Taiwan, it seems probable that grassy stunt as well as ragged stunt occurs in China. This suggestion is strengthened by the presence of both diseases in Japan, a country invaded annually by brown planthoppers from China.

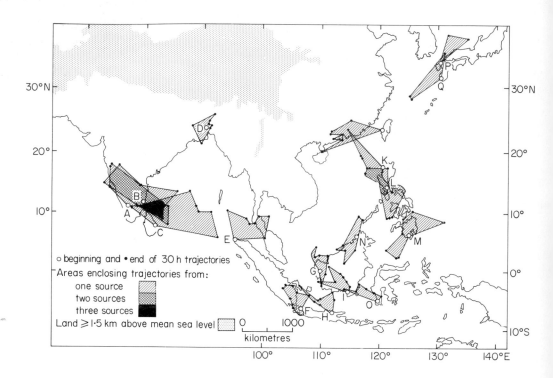

Figure 4. Areas within which brown planthoppers were dispersed from Points A to Q (cf. Fig. 2), 16-23 September 1979. (Cool conditions precluded take-off at P on 5 nights and at Q on 4 nights).

DISCUSSION

Trajectory analysis has generally been used retrospectively to locate distant sources of outbreaks (Johnson, 1969). Here, it has been used to predict the spread of grassy stunt and ragged stunt by modelling displacements of the vector.

The analyses presented here used weather data which, because they are transmitted daily, can be readily obtained from national meteorological services. Three sources of error arose from using these data. First, they allowed synoptic but not smaller weather systems to be identified. These latter features, such as land and sea breezes and thunderstorms lasting only a few hours, would have affected the track of insects flying nearby. Secondly, because wind-fields change frequently, errors were introduced by assuming that 0000 h and 1200 h GMT observations represent 12 h periods. Thirdly, where upper air and ships' observations were sparse, wind speed and direction were estimated by using the wind-fields expected from the weather systems that occurred.

Trajectory analysis can be used for other diseases that are acquired before and retained throughout migration provided that the vector's air speed is slow enough for its track to closely resemble that of the wind. The accuracy of predictions based on such a model will, of course, be positively correlated with knowledge of the vector's migratory behaviour. Trajectories based on changes in wind speed and direction, between and within a number of seasons, would be needed to determine the spread of a disease.

REFERENCES

Anon. (1977) *The Nautical Almanac for the year 1979*. Her Majesty's Stationery Office, London.

Anon. (1981) Test on the releasing and recapturing of marked plant-hoppers, *Nilaparvata lugens* and *Sogatella furcifera*. *Acta Ecologica Sinica* 1, 49-53.

Cheng S.-N., Chen J.-C., Si H., Yan L.-M., Chu T.-L., Wu C.-T., Chien J.-K. & Yan C.-S. (1979) Studies on the migrations of brown plant-hopper *Nilaparvata lugens* Stål. *Acta Entomologica Sinica* 22, 1-21.

Dyck V.A., Misra B.C., Alam S., Chen C.N., Hsieh C.Y. & Rejesus R.S. (1979) Ecology of the brown planthopper in the tropics. *Brown Planthopper: a Threat to Rice Production in Asia*. (Ed. by International Rice Research Institute), pp. 61-98. International Rice Research Institute, Los Baños.

Hibino H. (1979) Rice ragged stunt, a new virus disease occurring in tropical Asia. *Review of Plant Protection Research* 12, 98-110.

Johnson C.G. (1969) *Migration and Dispersal of Insects by Flight*. Methuen, London.

Kisimoto R. (1976) Synoptic weather conditions inducing long-distance immigration of planthoppers, *Sogatella furcifera* Horváth and *Nilaparvata lugens* Stål. *Ecological Entomology* 1, 95-109.

Kisimoto R. & Dyck V.A. (1976) Climate and rice insects. *Proceedings of the Symposium on Climate and Rice* (Ed. by International Rice Research Institute), pp. 367-91. International Rice Research Institute, Los Baños.

Kusukabe S.-I. & Hirao J. (1976) Tolerance for starvation in the brown planthopper, *Nilparvata lugens* Stål. (Hemiptera : Delphacidae). *Applied Entomology and Zoology* 11, 369-71.

Lee J.O. & Park J.S. (1977) Biology and control of the brown plant-hopper (*Nilaparvata lugens*) in Korea. *The Rice Brown Planthopper*. (Compiled by Food and Fertilizer Technology Center for the Asian and Pacific Region), pp. 199-213. Asian and Pacific Council, Food and Fertilizer Technology Center, Taipei.

Lewis T. & Taylor L.R. (1967) *Introduction to Experimental Ecology*. Academic Press, London.

Ling K.C. (1972) *Rice Virus Diseases*. International Rice Research Institute, Los Baños.

MacQuillan M.J. (1975) Seasonal and diurnal flight activity of *Nilaparvata lugens* Stål. (Hemiptera : Delphacidae) on Guadalcanal. *Applied Entomology and Zoology* 10, 185-8.

Nasu S. (1969) Vectors of rice viruses in Asia. *The Virus Diseases of the Rice Plant. Proceedings of a Symposium at the International*

Rice Research Institute, April 1967 (Ed. by International Rice Research Institute), pp. 93-109. Johns Hopkins University Press, Baltimore.

Ohkubo N. (1973) Experimental studies on the flight of planthoppers by the tethered flight technique. I. Characteristics of flight of the brown planthopper *Nilaparvata lugens* (Stål.) and effects of some physical factors. *Japanese Journal of Applied Entomology and Zoology* <u>17</u>, 10-8.

Ohkubo N. & Kisimoto R. (1971) Diurnal periodicity of flight behaviour of the brown planthopper, *Nilaparvata lugens* Stål. in the 4th and 5th emergence periods. *Japanese Journal of Applied Entomology and Zoology* <u>15</u>, 8-16.

Palmer C.E., Wise C.W., Stempson L.J. & Duncan G.H. (1955) The practical aspect of tropical meteorology. *Air Force Surveys in Geophysics*. No <u>76</u>, 195 pp.

Saxena R.C. & Justo H.D. (1980) Long distance migration of brown planthopper in the Philippine Archipelago. *Paper presented at the eleventh Annual Conference of the Pest Control Council of the Philippines held at Cebu City, 23-26 April 1980,* 21 pp.

Tu C.-W. (1980) The brown planthopper and its control in China. *Rice improvement in China and other Asian countries* (Ed. by T.R. Hargrove), pp. 149-56. International Rice Research Institute, Los Baños.

Changes in cropping practices and the incidence of hopper-borne diseases of rice in Japan

K. KIRITANI

National Institute of Agricultural Science, Yatabe,
Tsukuba, Ibaraki 305, Japan

INTRODUCTION

Dwarf and stripe were the only economically important insect-borne
virus diseases of rice known in Japan at the end of the Second World
War but six more insect-borne diseases have since been recognized. The
causal agents of these diseases are either viruses or mycoplasmas and
they are all transmitted by leafhoppers or planthoppers.

Rice dwarf virus was the first disease agent proved to be transmitted by a leafhopper, *Recilia dorsalis* (Takada, 1896). Fukushi
(1934) later reported that the green rice leafhopper, *Nephotettix
cincticeps*, is the main vector of rice dwarf and that the virus is
transmitted transovarially. Some characteristics of the hopper-borne
diseases of rice reported in Japan are summarized in Table 1.

Essential requirements for epidemics of rice virus diseases are an
abundance of virus sources, the presence of susceptible hosts and an
adequate population density of vectors (Kiritani, 1979). However, even
if these requirements are satisfied, epidemics are not inevitable as
environmental factors are also important and these include the cropping
system and climate.

RICE VIRUS DISEASE EPIDEMICS AND RICE CROPPING SYSTEMS

Although research on virus diseases of rice in Japan started in the
late 19th century, serious outbreaks of rice virus diseases before
the Second World War were reported only from limited areas and seldom
occurred for more than two years in succession. By contrast, later
outbreaks of dwarf and stripe have persisted longer and have affected
larger areas.

Since 1950, rice cropping systems have changed greatly in several
ways that influence virus spread. Firstly, the increased cultivation
of early-season (early) rice, transplanted in late April, has resulted
in an overlap with middle-season rice, transplanted in June. Secondly,

Plumb R.T. & Thresh J.M. (1983) *Plant Virus Epidemiology.*
Blackwell Scientific Publications, Oxford.

Table 1. The main hopper-borne virus (V) and mycoplasma (M)-induced diseases of rice in Japan.

Diseases	Main vectors	Status of disease in Japan	First record
Dwarf (V)	*Nephotettix cincticeps* *N. apicalis* *Recilia dorsalis*	Indigenous	1880
Stripe (V)	*Laodelphax striatellus*	Indigenous	1902
Waika (V)	*N. cincticeps* *N. virescens*	Exotic	1967
Yellow dwarf (M)	*N. cincticeps* *N. virescens* *N. nigropictus*	Indigenous	1919

since the early 1960s much middle-season rice has been sown earlier resulting in sequential plantings of rice within particular areas. Thirdly, the early planting of rice has reduced the area sown in the autumn with wheat and barley, which by 1965 was less than half that recorded at its maximum in 1950 (Fig. 1). As a result the area of paddy rice planted after a winter fallow has greatly increased. Fourthly, since 1965 the replacement of hand planting by machine planting has led to an increased requirement for seedlings raised under protected conditions. Fifthly, since 1970 the withdrawal of some paddy fields from rice cultivation has been encouraged by government subsidy and this has resulted in an increase in the area of uncultivated land, although since 1973 more of this land has been used for wheat and barley.

The introduction of early rice and the earlier planting of middle-season crops created favourable conditions for disease epidemics by producing a sequence of crops. The incidence of stripe and dwarf was affected differently by the reduction in the areas of wheat and barley. These crops provide breeding sites for the first generation of the smaller brown planthopper, *Laodelphax striatellus*, the vector of rice stripe and rice black-streaked dwarf viruses. By contrast, winter cultivations greatly reduce the overwintering population of *N. cincticeps*, the vector of rice dwarf virus. This is because overwintering nymphs of *N. cincticeps* develop on grass weeds such as *Alopecurus aequalis* that grow abundantly in fallow paddy fields but are destroyed by winter ploughing.

A prophylactic insecticide treatment applied to the nursery trays in which plants are produced for machine planting is now commonly used

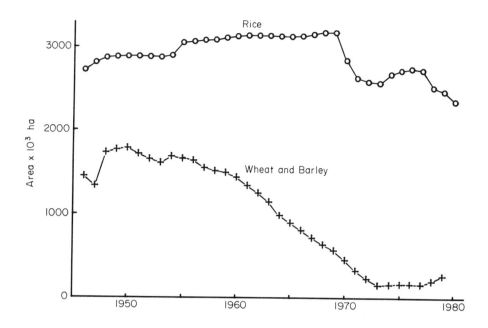

Figure 1. Areas of rice, wheat and barley in Japan, 1946-79.

to control insect pests of young rice. This treatment also reduces
vector populations (Table 2) and disease spread. Apart from these
cultural practices, weather during the winter determines the survival
of rice ratoons (regenerated tillers) which act as virus sources the
following spring. Therefore, severe winters are often followed by less
infection with persistently transmitted viruses (such as yellow dwarf
virus) that are not transmitted transovarially, and the semi-
persistently (or non-persistently) transmitted waika disease.

Changes in the areas of rice, wheat and barley are shown in Fig. 1
and changes in the areas affected by dwarf, yellow dwarf and stripe
are given in Fig. 2. The following sections provide detailed case
studies of these three important diseases and of waika disease, which
resembles tungro (Ling *et al.*, this volume).

RICE DWARF DISEASE

Planting rice early avoids damage caused by autumn typhoons and has
been rapidly accepted by growers in southern Japan. Where crops were
sown sequentially, five times as many *N. cincticeps* were caught in light
traps as in localities where crops were sown more or less concurrently
(Kiritani *et al.*, 1969). The main reason for this difference was that

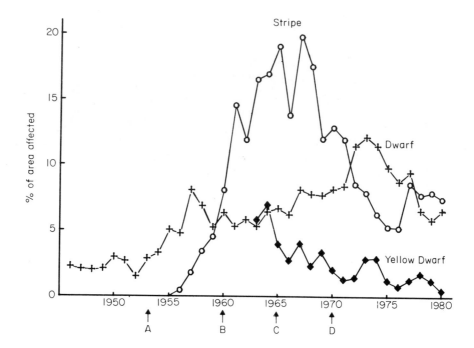

Figure 2. Percentages of rice areas affected by insect-borne diseases 1949-1980. Arrows indicate start of: A early-season rice cultivation, B early planting of middle-season rice, C machine transplanting, D restrictions on rice cultivation.

the early planting of rice provided favourable and abundant host plants for the first generation of *N. cincticeps*, which would otherwise have developed on graminaceous weeds in fallow paddy fields. Moreover, the resulting infection of young rice plants provided an ample source of dwarf virus for the first generation of vectors and led to an increase in the percentage of infective insects in that generation (Nakasuji & Kiritani, 1971). By contrast, the percentage of infective insects decreases from the overwintering to the first generation in the exclusively middle-season rice areas because acquisition of dwarf virus seldom occurs when the first generation develops on graminaceous weeds. Indeed, no measurable infection by dwarf occurs in the areas where only early or middle-season rice is grown (Nakasuji, 1974).

Since 1945 the following periods of rice dwarf incidence can be identified:-

1.   an endemic period 1945-52,
2.   the first epidemic period 1953-66,
3.   the second epidemic period 1967-74,
4.   the decline period since 1975.

The first epidemic period coincided with an extension of early rice growing in southern Japan, while the second started when the area planted to wheat and barley was less than half the previous maximum. The second epidemic period was characterized by the spread of dwarf disease to middle-season rice areas.

Various factors are involved in the recent decline in the incidence of disease during the fourth period:-

1. the decrease in the number of uncultivated paddy fields as rice was replaced by other crops,
2. spring and winter ploughing of fallow paddy fields (Fig. 3),
3. the use of insecticide in nursery trays (Table 2).

Table 2. Light trap catches of *N. cincticeps* in relation to rice transplanting methods in a locality of Saga Prefecture, Kyushu (Miyahara, unpublished)

Period	Planting method	Fields machine-planted (%)	Annual catches of *N. cincticeps*
1969–71	Mature seedlings by hand	9	219 000
1972–73	Young seedlings by machine	67	31 300
1974–76	Insecticidal treatment introduced	83	1900

RICE YELLOW DWARF DISEASE

The causal agent of yellow dwarf is a mycoplasma that is transmitted, but not transovarially, by *N. cincticeps*. Consequently, adults of the first generation which develop on graminaceous weeds are rarely infective because these plants are not suitable sources of the pathogen. However, a continuous transmission cycle of yellow dwarf is ensured when there are early crops of rice in areas mainly sown with middle-season crops. The early sowing of rice over large areas, however, causes a drastic decrease in the proportion of infective insects because the rice is harvested early thus destroying the main sources of infection for the overwintering generation. Infected ratoons from the stubble provide the only disease source for this generation and

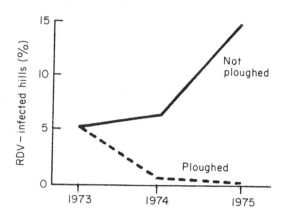

Figure 3. The effects of winter ploughing on the incidence of rice dwarf in Kochi Prefecture. Data from surveys in an area of 300 ha where winter ploughing was practised throughout and for a comparable area 2 km away where there was no winter ploughing.

they only survive in mild winters. The epidemics of yellow dwarf after the Second World War coincided with an increase in the area of middle-season rice sown early; this began about 1961 and infection with yellow dwarf reached a maximum during 1962-64. Concurrently the disease, which was previously limited to the Pacific coastal area of southern Japan, appeared in northern Kanto in central Japan. As for rice dwarf, outbreaks of yellow dwarf were triggered by the sequential cropping resulting from the early planting of rice.

There has been a general decline in the incidence of yellow dwarf since 1966 (Fig. 2). This has occurred despite the continuing decrease in the areas of barley and wheat, which might have been expected to lead to an increase in overwintering populations of *N. cincticeps* on weeds in the uncultivated areas. The reasons for this apparent anomaly are not known, but it is thought that the lack of transovarial transmission makes it difficult for yellow dwarf to complete its year-round transmission cycle in temperate conditions.

RICE STRIPE DISEASE

Stripe, which is transmitted by *L. striatellus*, causes sudden epidemics, in striking contrast to dwarf which tends to be endemic (Fig. 2). Stripe was prevalent only in central Japan before the Second World War but spread south to the whole of southern Japan, in parallel with the trend to earlier planting. Around 1960 local outbreaks were seen everywhere in southern and central Japan. The outbreaks were most serious in 1968 when 20% of the total area of rice in Japan was affected to some

extent. The decline in incidence during the last decade (Fig. 2) has
coincided with the decrease in the area of wheat and barley (Fig. 1).
Since wheat and barley are rarely infected with stripe, the reduced
incidence in rice is mainly attributed to a decrease in the number of
L. striatellus produced in the first generation.

There was more stripe in 1977 than for several years (Fig. 2) in the
Kanto district, where vectors were unusually abundant on wheat in 1977.
The percentage of infective insects in this district increased several
fold to 20% in 1980 after two epidemic years. Kiritani (1981) postulates that there will be an upper limit to the percentage of infective
insects because of hereditary and physiological factors affecting the
vector population, so that the decline in the incidence of stripe will
continue until there is a further outbreak of *L. striatellus*. As a
precautionary measure, growers in areas where rice is vulnerable to
severe stripe infection are recommended to delay transplanting to escape
an influx of adults of the first generation. The adoption of such
practices may have contributed to the recent decline in stripe incidence.

RICE WAIKA DISEASE

Waika disease resembles tungro disease which is prevalent in other rice-
growing areas of south-east Asia (Ling *et al.*, this volume). The first
serious outbreaks of waika in Japan were reported in western Kyushu in
1972 and 1973. The causal virus is transmitted semi-persistently by
N. cincticeps (Table 1) and the virus is dependent on diseased plants to
complete the transmission cycle.

Virus disease epidemics are often triggered by introducing new,
highly susceptible cultivars. The epidemics of waika seemed to be
associated with the rapid adoption of the recommended cultivar Reiho
which was first introduced in 1969 (Table 3). Since the virus causing
waika is unable to persist over winter in its vectors, it must survive
in host plants, but only rice was found to be susceptible in tests on
over 50 plant species (Shinkai, 1979). This suggests that the virus
survives the winter in ratoons from stubbles of diseased rice plants
and that the survival of ratoons until the following spring is essential for completion of the infection cycle. The winters preceding 1972
and 1973, the worst years for the disease, were both mild (Table 3).
The rapid decline in waika since then can be attributed to the replacement of Reiho by more tolerant cultivars, to improved chemical control
of *N. cincticeps* and to the cold winter of 1973/74.

CONCLUSIONS

Rice cultivation in Japan since 1945 has been characterized by widespread epidemics of insect-borne diseases. A common feature of rice
cultivation was that all the lowland rice cultivars were susceptible to
virus diseases and heavy applications of nitrogenous fertilizers were
used to obtain high yields. There was also an increase in the

Table 3. Annual incidence of rice diseases transmitted by
Nephotettix spp. and factors influencing waika disease in Kyushu
district (Shinkai, 1979).

Year	Percentage area affected*			Area affected by waika* (ha)	cv.Reiho (% total area)	Winter climate†
	Dwarf	Yellow dwarf	Waika			
1966	33.1	17.7	0.0	0	0.0	rather mild
1967	39.6	11.9	trace	2	0.0	average
1968	34.2	4.5	trace	50	0.0	cold
1969	30.7	3.3	trace	150	5.7	mild
1970	37.3	1.9	trace	23	27.1	average
1971	39.9	3.7	0.5	1647	46.2	average
1972	57.2	2.1	3.6	11 924	51.9	mild
1973	60.6	1.4	7.4	24 825	49.8	mild
1974	58.2	0.9	0.2	612	36.8	cold
1975	46.8	0.4	0.0	24	26.0	average
1976	38.8	0.3	0.0	17	22.0	rather mild
1977	38.9	0.5	0.0	0	17.1	cold
1978	--	--	0.0	0	0.0	mild

* Areas with at least some diseased plants
† December-February

cultivation of early rice and some middle-season rice was planted
earlier. These changes facilitated disease spread and led to serious
epidemics. The duration and frequency of these epidemics has been
influenced by subsequent changes in cultural practices including the
decreased cultivation of wheat and barley, the use of new rice cultivars
and insecticides, and the relative amounts of winter and spring
ploughing. This emphasises the importance of comprehensive studies of
the whole agroecosystem in developing effective strategies of pest and
disease control.

REFERENCES

Fukushi T. (1934) Studies on dwarf disease of rice plant. *Journal of
    Faculty of Agriculture, Hokkaido Imperial University* 37, 41-164.
Kiritani K. (1979) Pest management in rice. *Annual Review of
    Entomology* 24, 279-312.
Kiritani K. (1981) Spacio-temporal aspects of epidemiology in insect
    borne rice virus diseases. *Japan Agricultural Research Quarterly*
    15, 92-9.

Kiritani K., Sasaba T. & Inoue T. (1969) Studies on the resistance mechanism to malathion in the green rice leafhopper, *Nephotettix cincticeps*, in Kochi Prefecture. *Bulletin of Kochi Institute of Agricultural & Forest Science* 2, 39-46.

Nakasuji F. (1974) Epidemiological study on rice dwarf virus transmitted by the green rice leafhopper, *Nephotettix cincticeps*. *Japan Agricultural Research Quarterly* 8, 84-91.

Nakasuji F. & Kiritani K. (1971) Inter-generational changes in relative abundance of insects infected with rice dwarf virus in populations of *Nephotettix cincticeps* Uhler. *Applied Entomology and Zoology* 6, 75-83.

Shinkai A. (1979) Recent topics in rice cropping; control of waika disease. *Nōgyō oyobi Engei* 54, 25-8; 302-6.

Takada K. (1895) Report on rice dwarf disease. *Dai-Nippon Nōkaihō* 171, 1-4

Takada K. (1896) Report on rice dwarf disease. *Dai-Nippon Nōkaihō* 172, 13-32.

Epidemiological studies of rice tungro

K.C. LING*, E.R. TIONGCO & ZENAIDA M. FLORES
International Rice Research Institute, Los Baños,
Laguna, Philippines

INTRODUCTION

"Tungro" means degenerated growth in a Philippine language. Although
there have been reports of symptoms which suggest that tungro has been
present at least since 1940 the first identification of rice tungro
disease was by Rivera & Ou (1965). Tungro also occurs in Bangladesh
(Nuque & Miah, 1969), India (John, 1968), Indonesia (Rivera et al.,
1968), Malaysia (Ting & Paramsothy, 1970), Nepal (John et al., 1979) and
Thailand (Lamey et al., 1967). However several diseases described by
different names, including cella pance and penyakit habang in Indonesia
(Tantera, 1973), leaf yellowing in India (Raychaudhuri et al., 1967),
penyakit merah in Malaysia (Lim, 1969; Ou et al., 1965; Singh, 1969)
and yellow-orange leaf in Thailand (Wathanakul & Weerapat, 1969), seem
indistinguishable from tungro, which therefore appears to be widely
distributed in south and south-east Asia.

Rice tungro is caused by a virus with isometric particles 30-33 nm in
diameter (Gálvez, 1968, 1971). Saito et al. (1975) found both isometric and small bacilliform particles 25 x 140 nm in diseased plants
and their results have been confirmed from different specimens (Saito,
(1977), Hibino et al. (1978) and Mishra et al. (1979)). However, the
precise relationship between the kinds of particles and the disease
symptoms they produce has not been determined due to the lack of a
suitable bioassay. The main symptoms of tungro are stunting of plants,
yellowing of leaves, delayed flowering, and production of fewer filled
grains.

Rice tungro virus is transmitted only by leafhoppers; those species
known to be vectors are:

---

*This paper is the last written by Dr. K.C. Ling who died on 12 February
1982. Dr. Ling joined the International Rice Research Institute in 1965
as plant pathologist specializing in virus diseases. When he died he
was a world authority on rice virus diseases and had published more than
150 papers.

Plumb R.T. & Thresh J.M. (1983) *Plant Virus Epidemiology.*
Blackwell Scientific Publications, Oxford.

Nephotettix malayanus (Anon., 1973)
N. nigropictus (Ling, 1970; Rivera & Ling, 1968)
N. parvus (Rivera *et al.*, 1972)
N. virescens (Ling, 1966; Rivera & Ou, 1965)
Recilia dorsalis (Rivera *et al.*, 1969)
Hybrids of *N. virescens* & *N. nigropictus* (Ling, 1968a)

N. virescens is the most important and efficient vector.

Tungro virus was the first leafhopper-transmitted virus shown not to persist in its vector (Ling, 1966, 1972). However, as the period for which vectors retain the virus is prolonged in cool conditions, the term "transitory" is used to describe the virus-vector relationship rather than "non-persistent" (Ling & Tiongco, 1979).

Tungro is endemic in south and south-east Asia and is the most important virus disease in the region; the disease occasionally becomes epidemic and destroys large areas of rice. In 1971 yield losses due to tungro in the Philippines were estimated as 456 000 tonnes of rough rice and in the last 15 years the disease has caused severe damage in many parts of the region. The occurrence of these outbreaks is unpredictable and recently much effort has been put into epidemiological studies (Mishra *et al.*, 1971; Lim, 1972; Mukhopadhyay & Chawdury, 1973; Anjaneyulu, 1974; Hino *et al.*, 1974; Lim *et al.*, 1974; Kondaiah *et al.*, 1976; John & Mishra, 1979) in an attempt to develop more effective control of the disease. This chapter considers the epidemiological work done at the International Rice Research Institute.

EXPERIMENTAL EPIDEMIOLOGY

The concept of a disease triangle (host-pathogen-environment) must be modified when a vector is also involved. Disease incidence then becomes a function of various aspects of the vector, virus source, host, environment and time. To investigate the effect of each of these variables on tungro incidence a glasshouse method using cages has been developed (Ling, 1974b).

The metal cages measure 52 x 52 x 71 cm and they are covered by 28-mesh metal screens. The cages are used to simulate field conditions and as a convenient substitute for outdoor plots in studying the effects of insect vector, virus source, and host plant on disease spread. These factors can be changed either individually or collectively as required. Furthermore, the cages can be kept at different temperatures or under various environmental conditions to determine effects on spread.

The effects of the vector

The number (x) of *N. virescens* capable of transmitting tungro determines the percentage of seedlings infected ($\hat{Y}$) and the curvilinear relationship can be described as

$$\hat{Y} = 100 - 100e^{-2.8x} \tag{i}$$

where the constant $e = 2.719$. The percentage of infected seedlings increases rapidly up to 1 N. virescens/seedling but only slightly when the number of vectors increases from 1/seedling to 3/seedling (Anon., 1976).

An increase in the number of N. virescens (x) in an area or in a cage will increase the proportion ($\hat{Y}$) of plants infected, provided all other factors remain constant. This relationship can be expressed by

$$\hat{Y} = 100 - 100(1 + ax + bx^2) \tag{ii}$$

where a and b are variables influenced by the rate of mortality of the leafhoppers during the test period (Ling, 1975a).

When the number of seedlings and N. virescens in an area vary but the number of insects/seedling is the same at the beginning of each experiment, the proportion of plants infected ($\hat{Y}$) initially increases with an increase in the number (x) of insects and the number of seedlings but further increases in insect and seedling number result in a decrease in the proportion of seedlings infected as described by the equation

$$\hat{Y} = 32 + 0.68x - 0.003x^2 \tag{iii}$$

The time insects stay in an area also affects the incidence of tungro (Ling, 1975a). The proportion of seedlings infected ($\hat{Y}$) is related to the number of days (t) insects remain by

$$\hat{Y} = 100(1 - e^{-0.14t}) \tag{iv}$$

However, adult insects are three times more efficient vectors than nymphs (Ling, 1975a). Based on the shortest recorded inoculation access time of 5 min for successful transmission, a single infective N. virescens can, theoretically, infect 288 plants/day but in undisturbed cage experiments at 27°C the maximum number of infections was 11/day (Ling & Carbonell, 1975). When the insect was disturbed up to 30 plants/day were infected (Ling, 1974c) and when the experiment was repeated at 34°C a maximum of 40 plants/day were infected (Ling & Tiongco, 1975). However, the rate of spread is determined not only by the frequency of movement of viruliferous insects but also by the duration of their feeding period (Ling & Carbonell, 1975) which must be sufficient to transmit the virus.

Vector dispersal as measured at different numbers of plant spaces (n) from the point of release can be expressed by

$$\hat{Y} = a(1 + n)^{-b} \tag{v}$$

where $\hat{Y}$ is the proportion of the total number of insects released/hill.

Dispersal as a function of distance (n) (0-5 hills) and time (t) (1-11 days) can be expressed by

$$\hat{Y} = 69.8e^{-0.23t}(1 + n)^{-3.6e^{-0.17t}}$$ (vi)

Dispersal was not affected by the number of insects released or by the proportion of healthy or tungro-infected plants but was affected by cultivar and plant size. As plant height increased, the values for a and b (equation v) increased (Ling & Tiongco, 1977).

The effects of virus source

If the number of test plants in an area remains constant but the number of sources of virus is increased, more plants are infected. The relation between the proportion of plants infected in an area and the size of the virus source (s as a percentage of the total stand) can be described as:

$$\hat{Y} = as + b\sqrt{s}$$ (vii)

where the values for a and b are determined by the presence of nymphs or adult insects (Ling, 1975b). The distribution of source plants also affects disease incidence. More virus spread occurs when source plants are scattered than when they are clustered (Ling, 1975b). The furthest distance from a source that a hopper will transmit virus has been estimated to be about 250 m (Ling, 1975b) because N. cincticeps can travel up to 41 m/day in a paddy field (Miyashita et al., 1964) and the longest retention period of tungro virus by N. virescens in glasshouse conditions is 6 days (Ling, 1972).

Not all diseased plants are equally good acquisition sources for tungro virus. More plants were infected when the cultivars TN1 and IR22 were used as sources than when the cultivars C4-63G and IR20 were used (Ling, 1975b).

The effects of cultivar

Rice cultivars differ in their susceptibility to tungro but all cultivars become more difficult to infect and are less affected by infection as they age (Ling & Palomar, 1966).

The effects of environment

Although temperature influences virus transmission (Ling & Tiongco, 1979) the effect is probably small in the tropics, especially when virus sources are common (Ling & Tiongco, 1975). However, any factor that disturbs the vector may increase tungro incidence (Ling, 1975a).

FIELD EPIDEMIOLOGY

The epidemiology of tungro was studied in 37 fields in five provinces

Table 1. Tungro vector assessments May-July, and incidence of rice tungro disease June-October, in study fields in Luzon.

| | Tungro vectors caught May-July | | | | Tungro surveys June-October | | |
Year	Collections (no.)	Number*	Tested (no.)	Infective (%)	Observations (no.)	Affected fields (%)	Mean incidence (%)
1973	76	10.7	1551	0	151	1†	0.60
1974	172	10.9	3461	0.90	244	45¶	2.38¶
1975	173	10.2	3118	0.80	234	72	3.32
1976	140	3.0	2014	1.40	267	25	0.33
1977	178	2.7	2180	0.05	236	17	0.03
1978	163	3.1	1747	0	241	0	0
1979	165	2.0	1455	0.34	234	18	2.01§
1980	161	5.8	2135	0.23	203	11	0.06

*Mean numbers of *Nephotettix virescens, N. nigropictus,* and *Recilia dorsalis* per 10 sweeps made biweekly.
†Rice fields with only a few tungro-infected plants were excluded.
¶The Philippine Government launched a green leafhopper control programme in the area in July and August.
§About 95% of the incidence was on cultivars C1 and C21 in three study fields.

in Luzon, Philippines. Vectors were collected and tested for infectivity, and tungro incidence and plant growth stage were recorded biweekly in 1972-80. The principal cultivars grown during the study were IR20 in 1973, IR1561-228-3-3 and IR20 in 1974, IR1561-228-3-3 and IR26 in 1975, IR26 and IR30 in 1976 and IR36 in 1977-80.

The number of tungro vectors (*N. virescens, N. nigropictus* and *R. dorsalis*), the proportion of infective vectors, and tungro incidence varied from field to field and with time. However, the incidence of tungro from June to October in the wet season rice crop was correlated with the number of tungro vectors and the proportion that transmitted tungro from May to July (Table 1). The May-July period is either immediately before seedbeds are prepared or when rice seedlings are at very early stages of growth, either in the seed beds or soon after transplanting.

FORECASTING TUNGRO

Forecasting tungro is very difficult as the factors that influence virus occurrence are very variable. It is easier to forecast when no, or only a little, disease will occur, as only one important factor needs to be unfavourable to prevent a serious outbreak, than it is to accurately predict the full combination of factors required for an epidemic. Results from the experimental epidemiology studies showed that if only few vectors are present in May-July it is unlikely that there will be much tungro even if other factors are favourable. If vectors are numerous then chemical treatment is justified to prevent an outbreak.

The results of the experimental studies also demonstrated that the proportion of infected plants in an area has an important influence on virus spread. During April-June there are few rice crops in the study area and the regrowth of infected rice stubble remaining from previous crops is the main source of virus and vectors. When the stubbles are ploughed the vectors are forced to disperse and may move to seedbeds where they lay eggs and transmit tungro. The seedlings and the eggs they carry are then transplanted into fields and provide sources of virus from which the newly hatched nymphs can acquire virus. Virus is then rapidly dispersed to the young crop as more eggs hatch and nymphs become adults. To determine whether an outbreak of tungro is likely, both the stubble regrowths and young standing crops must be examined for infection and the infectivity of leafhoppers has to be tested by confining them in cages on test seedlings. If there are both few infected plants and a small proportion of infective vectors tungro cannot become epidemic because by the time (3-4 months) both sources and vectors of virus have become numerous the rice crop is ready for harvest. If the cultivar of rice is resistant to both tungro and the vector then a disease outbreak is also unlikely to occur.

STRATEGIES FOR CONTROL

There are many possible ways of controlling rice tungro disease; the most practicable methods at present are the use of resistant cultivars and chemical control of the vectors. The resistance of cultivars to tungro can be tested by the mass screening method in the glasshouse (Ling, 1974a) and evaluated in crops using the tungro propagation method (Ling et al., 1979). While resistance to tungro has been introduced into commercial cultivars (Khush, 1980) this resistance is not always associated with resistance to the vector. Resistance to vectors is of two types, antibiosis (Ling, 1968b) and non-preference (Ling & Carbonell, 1975).

It is important that farmers are taught to monitor vector insect populations, especially at the early stages of crop growth, so that appropriate insecticides can be applied at the right time. At present tungro is controlled best by growing resistant cultivars and using knowledge of the epidemiology of the disease to decrease both virus sources and vector populations.

REFERENCES

Anjaneyulu A. (1974) Epidemiological studies of rice tungro virus
    disease in India. *International Rice Research Conference.*
    International Rice Research Institute, Los Baños, Philippines
    (mimeograph).
Anon. (1973) *Annual Report for 1972.* International Rice Research
    Institute, Los Baños, Philippines.
Anon. (1976) *Annual Report for 1975.* International Rice Research
    Institute, Los Baños, Philippines.
Gálvez G.E. (1968) Purification and characterization of rice tungro
    virus by analytical density-gradient centrifugation. *Virology* 35,
    418-26.
Gálvez G.E. (1971) Rice Tungro Virus. *C.M.I./A.A.B. Descriptions of
    Plant Viruses* No. 67, 3 pp.
Hibino H., Roechan M. & Sudarisman S. (1978) Association of two types
    of virus particles with penyakit habang (tungro disease) of rice in
    Indonesia. *Phytopathology* 68, 1412-6.
Hino T., Wathanakul L., Nabheerong N., Surin P., Chaimongkil U.,
    Disthaporn S., Putta M., Kerdchokchai D. & Surin A. (1974) Studies
    on rice yellow orange leaf virus disease in Thailand. *Tropical
    Agriculture Research Center Technical Bulletin (Japan)* 7, 67 pp.
John V.T. (1968) Identification and characterization of tungro, a
    virus disease of rice in India. *Plant Disease Reporter* 52, 871-5.
John V.T., Freeman W.H. & Shahi B.B. (1979) Occurrence of tungro
    disease in Nepal. *International Rice Research Newsletter* 4(3), 16.
John V.T. & Mishra M.D. (1978) Epidemiology and control of rice tungro
    virus. *National Symposium on Increasing Rice Yield in Kharif.*
    Indian Council of Agricultural Research (mimeograph).
Khush G.S. (1980) Breeding for multiple disease and insect resistance
    in rice. *Biology and Breeding for Resistance to Arthropods and
    Pathogens in Agricultural Plants* (Ed. by M.K. Harris), pp. 341-54.
    Texas Agricultural Experiment Station, College Station, Texas.
Kondaiah A., Rao A.V. & Srinivasan T.E. (1976) Factors favoring
    spread of rice "tungro" disease under field conditions. *Plant
    Disease Reporter* 60, 803-6.
Lamey H.A., Surin P. & Leeuwangh J. (1967) Transmission experiments on
    the tungro virus in Thailand. *International Rice Commission News-
    letter* 16(4), 15-9.
Lim G.S. (1969) The bionomics and control of *Nephotettix impicticeps*
    Ishihara and transmission studies on its associated viruses in West
    Malaysia. *Ministry of Agriculture and Co-operatives (Malaysia)
    Bulletin* 121, 62 pp.
Lim G.S. (1972) Studies on penyakit merah disease of rice. III.
    Factors contributing to an epidemic in North Krian, Malaysia.
    Malaysian Agricultural Journal 48, 278-94.
Lim G.S., Ting W.P. & Heong K.L. (1974) Epidemiological studies of
    tungro virus in Malaysia. *Lapuran MARDI Report* 21, 12 pp.
Ling K.C. (1966) Nonpersistence of the tungro virus of rice in its
    leafhopper vector, *Nephotettix impicticeps*. *Phytopathology* 56,
    1252-6.
Ling K.C. (1968a) Hybrids of *Nephotettix impicticeps* Ish. and *N.*

UNIVERSITY OF GREENWICH LIBRARY

apicalis (Motsch.) and their ability to transmit the tungro virus of rice. *Bulletin of Entomological Research* 58, 393-8.

Ling K.C. (1968b) Mechanism of tungro resistance in rice variety Pankhari 203. *Philippine Phytopathology* 4, 21-38.

Ling K.C. (1970) Ability of *Nephotettix apicalis* to transmit the rice tungro virus. *Journal of Economic Entomology* 63, 582-6.

Ling K.C. (1972) *Rice Virus Diseases.* International Rice Research Institute, Los Baños.

Ling K.C. (1974a) An improved mass screening method for testing the resistance of rice varieties to tungro disease in the greenhouse. *Philippine Phytopathology* 10, 19-30.

Ling K.C. (1974b) A cage method for studying experimental epidemiology of rice tungro disease. *Philippine Phytopathology* 10, 31-41.

Ling K.C. (1974c) The capacity of *Nephotettix virescens* to infect rice seedlings with tungro. *Philippine Phytopathology* 10, 42-9.

Ling K.C. (1975a) Experimental epidemiology of rice tungro disease I. Effect of some factors of vector (*Nephotettix virescens*) on disease incidence. *Philippine Phytopathology* 11, 11-20.

Ling K.C. (1975b) Experimental epidemiology of rice tungro disease II. Effect of virus source on disease incidence. *Philippine Phytopathology* 11, 21-31.

Ling K.C. & Carbonell M.P. (1975) Movement of individual viruliferous *Nephotettix virescens* in cages and tungro infection of rice seedlings. *Philippine Phytopathology* 11, 32-45.

Ling K.C. & Palomar M.K. (1966) Studies on rice plants infected with the tungro virus at different ages. *Philippine Agriculturist* 50, 165-77.

Ling K.C. & Tiongco E.R. (1975) Effect of temperature on the transmission of rice tungro virus by *Nephotettix virescens*. *Philippine Phytopathology* 11, 46-57.

Ling K.C. & Tiongco E.R. (1977) Dispersal of *Nephotettix virescens*. *Philippine Phytopathology* 13, 2.

Ling K.C. & Tiongco E.R. (1979) Transmission of rice tungro virus at various temperatures: a transitory virus-vector interaction. *Leafhopper Vectors and Plant Disease Agents* (Ed. by K. Maramorosch & K.F. Harris), pp. 349-66. Academic Press, New York.

Ling K.C., Tiongco E.R. & Daquioag R.D. (1979) Tungro propagation. *International Rice Research Newsletter* 4(5), 8-9.

Mishra M.D., Raychaudhuri S.P., Everett T.R. & Basu A.N. (1971) Possibilities of forecasting outbreaks of tungro and yellow dwarf of rice in India and their control. *Proceedings of the Indian National Science Academy* 37B, 352-6.

Mishra M.D., Saito Y., Basu A.N. & Niazi F.R. (1979) Etiology of two strains of rice tungro virus in India. *Indian Phytopathology* 32, 260-3.

Miyashita K., Ito Y., Yasuo S., Yamaguchi T. & Ishii M. (1964) Studies of the dispersal of plant- and leafhoppers. II. Dispersals of *Delphacodes striatella* Fallen, *Nephotettix cincticeps* Uhler, and *Deltocephlus dorsalis* Motschulsky in nursery and paddy field. *Japan Journal of Ecology* 14, 233-41.

Mukhopadhyay S. & Chawdhury A.K. (1973) Some epidemiological aspects of tungro virus disease of rice in West Bengal. *International Rice*

Commission Newsletter 22(4), 44-57.

Nuque F.L. & Miah S.A. (1969) A rice virus disease resembling tungro in East Pakistan. *Plant Disease Reporter* 53, 888-90.

Ou S.H., Rivera C.T., Navaratham S.J. & Goh K.G. (1965) Virus nature of "penyakit merah" disease of rice in Malaysia. *Plant Disease Reporter* 49, 778-82.

Raychaudhuri S.P., Mishra M.D. & Ghosh A. (1967) Preliminary note on the transmission of a virus disease resembling tungro of rice in India and other virus-like symptoms. *Plant Disease Reporter* 51, 300-1.

Rivera C.T., Aguiero V.M., Dimasuay D.F. & Ling K.C. (1972) New vector of rice tungro and yellow dwarf. *Philippine Phytopathology* 8, 10.

Rivera C.T. & Ling K.C. (1968) Transmission of rice tungro virus by a new vector, *Nephotettix apicalis*. *Philippine Phytopathology* 4, 16.

Rivera C.T., Ling K.C., Ou S.H. & Aguiero V.M. (1969) Transmission of two strains of rice tungro virus by *Recilia dorsalis*. *Philippine Phytopathology* 5, 17.

Rivera C.T. & Ou S.H. (1965) Leafhopper transmission of "tungro" disease of rice. *Plant Disease Reporter* 49, 127-31.

Rivera C.T., Ou S.H. & Tantere D.M. (1968) Tungro disease of rice in Indonesia. *Plant Disease Reporter* 52, 122-4.

Saito Y. (1977) Interrelationship among waika disease, tungro and other similar diseases of rice in Asia. *Tropical Agriculture Research (Japan) Series* No. 10, 129-35.

Saito Y., Roechan M., Tantera D.M. & Iwaki M. (1975) Small bacilliform particles associated with penyakit habang (tungro-like) disease of rice in Indonesia. *Phytopathology* 65, 793-6.

Singh K.G. (1969) Virus vector relationship in penyakit merah of rice. *Annals of the Phytopathological Society of Japan* 35, 322-4.

Tantera D.M. (1973) Studies on rice virus/mycoplasma diseases in 1972. *July Staff Meeting*. Central Research Institute for Agriculture, Bogor, Indonesia (mimeograph).

Ting W.-P. & Paramsothy S. (1970) Studies on penyakit merah disease of rice I. Virus-vector interaction. *Malaysian Agricultural Journal* 47, 290-8.

Wathanakul L. & Weerapat P. (1969) Virus diseases of rice in Thailand. *Virus Diseases of the Rice Plant,* pp. 79-85. Johns Hopkins Press, Baltimore.

Origins in Mesoamerica of maize viruses and mycoplasmas and their leafhopper vectors

L.R. NAULT
Department of Entomology, Ohio Agricultural Research and
Development Center, Wooster, Ohio 44691, USA

In searching for the original insect-borne viruses and mycoplasmas that infect maize, I have narrowed the possibilities to three pathogens that cause stunting: the maize rayado fino virus, the maize bushy stunt mycoplasma and the corn stunt spiroplasma. All three are transmitted by species of neotropical leafhoppers in the genus *Dalbulus*. In this chapter evidence is presented to support the hypothesis that leafhoppers in the genus *Dalbulus*, and the closely related genus *Baldulus*, have a long co-evolutionary history with maize and its wild relatives. It is also speculated that pathogens that cause stunting restrict the distribution of extant maize relatives as well as their leafhopper vectors.

MAIZE AND ITS RELATIVES

Maize was probably first domesticated in central Mexico 8-10 000 years ago (Wilkes, 1972; Galinat, 1978; Beadle, 1980). It is here and in Guatemala that the closest living relatives of maize occur. These are the teosintes, of which there are both annual and perennial forms. The teosinte habitat is characterized by dry winters and wet summers, limestone soils and elevations of 800-2250 m (Wilkes, 1972, 1977, 1979). The teosintes and maize comprise the genus *Zea*. Several species, subspecies and races are recognized (Iltis *et al.*, 1979; Doebley & Iltis, 1980; Iltis & Doebley, 1980). They include the recently discovered diploid perennial, *Z. diploperennis*, and its probable descendant, the autotetraploid *Z. perennis*. Both species are found in a few remnant populations at intermediate to high elevations in southern Jalisco, Mexico. *Z. diploperennis* is considered to be the most primitive *Zea* sp. The annual forms include *Z. luxurians*, which is indigenous to southern Guatemala. All other annuals, including maize, are *Z. mays*. The three subspecies are:

> *Z. mays* ssp. *parviglumis*, which includes the extensive populations of Balsas teosinte that grow along the western escarpment of central Mexico and a population that occurs in western Guatemala,

Plumb R.T. & Thresh J.M. (1983) *Plant Virus Epidemiology.*
Blackwell Scientific Publications, Oxford.

Z. *mays* ssp. *mexicana*, which includes the more maize-like
races, Nobogame, Central Plateau and Chalco,
Z. *mays* ssp. *mays*, which is cultivated maize.

The natural distribution of the teosintes in Mexico falls within some
of the best agricultural land.   In the states of Jalisco, Guanajuato and
Michoacan, the Balsas teosintes only occur in semi-isolated populations,
mostly along stone fences bordering maize fields or steep banks (Wilkes,
1972, 1977).   All of the teosintes, with the exception of Z. *perennis*,
are known to hybridize naturally with maize, but the Central Plateau and
Chalco races appear to do so most frequently (Wilkes, 1977).   Chalco,
the most maize-like of the teosintes, invades corn fields in the Valley
of Mexico.   Here teosinte is able to survive by becoming a "maize-
mimetic weed" (Wilkes, 1977, 1979).

The only other group of maize relatives are grasses in the genus
Tripsacum.   Together, *Tripsacum* and *Zea* make up the Tripsacinae of the
Andropogoneae (de Wet *et al.*, 1976).   *Tripsacum* is the only genus which
has been experimentally hybridized with maize (Mangelsdorf, 1974; de
Wet & Harlan, 1978) and natural hybrids are unknown.   Thirteen
Tripsacum spp. are recognized, and additional studies will probably
reveal more (de Wet *et al.*, 1976; de Wet *et al.*, 1981).   The centre of
variation of *Tripsacum*, of which there are both diploid and polyploid
species, is the western escarpment of central Mexico and it almost
exactly overlaps that of the Balsas teosinte populations (Wilkes, 1972).
The presumed hybridization of diploids to produce allotetraploids and
the continual hybridization of the polyploids makes distinguishing
species difficult.   The preferred habitats of *Tripsacum* spp. in Mexico
are almost identical to those of teosinte (Wilkes, 1972, 1979), and
they too are now restricted to a shrinking area of untilled, ungrazed
land near fences and on steep embankments.   Most *Tripsacum* spp. are
perennial and are morphologically similar to the perennial *Zea* spp.,
their putative descendants (Iltis *et al.*, 1979; Doebley & Iltis, 1980).

The origin of maize is still disputed.   The oldest theory suggests
prehistoric human selection of teosinte (Ascherson, 1875), but this
theory was challenged by Mangelsdorf & Reeves (1939), who assumed that
maize was domesticated from an extinct pod/pop corn and that teosinte
resulted from hybridization between *Tripsacum* spp. and domesticated
maize.   However, it has since been shown that teosinte is not a
Tripsacum spp. x maize hybrid (de Wet & Harlan, 1972).   More recently,
several authors (Beadle, 1972, 1980; Iltis, 1972; Harlan *et al.*, 1973;
Galinat, 1978; Iltis *et al.*, 1979) have returned to teosinte as the
ancestor of maize.   This theory is strongly supported by thorough
taxonomic studies (Doebley & Iltis, 1980; Iltis & Doebley, 1980).
Wilkes (1979), however, has resurrected the theory of Mangelsdorf &
Reeves (1939), but has substituted Z. *diploperennis* for *Tripsacum* as
one parent and a primitive maize such as Chapalote/Nal Tel as the
other.   He postulates that the progeny of this cross are the annual
teosintes and teosinte-introgressed maize.

MAIZE-FEEDING LEAFHOPPERS

Deltocephaline leafhoppers of the genus *Dalbulus* (DeLong, 1950) are small (4 mm long), pale yellow or cream and inconspicuous. They would have gone largely unnoticed and unstudied if several species had not been shown to transmit pathogens that stunt maize.

Before starting detailed studies in 1979, I had noted that *Dalbulus* spp. and those in the closely related genus, *Baldulus*, appeared to have a close association with maize and its relatives. The most studied species, the corn leafhopper, *D. maidis*, is aptly named. It is common on maize in all neotropical regions from the southern United States of America (Nault & Knoke, 1981) to Argentina (Oman, 1948), and can seriously damage maize, even in the absence of corn-stunting pathogens (Bushing & Burton, 1974). *D. maidis* also feeds on the annual teosintes in Mexico (Barnes, 1954). Pitre (1966, 1970a, 1970b) reared *D. maidis* experimentally on *Tripsacum dactyloides*, but they took longer to develop, died sooner and weighed less than those reared on maize.

A second species, *D. elimatus*, which also feeds preferentially on maize, has tentatively been designated as the "Mexican corn leafhopper" (Nielsen, 1968). Barnes (1954) provided ample evidence for the appropriateness of this name and also noted that while both *D. elimatus* and *D. maidis* co-exist on maize, the former is the most frequent above 750 m while the latter is the most numerous below this altitude. Barnes concluded that *D. maidis* utilizes only maize and teosinte as hosts, whereas *D. elimatus* is less fastidious. After harvest, the latter species disperses to wild and cultivated plants on which it feeds during the winter when maize is not present. Moreover, leafhoppers lay eggs in wheat and barley where nymphs hatch and mature to adults. Of 15 wild and introduced grasses tested, two *Bromus* species are hosts. Nevertheless, maize is still the much preferred host of *D. elimatus*. Barnes did not test or collect from *Tripsacum* spp.

Barnes also collected a few *D. gelbus* and *D. longulus* from maize. C. DeLeon (personal communication in Nault & DeLong, 1980) reported that Ramírez *et al.* (1975) also collected these two species as well as a third, *D. guevarai*, from maize at the Morelos Research Station of the International Center for Improvement of Maize and Wheat (CIMMYT). However, compared to *D. maidis*, very few of these three species were caught.

Leafhoppers in the genus *Baldulus* are morphologically similar to *Dalbulus* spp. At one time *D. maidis* was placed in the genus *Baldulus* until DeLong (1950) separated the two genera by differences in the shapes of the crown and male genitalia. The similarity between the two genera may be more than morphological. A species from the eastern USA, *B. tripsaci*, was named from its host, *T. dactyloides* (Kramer & Whitcomb, 1968). Although *B. tripsaci* has never been collected from maize, Granados & Whitcomb (1971) reared it on sweetcorn for 8 months in the laboratory and used it to transmit the corn stunt spiroplasma.

Before 1979, little was known of the host preferences of most *Dalbulus* and *Baldulus* spp. except that their distributions appeared to overlap those of *Zea* and *Tripsacum* in Mexico (Nault & DeLong, 1980). Visits to four Mexican states in 1979 and 1980 confirmed the evidence obtained from limited host range tests and distribution records, that these leafhoppers feed on wild *Zea* and *Tripsacum* spp. (Nault & DeLong, 1980; Nault, unpublished). *D. gelbus, D. guevarai, D. longulus, D. maidis* and *D. elimatus*, were collected from *Tripsacum* spp. *D. maidis* and *D. elimatus* were also collected from annual and perennial *Zea* spp. In addition, three new leafhopper species were discovered on *Tripsacum* spp.: *D. tripsacoides, D. quinquenotatus* and *D. guzmani* (Nault & DeLong, 1980; Nault & DeLong, unpublished). The aedaegus of *D. tripsacoides* is long, slender and less elaborated than in other members of the genus, suggesting its primitive status.

Based on their natural occurrence on *Tripsacum* spp., I postulate that leafhoppers of the genera *Dalbulus* and *Baldulus* originated on these plants. The mountainous habitats of these grasses form natural barriers that favour their isolation and speciation and that of their resident leafhoppers. I further speculate that with the evolution of *Zea*, two leafhopper species, *D. maidis* and *D. elimatus*, specialized on these plants. When the pre-Columbian civilizations domesticated *Zea* some 8-10 000 years ago by developing and then dispersing maize throughout the Americas, *D. maidis* became dominant and is now a maize pest throughout the neotropics. The corn leafhopper may have been the first insect to plague man as he adopted an agrarian way of life in Mesoamerica.

THE STUNTING PATHOGENS

Maize rayado fino virus (MRFV), which has isometric particles *c*. 31 nm in diameter, is found from the southern USA (Bradfute *et al.*, 1980) to Uruguay (Gámez *et al.*, 1979; Gámez, 1980). It is efficiently transmitted by *D. maidis* and *D. elimatus*; several other deltocephaline leafhoppers are inefficient vectors (Nault *et al.*, 1980). The virus is transmitted in a persistent manner and probably multiplies in its leafhopper vectors (Nault *et al.*, 1980), but it is not known if the virus is also pathogenic to them. In maize, infection by MRFV causes fine chlorotic stripes and dots along leaf veins and, in certain cultivars, severe stunting. The virus has a limited host range which includes experimentally infected annual *Zea* spp., *Z. diploperennis, Tripsacum australe* and *Rottboellia exaltata* (Nault *et al.*, 1980). MRFV has also been found infecting Chalco teosinte from the Valley of Mexico.

The corn stunt spiroplasma (CSS) (Davis & Worley, 1973) has the same distribution as MRFV (Nault & Knoke, 1981) and is transmitted by several *Dalbulus* spp. and other deltocephaline leafhoppers (Nault, 1980). The spiroplasma is transmitted in a persistent manner and multiplies in its vectors, to which it is pathogenic (Granados & Meehan, 1975). The spiroplasma has helical filaments observable microscopically (Davis, 1977) and it has been cultured *in vitro* (Chen & Liao, 1975; Williamson & Whitcomb, 1975). Symptoms produced by CSS infection are chlorotic spots and stripes at the bases of young leaves, reddening at the tips of older leaves and plant stunting (Nault, 1980). Symptoms are more severe when plants are grown in warm (30°C), rather than cool (22°C), conditions. No natural alternate host for CSS is known, although annual and perennial *Zea* spp. (Nault, 1980) and two dicotyledonous species (Markham *et al.*, 1977) have been infected experimentally.

Several strains of corn stunt have been described; however, only the Rio Grande strain is associated with a spiroplasma and properly called CSS (Nault & Bradfute, 1979; Nault, 1980). Other "strains", such as Mesa Central and Louisiana, are considered to be isolates of the maize bushy stunt mycoplasma (MBSM) (Nault & Bradfute, 1979; Nault, 1980). MBSM is not helical and has not been cultured. It is transmitted by *Dalbulus* spp., but not by two other deltocephaline leafhoppers that transmit CSS (Nault, 1980), in a persistent manner and its long incubation period suggests propagation in the vector, although there is no evidence that it is pathogenic to the leafhopper. The symptoms produced by MBSM are distinct from those produced by CSS on maize and are more severe (Nault, 1980). In Mexico, CSS is more prevalent below 1000 m and MBSM above 1000 m (Maramorosch, 1955; Davis, 1974, 1977). MBSM occurs in the southern USA, Mexico, Peru and Colombia and probably has the same distribution as MRFV and CSS.

It is reasonable to suggest that *Dalbulus* spp. and *Baldulus* spp. have co-evolved with maize and its ancestors, but it is more difficult to make such assumptions for the stunting pathogens. Whitcomb (1981) considers CSS to be a "maize specialist" that originated from *Spiroplasma citri*, which has a broad host range. Pathogens related to MBSM and MRFV are unknown and it is not known how long these pathogens have been associated with maize and teosinte. However, these pathogens may have a profound influence on the distribution of present races of teosinte and maize. For example, the distribution of teosinte species above, but not below, 800 m, may be due, in part, to these diseases. Experiments in controlled environments showed that MBSM and CSS kill maize (Nault, 1980) and teosinte at temperatures characteristic of the lowland tropics, but the plants survive at the cooler temperatures typical of the teosinte habitat. The distributions of *Tripsacum* spp., which are similar to the distribution of teosinte, cannot be explained by the effects of these pathogens, since the species are relatively unaffected by the stunting pathogens.

Assuming that the stunting pathogens originated at least as early as the domestication of maize, it is more reasonable to suggest that they, and not maize mosaic virus as speculated by Brewbaker (1979), contributed to the collapse of the Mayan civilization*. Corn stunt has been designated a disease of principal concern in the lowlands of tropical Central America by CIMMYT and is also important in South America, even

---

Peregrinus maidis, the vector of maize mosaic virus, and probably MMV, has an African origin on *Sorghum* spp. and was probably introduced into the New World less than 200 years ago.

at high altitude. Puca poncho disease, which features in Indian folklore in the Peruvian Andes, may be associated with CSS and MBSM infection (Nault *et al.*, 1981).

The story of corn stunting diseases remains incomplete. Their distribution from the southern USA to South America overlaps that of their principal leafhopper vector, *D. maidis*, but it is not understood why *D. maidis* has spread beyond its centre of origin in Mexico, whereas other *Dalbulus* species, particularly *D. elimatus*, have not. It is possible that the stunting pathogens are determining the geographical ranges of their leafhopper vectors. It has been noted that CSS is differentially pathogenic to its vectors, *D. maidis* being much less affected than other members of the genus (Granados & Meehan, 1975; Nault, unpublished). This pathogen, which is most common at low altitudes, may restrict the distribution of other *Dalbulus* spp. to higher regions or to *Tripsacum* spp. not susceptible to CSS. Perhaps the most important question is why has *D. maidis* not expanded its distribution to the northern Corn Belt states of the USA. If this were known it could help to keep corn stunting pathogens out of the major maize-producing regions of the world.

Acknowledgement

Published with the approval of the Associate Director, Ohio Agricultural Research and Development Center, as journal article no. 100-81.

REFERENCES

Ascherson P. (1875) Ueber Euchlaena Mexicana Schrad. *Botanische Vereinschaften Provance Brandenburg* 17, 76-80.

Barnes D. (1954) Biología, ecología y distribución de las chicharritas, *Dalbulus elimatus* (Ball) y *Dalbulus maidis* (DeL. & W.) *Folleto Técnico, Oficina de Estudios Especiales, Secretaría de Agricultura y Ganadería, México DF* No. 11, 112 pp.

Beadle G.W. (1972) The mystery of maize. *Field Museum Natural History Bulletin* 43, 241.

Beadle G.W. (1980) The ancestry of corn. *Scientific American* 242, 112-9.

Bradfute O.E., Nault L.R., Gordon D.T., Robertson D.C., Toler R.W. & Boothroyd C.W. (1980) Identification of maize rayado fino virus in the United States. *Plant Disease* 64, 50-3.

Brewbaker J.L. (1979) Diseases of maize in the wet lowland tropics and the collapse of the classic Maya civilization. *Economic Botany* 33, 101-18.

Bushing R.W. & Burton V.E. (1974) Leafhopper damage to silage corn in California. *Journal of Economic Entomology* 67, 656-8.

Chen T.A. & Liao C.H. (1975) Corn stunt spiroplasma: isolation, cultivation and proof of pathogenicity. *Science* 188, 1015-7.

Davis R.E. (1974) Occurrence of spiroplasma in corn stunt-infected plants in Mexico. *Plant Disease Reporter* 57, 333-7.

Davis R.E. (1977) Spiroplasma: Role in the diagnosis of corn stunt disease. *Proceedings Maize Virus Disease Colloquium and Workshop,*

16-9 August 1976, pp. 92-8. Ohio Agricultural Research and Development Center, Wooster.

Davis R.E. & Worley J.R. (1973) Spiroplasma: motile, helical micro-organism associated with corn stunt disease. *Phytopathology* 63, 403-8.

DeLong D.M. (1950) The genera *Dalbulus* and *Baldulus* in North America including Mexico. (Homoptera : Cicadellidae.) *Bulletin Brooklyn Entomological Society* 45, 105-16.

de Wet J.M.J. & Harlan J.R. (1972) Origin of maize: the tripartite hypothesis. *Euphytica* 21, 271-9.

de Wet J.M.J. & Harlan J.R. (1978) *Tripsacum* and the origin of maize. *Maize Breeding and Genetics* (Ed. by D.B. Walden), pp. 129-41. Wiley Interscience, New York.

de Wet J.M.J., Gray J.R. & Harlan J.R. (1976) Systematics of *Tripsacum* (Gramineae). *Phytologia* 33, 203-27.

de Wet J.M.J., Timothy D.H., Hilu K.W. & Fletcher G.B. (1981) Systematics of South American *Tripsacum* (Gramineae). *American Journal of Botany* 68, 269-76.

Doebley J.F. & Iltis H.H. (1980) Taxonomy of *Zea* (Gramineae). I. A subgeneric classification with key to taxa. *American Journal of Botany* 67, 982-93.

Galinat W.C. (1978) The inheritance of some traits essential to maize and teosinte. *Maize Breeding and Genetics* (Ed. by D.B. Walden), pp. 93-112. Wiley Interscience, New York.

Gámez R. (1980) Rayado fino virus disease of maize in the American tropics. *Tropical Pest Management* 26, 26-33.

Gámez R., Kitajima E.W. & Lin M.T. (1979) The geographical distribution of maize rayado fino virus. *Plant Disease Reporter* 63, 830-3.

Granados R.R. & Meehan D.J. (1975) Pathogenicity of the corn stunt agent to an insect vector, *Dalbulus elimatus*. *Journal of Invertebrate Pathology* 26, 313-20.

Granados R.R. & Whitcomb R.F. (1971) Transmission of corn stunt mycoplasma by the leafhopper, *Baldulus tripsaci*. *Phytopathology* 61, 240-1.

Harlan J.R., de Wet J.M.J. & Price E.G. (1973) Comparative evolution of cereals. *Evolution* 27, 311-25.

Iltis H.H. (1972) The taxonomy of *Zea mays* (Gramineae). *Phytologia* 23, 248-9.

Iltis H.H. & Doebley J.F. (1980) Taxonomy of *Zea* (Gramineae). II. Subspecific categories in the *Zea mays* complex and a generic synopsis. *American Journal of Botany* 67, 994-1004.

Iltis H.H., Doebley J.F., Guzman J.R. & Pazy B. (1979) *Zea diploperennis* (Gramineae): A new teosinte from Mexico. *Science* 203, 186-8.

Kramer J.P. & Whitcomb R.F. (1968) A new species of *Baldulus* from gamagrass in Eastern United States with its possible implications in the corn stunt virus problem (Homoptera : Cicadellidae : Deltocephalinae). *Proceedings of the Entomological Society of Washington* 70, 88-92.

Mangelsdorf P.C. (1974)· *Corn, Its Origin, Evolution and Improvement*. Belknap Press, Cambridge, Mass.

Mangelsdorf P.C. & Reeves R.G. (1939) The origin of Indian corn and its relatives. *Texas Agricultural Experiment Station Bulletin* 574, 1-315.

Maramorosch K. (1955) The occurrence of two distinct types of corn stunt in Mexico. *Plant Disease Reporter* 39, 896-8.

Markham P.G., Townsend R., Plaskitt K. & Saglio P. (1977) Transmission of corn stunt to dicotyledonous plants. *Plant Disease Reporter* 61, 342-5.

Nault L.R. (1980) Maize bushy stunt and corn stunt: a comparison of disease symptoms, host range and vectors. *Phytopathology* 70, 659-62.

Nault L.R. & Bradfute O.E. (1979) Corn stunt: involvement of a complex of leafhopper-borne pathogens. *Leafhopper Vectors and Plant Disease Agents* (Ed. by K. Maramorosch & K. Harris), pp. 561-86. Academic Press, New York.

Nault L.R. & DeLong D.M. (1980) Evidence for co-evolution of leaf-hoppers in the genus *Dalbulus* (Cicadellidae : Homoptera) with maize and its ancestors. *Annals of the Entomological Society of America* 73, 349-53.

Nault L.R. & Knoke J.K. (1981) Maize vectors. *Virus and viruslike diseases of maize in the United States* (Ed. by D.T. Gordon, J.K. Knoke and G.E. Scott), pp. 77-84. *Southern Cooperative Series Bulletin* 247.

Nault L.R., Gingery R.E. & Gordon D.T. (1980) Leafhopper transmission and host range of maize rayado fino virus. *Phytopathology* 70, 709-12.

Nault L.R., Gordon D.T. & Castillo Loayza J. (1981) Maize virus and mycoplasma diseases in Peru. *Tropical Pest Management* 27, 363-9.

Nielsen M.W. (1968) The leafhopper vectors of phytopathogenic viruses (Homoptera, Cicadellidae) taxonomy, biology and virus transmission. *USDA Technical Bulletin* 1382, 386 pp.

Oman P.W. (1948) Distribution of *Baldulus maidis* (DeLong & Wolcott). *Proceedings of the Entomological Society of America* 50, 34 pp.

Pitre H.N. (1966) Gamagrass, *Tripsacum dactyloides*: a new host of *Dalbulus maidis*, vector of corn stunt virus. *Plant Disease Reporter* 50, 570-1.

Pitre H.N. (1970a) Observations on the life cycle of *Dalbulus maidis* on three plant species. *Florida Entomologist* 53, 33-7.

Pitre H.N. (1970b) Notes on the life history of *Dalbulus maidis* on gamagrass and plant susceptibility to the corn stunt disease agent. *Journal of Economic Entomology* 63, 1661-2.

Ramírez J.L., DeLeon C., Garcia C. & Granados G. (1975) *Dalbulus guevarai* (DeL.) Nuevo vector del achaparramiento del maíz en México. Incidencia de la enfermedad y su relación con el vector *Dalbulus maidis* (DeL. & W.) en Juna, Yucatan. *Agrociencia* 22, 39-49.

Whitcomb R.F. (1981) The biology of spiroplasmas. *Annual Review of Entomology* 26, 397-425.

Wilkes H.G. (1972) Maize and its wild relatives. *Science* 177, 1071-7.

Wilkes H.G. (1977) Hybridization of maize and teosinte in Mexico and Guatemala and the improvement of maize. *Economic Botany* 31, 254-93.

Wilkes H.G. (1979) Mexico and Central America as a centre for the evolution of agriculture and the evolution of maize. *Crop Improvement* 6, 1-18.

Williamson D.L. & Whitcomb R.F. (1975) Plant mycoplasmas: a cultivable spiroplasma causes corn stunt disease. *Science* 188, 1018-20.

The ecology of maize rayado fino virus in the American tropics

RODRIGO GÁMEZ

Centro de Investigación en Biología Celular y Molecular, Universidad de
Costa Rica, Ciudad Universitaria, Costa Rica

INTRODUCTION

Maize rayado fino virus (MRFV) (Gámez, 1980a), which appears to be
restricted to the Americas, is widespread and increasingly important in
maize in tropical areas (Gámez et al., 1979; Gámez, 1980b). It is
transmitted in the persistent manner by leafhoppers of the genus
Dalbulus; D. maidis, the principal vector, is abundant on maize (Zea
mays ssp. mays) in all tropical areas of the continent (Gámez, 1980b).
The symptoms of the disease are distinct, conspicuous, small chlorotic
spots that develop at the base and along the veins of young leaves in
characteristic stippled stripes. As the virus is spread only by leafhoppers, the epidemiology of MRFV is determined largely by the biology
and phenology of its vector species.

This chapter considers the ecology and epidemiology of MRFV and aims
to demonstrate the close biological interaction between the virus, its
leafhopper vector and the maize host, and how this interaction has been
successfully established in widely different ecosystems and geographical locations.

MAIZE RAYADO FINO VIRUS

The virus

MRFV is an isometric virus *c*. 31 nm in diameter containing single-
stranded RNA. Strains of MRFV from Colombia and Brazil are serologically different and differ in the severity of the disease they induce
in maize (Gámez, 1980a).

MRFV multiplies in *D. maidis* and, in enzyme-linked immunosorbent
assays, an increase in virus titre has been detected in vectors. Virus
particles have also been detected, by electron microscopy, in the
internal organs of viruliferous *D. maidis* (Rivera *et al*., 1981; Gámez
et al., 1981).

Plumb R.T. & Thresh J.M. (1983) *Plant Virus Epidemiology.*
Blackwell Scientific Publications, Oxford.

Transmission

MRFV is transmitted by leafhopper vectors but not by sap inoculation or through the seed of infected plants (Gámez, 1980a, 1980b). Transmission of MRFV by *D. maidis* is typical of propagative viruses (Gámez, 1969, 1973; González & Gámez, 1974; Paniagua & Gámez, 1976; Wolanski & Maramorosch, 1979; Nault *et al.*, 1980). The virus can be acquired during a 6 h acquisition feed and inoculated to healthy plants during an 8 h infection feed. However, there is a long incubation period (8-37 days) in the vector; leafhoppers remain infective for 1-20 days. The ability to transmit MRFV is retained after moulting but the virus is not transmitted transovarially. Individual insects differ in their ability to transmit and the percentage of transmitters in a colony ranges from 10 to 34%.

There is no evidence that MRFV harms *D. maidis* and the longevity of transmitters and non-transmitters is the same (González & Gámez, 1974). MRFV and the corn stunt spiroplasma can be transmitted simultaneously by *D. maidis* and *D. elimatus* (Gámez, 1973; Wolanski & Maramorosch, 1979). Species of *Dalbulus* have probably co-evolved with maize in Mesoamerica (Nault & Delong, 1980; Nault, this volume) and it is possible that both MRFV and its vectors also co-evolved to attain the present close and non-deleterious interrelationship.

Hosts of the virus

Host range studies (Gámez, 1973; Paniagua & Gámez, 1976; Nault *et al.*, 1980) have tested 54 species and three subspecies in 30 genera representing the major Gramineae assemblages (Watson & Gibbs, 1974). Wild grasses common in Central America and the southern and central United States of America, the major cultivated cereal grains and those species considered as close, wild relatives of maize have been tested under experimental conditions. The virus has a narrow host range within the Andropogonoids (Table 1), mainly in the genus *Zea*. One of the perennial teosintes of Mesoamerica, *Z. diploperennis*, the annuals *Z. luxurians, Z. mays* ssp. *mexicana*, races Mesa Central and Chalco, *Z. mays* ssp. *parviglumis* and ssp. *huehuetenagensis*, numerous maize races from all the American continent and several hundred maize cultivars and inbreds, are susceptible to MRFV. Maize and the annual teosintes are highly sensitive to MRFV but the perennial diploid *Z. diploperennis* is tolerant and the tetraploid perennial *Z. perennis* is immune.

Of the teosintes, the two perennial species are most distantly related to maize and most closely related to *Tripsacum* (Doebley & Iltis, 1980; Iltis & Doebley, 1980). Although even more distantly related to maize, *T. australe* and *Rottboellia exaltata* are susceptible to the virus (Nault *et al.*, 1980).

Maize cultivars differ in their susceptibility and sensitivity to MRFV (Martínez-López, 1977; Gámez, 1980b). Losses of 40-50% of the weight of mature ears have been recorded on individual plants of locally adapted Central American cultivars, but losses may reach 100%

Table 1. Taxonomic analysis* of the species tested for susceptibility to maize rayado fino virus†

| | Grass species (and genera) tested | | |
	Infected, with symptoms	Infected, symptomless	Not infected
Bamboos	0	0	1 (1)
Oryzoids	0	0	1 (1)
Festucoids	0	0	15 (10)
Chloridoids	0	0	5 (3)
Panicoids	0	0	12 (3)
Andropogonoids	3 (2)	2 (1)	15 (10)

* As proposed by Watson & Gibbs (1974).
† Data from Gámez (1973, and unpublished), Nault *et al.* (1980) and
Paniagua & Gámez (1976).

in some introduced and some newly developed maize genotypes in Central
America (Gámez, 1980b), El Salvador (A. Díaz, personal communication),
Mexico (Rocha-Peña, 1981) and Colombia (Martínez-López, 1977). The
different susceptibilities of maize cultivars indicate that MRFV may
play an important role in the adaptation of maize genotypes in the
American tropics and, as previously suggested (Nault & DeLong, 1980;
Nault, this volume), was probably a selective force in the evolution of
maize from its wild teosinte ancestors.

VECTORS OF MAIZE RAYADO FINO VIRUS

Identity, distribution and hosts

The natural vectors of MRFV are deltocephaline leafhoppers of the neotropical genus *Dalbulus*. *D. maidis* occurs everywhere that MRFV has
been recorded, from the southern USA to Uruguay in South America
(Gámez *et al.*, 1979).

D. *maidis* has a narrow host range; apart from maize and the annual
teosintes (Barnes, 1954), only Z. *diploperennis*, Z. *perennis* and T.
lanceolatum are natural hosts (Nault & DeLong, 1980). *D. maidis* may
feed on a few other grass species but survival of the leafhopper on
common wild and cultivated grasses is poor (Barnes, 1954; Pitre, 1967;
Gámez, 1973; León-Trochez *et al.*, 1976; Panigua & Gámez, 1976; Nault
et al., 1980).

D. *elimatus* is also a vector of MRFV (Wolanski & Maramorosch, 1979;

Nault *et al.*, 1980; Rocha-Peña, 1981). It is closely related to *D. maidis* but is more restricted in distribution and occurs mainly in Mexico and sporadically in the highlands of Central America above 750 m. *D. elimatus* has a relatively wide host range, which includes maize, annual and perennial teosintes, wheat, barley and two *Bromus* spp. (Barnes, 1954; Nault & DeLong, 1980). Other species of deltocephaline leafhoppers reported as vectors of MRFV under experimental conditions in the USA, are *Baldulus tripsaci, Graminella nigrifrons* and *Stirellus bicolor*. They are less efficient vectors than *Dalbulus* spp. (Nault *et al.*, 1980) and appear to be restricted to the southern USA, Mexico and the northern Caribbean. They are not found where MRFV is prevalent.

Field populations and dispersal

Little is known about the biology of *Dalbulus* spp. In central Mexico (Barnes, 1954), with an annual mean temperature of 16°C, both *D. maidis* and *D. elimatus* require *c*. 2 months to complete a generation. In the warmer areas of Latin America where MRFV is prevalent the generation time is 5-6 weeks. The number of generations/year depends on environment and host availability, but could be six to eight on the Pacific slopes of Central America where the mean annual temperature is $20-25^{\circ}$C and where several overlapping crops of maize are grown. Population densities are influenced by rainfall and temperature, and can be 170 leafhoppers/plant at times of maximum infestation on the Pacific plains of El Salvador and Nicaragua (Díaz-Chávez, 1969; Tapia & Sáenz, 1971) where the insect is a pest.

Little is known about the dispersal of *Dalbulus*; long distance flights or migrations may occur but were not observed in Mexico (Barnes, 1954).

DISTRIBUTION

Geographical

MRFV was first recognised in El Salvador and Costa Rica but has since been detected in all Central American countries including Guatemala, Honduras, Nicaragua and Panama (Gámez, 1969, 1973; Gámez *et al.*, 1979), Mexico (Gámez *et al.*, 1979; Rocha-Peña, 1981), the southern USA (Bradfute *et al.*, 1980), Colombia (Martínez-López, 1977), Peru (Gámez *et al.*, 1979), Venezuela (Lastra & de Uscategui, 1981), Brazil (Kitajima *et al.*, 1976) and Uruguay (Gámez *et al.*, 1979). The geographical range of MRFV (Fig. 1) extends from approximately latitude 30° N to 32° S (Gámez *et al.*, 1979).

The distributions of MRFV and *D. maidis* overlap, which suggests that this insect closely followed the dispersal of maize from its centre of origin in Mexico. If MRFV also originated in the southern Mexican highlands, it possibly spread with its maize host and leafhopper vectors.

Figure 1. Known occurrences (•) of maize rayado fino virus in the Americas. (Modified from Gámez *et al.*, 1979.)

Ecological

MRFV and *D. maidis* are adapted to a wide range of ecological conditions. Both occur in all maize-growing regions of Costa Rica, from sea level to nearly 3000 m. This includes eleven different vegetation zones in the basal, premontane, lower montane and montane altitudinal belts. These zones differ in altitude, annual mean temperature, amount and seasonal distribution of rainfall, light intensity, soil conditions and physiography (Holdridge, 1978). The virus is also present in environments as diverse as the Central Valley of Mexico (subtropical lower montane dry forest), the Savanna of Bogotá, Colombia (tropical lower montane dry forest), the Peruvian coast (subtropical basal desert), the Yucatán Peninsula of Mexico (subtropical basal very dry forest) and the Brazilian Cerrado (tropical premontane dry forest).

The diverse forms of maize, which allow it to thrive or at least grow in a wide range of environments, is greater than that exhibited by any other crop species (Mangelsdorf, 1974). MRFV and *D. maidis* have closely followed this successful adaptation of their host, so that they are now found wherever maize is grown in the continent.

EPIDEMIOLOGY

Disease incidence

The incidence of MRFV generally varies from 0 to 40%, but in some areas, such as Zapotitlán in the Pacific lowlands of El Salvador (Gámez *et al.*, 1979) or in the highlands of Colombia (Martínez-López, 1977), it reaches nearly 100%. Similar wide variations in virus incidence have been recorded in many parts of Mexico (Rocha-Peña, 1981).

Overseasoning and sources of MRFV

Grasses susceptible to MRFV under experimental conditions are not important natural perennating hosts of the virus or *D. maidis*. This may be because they are absent or rare where MRFV is endemic. For example, the annual and perennial teosintes are found only in certain regions of northern Honduras, Guatemala, and central and southern Mexico (Mangelsdorf, 1974; Iltis & Doebley, 1980) and although *Tripsacum* spp. are widely distributed in Central and South America, (Mangelsdorf, 1974) they are rare where MRFV is prevalent. *R. exaltata*, an introduced weed, is found mainly in rice fields in some regions of the warm lowlands of Central America and in similar tropical areas of South America.

It is possible that unidentified hosts and insect vectors of MRFV exist, but how they would survive during the dry season when maize is not grown is unknown. Nevertheless, both virus and vector quickly appear in the first crop sown in the early rainy season.

The typical overlapping sequence of crops during the rainy season allows almost continuous breeding of *D. maidis* and the rapid spread of MRFV, and mature crops appear to be the main source of viruliferous leafhoppers (F. Saavedra & R. Gámez, unpublished data).

Patterns and rates of disease spread

Only limited information is available on the patterns of dispersal and rates of increase of MRFV in maize fields. In the Central Valley of Costa Rica, where traditionally maize is planted throughout the rainy season and also, but less frequently, during the dry season, plots are 0.001-1 ha but are usually less than 0.1 ha. "Doublet count" analysis (Vanderplank, 1946) showed non-random aggregation of infected plants in five out of six fields. Disease incidence decreased linearly with distance from the source and increased with time, approximately following a sigmoid curve (R. Gámez, V. Quiroga & R. Pereira, unpublished data).

The interpretation of gradients in such small areas is complex. This is due largely to the effect of the multiple planting dates and the very variable environmental conditions on the phenology of *D. maidis*.

CONTROL

All maize cultivars screened for resistance in Colombia (Martínez-López, 1977), Costa Rica (Gámez, 1977; D. Mora & R. Gámez, unpublished data), El Salvador (A. Díaz-Cháves, personal communication) and Mexico (Rocha-Peña, 1981) are susceptible to the virus, but there are differences in tolerance. Insect control by systemic insecticides did not reduce the incidence of MRFV in Colombia but crop rotations and adjustment of planting dates did reduce virus infection (Martínez-López, 1977).

Acknowledgements

This work was supported by grants from the Consejo Nacional de Investigaciones Científicas y Tecnológicas de Costa Rica (CONICIT) and the Vicerrectoría de Investigación, Universidad de Costa Rica. R. Gámez is a scientific fellow of CONICIT. The comments of Drs. P. León, L.A. Fournier and J. Woolley are gratefully acknowledged. The collection of field and ecological data was possible through the collaboration of many maize pathologists and breeders throughout Latin America.

REFERENCES

Barnes D. (1954) Biología, ecología y distribución de las chicharritas *Dalbulus elimatus* (Ball) y *Dalbulus maidis* (DeL. & W.). *Folleto Técnico, Oficina de Estudios Especiales, Secretaría de Agricultura y Ganadería, México DF* No. 11, 112 pp.

Bradfute O.E., Nault L.R., Gordon D.T., Robertson D.C., Toler R.W. & Boothroyd C.W. (1980) Identification of maize rayado fino virus in the United States. *Plant Disease* 64, 50-3.

Díaz-Chavez A.J. (1969) Estudio de la población de *Dalbulus* sp., vector del virus causante del achaparramiento del maíz. *XV Reunión Anual, Programa Cooperativo Centroamericano para el Mejoramiento de Cultivos Alimenticios*, 2 pp. San Salvador.

Doebley J.F. & Iltis H.H. (1980) Taxonomy of *Zea* (Gramineae). I. A subgeneric classification with key to taxa. *American Journal of Botany* 67, 982-93.

Gámez R. (1969) A new leafhopper-borne virus of corn in Central America. *Plant Disease Reporter* 53, 929-32.

Gámez R. (1973) Transmission of rayado fino virus of maize (*Zea mays* L.) by *Dalbulus maidis* DeLong & Wolcott. *Annals of Applied Biology* 73, 285-92.

Gámez R. (1980a) Maize rayado fino virus. CMI/AAB *Descriptions of Plant Viruses* 220, 4 pp.

Gámez R. (1980b) Rayado fino virus disease of maize in the American tropics. *Tropical Pest Management* 26, 26-33.

Gámez R., Kitajima E.W. & Lin M.T. (1979) The geographical distribution of maize rayado fino virus. *Plant Disease Reporter* 63, 830-3.

Gámez R., Rivera C. & Kitajima E.W. (1981) The biological cycle of maize rayado fino virus in its insect vector *Dalbulus maidis*. *Abstracts, V International Congress of Virology*, Strasbourg, France, p. 213.

González V. & Gámez R. (1974) Algunos factores que afectan la transmisión del virus del rayado fino del maíz por *Dalbulus maidis* DeLong & Wolcott. *Turrialba* 24, 51-7.

Holdridge L.R. (1978) *Ecología Basada en Zonas de Vida*. Instituto Interamericano de Ciencias Agrícolas, San José, Costa Rica.

Iltis H.H. & Doebley J.F. (1980) Taxonomy of *Zea* (Gramineae). II. Subspecific categories in the *Zea mays* complex and a generic synopsis. *American Journal of Botany* 67, 994-1004.

Kitajima E.W., Yano T. & Costa A.S. (1976) Purification and intracellular location of isometric particles associated with the Brazilian corn streak virus infection. *Ciência e Cultura* 28, 427-30.

Lastra R. & de Uscategui R.C. (1980) El virus rayado fino del maíz en Venezuela. *Turrialba* 30, 405-8.

León-Tróchez A., Posada-Ochoa L. & Martínez-López G. (1967) Estudio sobre plantas hospedantes del saltahojas *Dalbulus maidis* (Homoptera : Cicadellidae) en la Sabana de Bogotá. *Revista Colombiana de Entomología* 2, 63-8.

Mangelsdorf P.C. (1974) *Corn. Its Origin, Evolution and Improvement*. Belknap Press, Cambridge, Mass.

Martínez-López G. (1977) New maize virus diseases in Colombia. *Proceedings International Maize Virus Disease Colloquium and Workshop* (Ed. by L.E. Williams, D.T. Gordon & L.R. Nault), pp. 20-9. Ohio Agricultural Research and Development Center, Wooster.

Nault L.R. & DeLong D.M. (1980) Evidence of co-evolution of leafhoppers in the genus *Dalbulus* (Cicadellidae : Homoptera) with maize and its ancestors. *Annals of the Entomological Society of America* 73, 349-53.

Nault L.R., Gingery R.E. & Gordon D.T. (1980) Leafhopper transmission and host range of maize rayado fino virus. *Phytopathology* 70, 709-12.

Paniagua R. & Gámez R. (1976) El virus del rayado fino del maíz: estudios adicionales sobre la relación del virus y su insecto vector *Turrialba* 26, 39-43.

Pitre H.N. (1967) Greenhouse studies of the host range of *Dalbulus maidis*, a vector of the corn stunt virus. *Journal of Economic Entomology* 60, 417-21.

Rivera C., Kozuka Y. & Gámez R. (1981) Rayado fino virus: detection in salivary glands and evidence of increase in virus titre in the leafhopper vector *Dalbulus maidis*. *Turrialba* 31, 78-80.

Rocha-Peña M.A. (1981) Algunos aspectos relacionados con el virus del rayado fino del maíz en México. *MSc Thesis, Colegio de Posgraduados, Chapingo*.

Tapia H. & Sáenz L. (1971) Información básica para el control del achaparramiento del maíz en Nicaragua. *XVIII Reunión Anual, Programa Cooperativo Centroamericano para el Mejoramiento de Cultivos Alimenticios. Maíz y Sorgo*, 14 pp. Panamá.

Vanderplank J.E. (1946) A method for estimating the number of random groups of adjacent diseased plants in a homogeneous field. *Transactions of the Royal Society of South Africa.* 31, 269-78.

Watson L. & Gibbs A.J. (1974) Taxonomic patterns in the host range of viruses among grasses and suggestions on generic sampling for host-range studies. *Annals of Applied Biology* 77, 23-32.

Wolanski B.S. & Maramorosch K. (1979) Rayado fino virus and corn stunt spiroplasma: phloem restriction and transmission by *Dalbulus elimatus* and *D. maidis*. *Fitopatología Brasileira* <u>4</u>, 47-54.

The comparative epidemiology of two diseases of maize caused by leafhopper-borne viruses in Mauritius

L.J.C. AUTREY & C. RICAUD

Sugar Industry Research Institute, Réduit, Mauritius

INTRODUCTION

Sugarcane has been the backbone of the economy of Mauritius for the last 250 years, contributing, in 1980, nearly 75% of all exports. It occupies 94% of the area under cultivation. To avoid the economic and biological hazards of monoculture and to achieve self-sufficiency in food, a large programme of agricultural diversification was started in the 1960s. Research was directed to crops that could be either inter-planted with sugarcane or grown in the period between successive cane crops. Maize is one of several crops being developed as an important part of this programme.

Although maize has been cultivated in Mauritius as long as sugarcane, it has never occupied a large area. The local, open-pollinated cultivar, though resistant to the main diseases present locally, proved unsuitable for the diversification programme in several respects: it has a marked tendency to lodge, over-abundant foliage and an erratic yield. Furthermore, owing to its excessive height and long, 140-day crop cycle, inter-planting between sugarcane caused reductions in cane yield. Short-cycle, dwarf maize hybrids from Europe and elsewhere suited the local conditions better but were found to be highly susceptible to important diseases and pests, which in some cultivars caused severe losses. The cultivation of these hybrids triggered a complete change in the pattern of infection of the pathogens known to infect maize in Mauritius. In particular, epidemics of virus diseases became apparent and some viruses were detected for the first time.

As maize and sugarcane are graminaceous crops with several diseases in common and as the vectors of some of the pathogens can feed on both species, the potential danger to sugarcane of the viruses of maize, especially as a result of inter-cropping, was appreciated. Studies of the virus diseases of maize in Mauritius were considered to be of utmost importance. The epidemiology of maize streak and maize mosaic diseases has been studied and some of the findings are discussed in this chapter.

Plumb R.T. & Thresh J.M. (1983) *Plant Virus Epidemiology.*
Blackwell Scientific Publications, Oxford.

VIRUSES AND VIRUS STRAINS IDENTIFIED IN MAURITIUS

The following viruses of maize have been identified in Mauritius
(Ricaud & Félix, 1976):-

> maize streak virus (MSV), which is transmitted by the leafhopper
> *Cicadulina mbila*,
> maize mosaic (MMV) and maize stripe viruses, which are transmitted
> by the leafhopper *Peregrinus maidis*,
> maize dwarf mosaic virus (MDMV), which has aphid vectors.

The diagnosis of these diseases was based on symptomatology, vector
transmission and serology. While the presence of MSV, MMV and MDMV has
also been confirmed by electron microscopy, attempts to isolate a virus
inducing maize stripe and examination of ultra-thin sections of infected
leaves have so far failed to confirm the presence of the 40 nm diameter
particles associated with the disease elsewhere (Kulkarni, 1973).

MSV is by far the most important of these viruses and causes severe
epidemics, especially in some exotic genotypes in which total infection
has sometimes been observed and serious losses have resulted (Anon.,
1976). The virus is known to infect various grass hosts but most of
the strains seem host-adapted and only a few can cause epidemics in
maize (Ricaud & Félix, 1978). Differences in the syndromes in the open-
pollinated, but fairly stabilized, local cultivar suggest the existence
of strains differing in virulence, but this variation has not been
studied.

MMV is second in importance and its incidence is usually low,
although occasionally infection is common. Three strains of the virus
have been identified, based on the syndromes induced in maize and other
hosts, and they have been designated MMV-fine, MMV-coarse and MMV-
broken (Autrey, 1980). The latter is believed to be identical to
maize line virus in Kenya (Autrey, 1980). These strains cross-protect
but differ in their effects on maize and in their vector relationships
(Autrey, 1980).

INCIDENCE AND PATTERN OF INFECTION OF MSV AND MMV IN MAIZE

Disease distribution

Field surveys in the early 1970s showed that streak disease was
prevalent in nearly all plantations throughout the island, infection
varying from negligible to 100%. It was more common in the east, west
and south-west, where successive plantings of the local cultivar are
made throughout the year. Mosaic was also encountered frequently in
these areas but in contrast to streak, it was present at a very low
level and has been reported in only five other localities in imported
hybrids. The incidence and spread of streak were in general greater in
imported hybrids than in the local cultivar, whereas for mosaic the
converse was true.

Incidence in the local cultivar

Records are scarce, but since 1924 there has been little mosaic and
streak in the local cultivar and no severe epidemic has been recorded.
In 1974, however, 100% infection by streak was found in the west of
Mauritius in plantations made late in the warm, wet season, while in
the dry, cool season infection was <5%. The incidence of mosaic was
negligible (Anon., 1975).

In 1977/78 the incidence of the two viruses was determined in 13
successive monthly plantings in an area prone to infection in the
vicinity of scrub land and near mountain slopes, in the east of
Mauritius. Most streak infection occurred in the March and October
plantings (Fig. 1) made in the mild periods at the beginning of the
cool/dry and warm/wet seasons, respectively. By contrast, mosaic
infection was greatest in the January crops planted at the middle of
the wet season, when temperatures are high (Fig. 1). The incidence of
mosaic was generally higher than that of streak except in March, April
and October. The greatest infection by mosaic was 36.5% while in the
March planting 38.0% of plants were infected by streak.

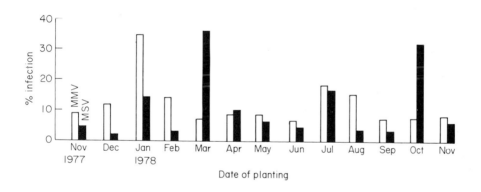

Figure 1. Comparative incidence of maize mosaic virus (MMV) and
maize streak virus (MSV) in monthly plantings made in the east of
Mauritius (November 1977-November 1978).

This trial also showed that for mosaic, irrespective of the month of
planting, there was little spread at the beginning or end of crop
growth, and infection was at a maximum when the plants were in the
middle of the vegetative cycle. This pattern is illustrated in Fig. 2
for the January planting. Rapid spread of streak occurred earlier in
the vegetative cycle in the local cultivar (Fig. 2).

Incidence in imported hybrids

The incidence of mosaic in plantings of imported hybrids since the 1970s

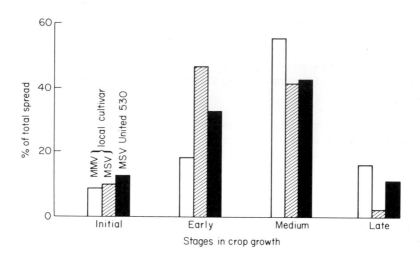

Figure 2. Spread of maize mosaic virus (MMV) and maize streak virus (MSV) in the local cultivar and the hybrid United 530 at different stages in crop growth.

was invariably low in monocultures and in inter-plantings with sugarcane. The greatest incidence was 5.9% in 1976 in the hybrid United 530 in the east of Mauritius, while in the west only 2% infection was found in the hybrid SSM20 in 1974.

The incidence of streak was high in the eastern and western sectors near hill slopes and, on one occasion, in an area in the south of the island at least 10 km from scrub land. At this latter site, United 530, established in December alongside a planting of the same hybrid made 6 weeks earlier, was totally infected although there was little infection in the earlier-sown crop. In general, however, infection in such localities is limited to a few plants and seldom exceeds 2%. In the eastern and western sectors, near scrub land, in maize inter-planted with sugarcane, 40-50% infection with streak occurred in the hybrids Anjou 360 and United 530.

Interference between MMV and MSV

On several occasions sequential infection occurred in the local cultivar and in foreign hybrids, so that the symptoms of streak masked those of initial mosaic infection. However, very rarely, in the open-pollinated local cultivar, symptoms of MMV masked those caused by MSV. This occurred in individual plants apparently resistant to MSV, since the symptoms of the latter disease before MMV infection were very mild. In a few instances, the symptoms of both diseases persisted throughout the growth of infected plants and neither of the viruses became predominant.

In glasshouse and field experiments, Autrey (1980) found that in imported hybrids MSV protected against MMV but not vice-versa. Furthermore, in highly susceptible hybrids, inoculation of MSV to MMV-infected plants proved lethal (Autrey, 1980).

FACTORS AFFECTING DISEASE INCIDENCE, SPREAD AND CARRY-OVER

Alternative hosts

The following members of the Gramineae have been found naturally infected with MSV in Mauritius: *Coix lachryma-jobi, Cenchrus echinatus, Brachiaria reptens, B. eruciformis, Panicum maximum, Paspalum conjugatum, Digitaria timorensis, D. horizontalis, D. didactyla* and *Saccharum* hybrids (Ricaud & Félix, 1978).

Distinct, host-adapted strains of MSV have been obtained from these species and only the first four grasses are likely to be of direct importance in the epidemiology of the virus in maize. The isolates from these grasses when transmitted to various hosts have proved very similar to the maize strain of the virus, and the latter strain can be readily acquired from and transmitted to them (Ricaud & Félix, 1978). Only wild sorghum (*Sorghum verticilliflorum*) has been found naturally infected with each of the three designated strains of MMV. However, virus transmission from and to that host by *P. maidis* is very inefficient (Autrey, 1980).

In Mauritius, the occurrence of grass weeds and the distribution of maize crops in relation to the weed reservoirs of viruses probably differs appreciably from that in other countries where maize is grown widely and where either or both viruses are important. Because Mauritius is a very small island and heavily dependent on sugar production, cane fields occupy almost all the suitable agricultural land and there is very little fallow or scrub land; the latter is mainly on hill slopes. Furthermore, weed control in cane fields is very efficient and only sparse growth of the smaller grass weed species occurs in and around plantings. Both *C. lachryma-jobi* and *S. verticilliflorum* are perennial grasses and constitute the main reservoirs of MSV and MMV, respectively, but their distribution is limited. They are usually only abundant in scrub land, near hills. The two species of *Brachiaria*, as well as *C. echinatus*, are annuals and are fairly well distributed over the island. They grow all year round with maximum growth during the warm, humid months of January-March. They are therefore important carry-over hosts of streak and help to bridge the gap between maize crops in the absence of the main perennial reservoir host.

Vectors

C. mbila is the only known vector of MSV in Mauritius. It is quite widespread in the island and at times high populations occur in maize. There is no severe dry season in Mauritius and green, actively growing grasses are always present. In addition, with the large areas under sugarcane, the insect has ample feeding hosts to maintain itself all the year round. It has been found, locally and elsewhere, to be a most efficient vector of MSV (Storey, 1928; Ricaud & Félix, unpublished).

P. maidis is the only known vector of MMV. Although large populations of the insect occur in maize plantations, in contrast to C. mbila, it has a sedentary habit clustering in the leaf axils and underneath the leaf sheaths of maize. It is a very inefficient vector of MMV, both in the field and in the glasshouse (Autrey, 1980). However, in the warm season populations increase and transmitting efficiency also improves appreciably.

Maize cropping

The epidemic build-up of both streak and mosaic in Mauritius has, until now, been governed largely by cropping practices. Plans for increasing maize production in Mauritius are now being pursued and there are currently two main cropping systems:-

1.   *Small peasant growers*. These are limited to particular localities in the eastern and western areas. They are the only growers cultivating maize in the same fields every year and they always grow two successive crops on the same land. They cultivate mainly the local cultivar which shows appreciable field resistance to both MSV and MMV. Such a cropping system would have failed if susceptible foreign hybrids had been used.
2.   *Cultivation on sugar estates*. This is a recent development to boost maize production and is the only extensive cropping. Most crops are grown from August to December between rows of sugarcane after the cane harvest or between successive cane plantings.

The following factors largely determine the severity of epidemics of streak or mosaic in Mauritius:-

1.   the maize genotypes grown,
2.   proximity to scrub land,
3.   crop sequence.

Most hybrids introduced to improve production in Mauritius are highly susceptible to MSV and/or MMV, with the exception of United 530 which is now the most popular and which shows some tolerance of both diseases. Severe losses have been encountered only in the susceptible foreign hybrids. The local cultivar, although susceptible and sometimes extensively infected, especially with MSV, generally shows appreciable field resistance and is very tolerant; yield losses have rarely been severe. The only severe epidemics recorded, especially of MSV, have been either in plantations established in the immediate vicinity of scrub land close to hill slopes, or elsewhere after successive cropping.

Streak infection can build up linearly or exponentially in the field (Rose, 1978) following the "simple interest" or "compound interest" models of Vanderplank (1963). When planting a susceptible maize crop in the immediate vicinity of scrub land in Mauritius, most of the initial infection may be due to extensive and rapid spread by vectors that have acquired the virus from the perennial grass reservoir hosts, although some transmission by originally virus-free vectors also takes place within plantings later in the season (Anon., 1976). This was observed in a field of United 530 in 1975 (Fig. 3a), when a linear build-up of infection occurred after fairly high initial infection, which accounted for the severity of the disease in the crop. However, in the local cultivar grown under similar ecological conditions, disease progress in several plots planted at intervals throughout the year always showed a marked lag phase (Fig. 3b) attributed to the greater resistance of the cultivar to infection.

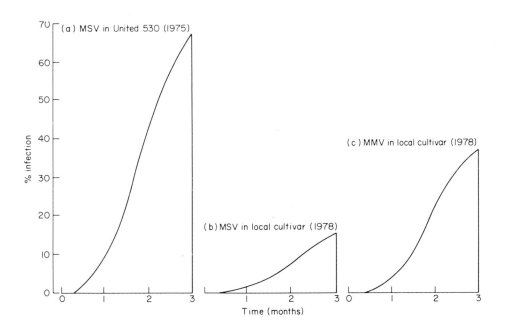

Figure 3. Spread of maize mosaic virus (MMV) and maize streak virus (MSV) in plantations established during the warm season close to scrub land in the east of Mauritius.

Although *C. mbila* can spread streak over long distances, the incidence of infection decreases markedly away from sources of virus and vector (Rose, 1978). When maize is first planted in an area away from scrub land, the build-up of streak infection is slow and, although such build-up has not been studied closely, it is inferred that spread within fields is relatively more important than from outside sources and that the disease progress curve is exponential. However, when a second crop follows in the same field or in the immediate vicinity, the first crop then serves as an important virus source, and disease build-up follows the same pattern as in fields of susceptible maize planted near scrub land, resulting in widespread infection. Spread can be by direct movement of vectors between fields or, when there is an interval between plantings, by a carry-over of the virus in vectors or in the widely distributed annual alternative grass hosts *Brachiaria* and *Cenchrus*. Such increases in the disease with successive cropping has often been reported elsewhere (Gorter, 1953) and indicates that infection builds up faster in maize than in the surrounding grass hosts. This may be explained by the vigorously growing stands of maize that constitute a better habitat for the insect than the sparse growth of the alternative grass hosts scattered among non-host plants.

In contrast to streak disease, infection curves of mosaic in Mauritius are invariably exponential (Fig. 3c). This indicates that large reservoirs of MMV in its alternative host are rare and that spread within maize plantings largely determines the build-up of infection. This pattern is due to the limited distribution of *S. verticilliflorum* which is the main reservoir of MMV, and to the inefficiency of transmission and sedentary habit of the vector. These features account for the generally low infection levels observed. However, it should be emphasized that these observations are valid only for the local cultivar and that masking of MMV symptoms by MSV makes it difficult to assess the spread of mosaic in foreign hybrids.

CONCLUSIONS

Studies at the Mauritius Sugar Industry Research Institute in the 1970s on the epidemiology of maize viruses have led to the identification of strains of MSV and MMV and have clarified their relative importance. Streak is evidently the most important maize virus disease in the island because its alternative hosts are widely distributed and its vector is very efficient. Although populations of the vector of MMV are large, especially in the warm season, mosaic is less important because the vector is sedentary and transmits inefficiently.

When the maize development programme was initiated and foreign hybrids were imported to suit local conditions, it was feared that intensive maize cultivation would lead to severe epidemics of virus diseases as has been encountered elsewhere. This has not occurred, largely because of the present pattern of maize cultivation. However, the crop will soon be grown on large areas and in sequence over a longer period. To avoid the problems that may eventually arise with sequential maize planting, it is essential that there should be continued emphasis on the use of hybrids with good field resistance to both MSV and MMV. These hybrids are being developed from crosses using the local genotypes which confer the desired disease resistance.

REFERENCES

Anon. (1975) *Report, Mauritius Sugar Industry Research Institute* 22,
     63-4.
Anon. (1976) *Report, Mauritius Sugar Industry Research Institute* 23,
     56-7.
Autrey L.J.C. (1980) Studies on maize mosaic virus, its strains and
     economic importance. *PhD Thesis, Exeter University.*
Gorter G.J.M.A. (1953) Studies on the spread and control of streak
     disease of maize. *Scientific Bulletin, South African Department of
     Agriculture* 341, 20 pp.
Kulkarni H.Y. (1973) Comparison and characterization of maize stripe
     and maize line viruses. *Annals of Applied Biology* 75, 205-16.
Ricaud C. & Félix S. (1976) Identification et importance relative de
     viroses du mais à l'Ile Maurice. *Revue Agricole et Sucrière de
     l'Ile Maurice* 55, 163-9.
Ricaud C. & Félix S. (1978) Sources and strains of streak virus
     infecting graminaceous hosts. *Proceedings, International Congress
     of Plant Pathology* 3, 23 (Abstract).
Rose D.J.W. (1978) The epidemiology of maize streak disease. *Annual
     Review of Entomology* 23, 259-82.
Shepherd E.F.S. (1929) Maize chlorosis. *Tropical Agriculture* 6, 320.
Storey H.H. (1928) Transmission studies of maize streak disease.
     Annals of Applied Biology 15, 1-25.
Vanderplank J.E. (1963) *Plant Diseases: Epidemics and Control.*
     Academic Press, New York.

Monitoring the Fiji disease epidemic in sugarcane at Bundaberg, Australia

B.T. EGAN & P. HALL
Bureau of Sugar Experiment Stations, P.O. Box 86,
Indooroopilly, Queensland 4068, Australia

INTRODUCTION

Fiji disease of sugarcane is caused by a reovirus, Fiji disease virus
(FDV), that is transmitted by planthoppers of the Delphacid genus
Perkinsiella. The disease was described initially in Fiji in the 1880s
(Hughes & Robinson, 1961). It occurs from Samoa through Fiji, Australia
and New Guinea to the Philippines, and is also present in the Malagasy
Republic, Thailand and Malaysia. It occurs naturally only in *Saccharum*
spp. and some of their hybrids. Yield losses can be quite severe and
infected planting material may produce very stunted plants of no value.
Plants of sensitive cultivars infected in one crop may be reduced to
stunted stools in the subsequent ratoon crop.

The Bundaberg cane-growing area is on the eastern coast of Queensland,
Australia at latitude 25°S. Sugarcane is planted in autumn (March/April)
or spring (August/September) and harvested in the following year between
July and November. It is then allowed to ratoon from the underground
stool portions for a further two, three or four annual ratoon crops
before it is ploughed out.

The main cane-growing area is outlined in Fig. 4. The canefields are
often contiguous in the localities north, east and south of Bundaberg
city, but tend to be more separated in the peripheral areas and to the
west. The total area of sugarcane harvested annually varied from
27 100 to 32 900 ha in 1970-80. During this period the Bingera,
Fairymead, Millaquin and Qunaba sugar mills processed crops of up to
3.04 million tonnes of sugarcane, to produce up to 440 000 tonnes of raw
sugar, with a maximum value of over A$130 000 000 in the best year.

Fiji disease caused problems in the district in the 1940s, but
appeared to have been eradicated after 1953 (Toohey & Nielsen, 1972).
However, it was rediscovered in 1969 and continued to spread despite all
efforts at control. The cultivar NCo310 was selected in South Africa
and has been planted widely in many countries. It was approved for
planting at Bundaberg in the mid-1950s, reached 80% of the total crop
by 1970, 85% by 1971, 90% by 1973, and 95% in 1975. However, in the

Plumb R.T. & Thresh J.M. (1983) *Plant Virus Epidemiology.*
Blackwell Scientific Publications, Oxford.

localities where Fiji disease was first discovered, and first caused epidemics, NCo310 exceeded 90% of the crop by 1970 and constituted 95-99% of the crop during 1971-77. Unfortunately, NCo310 proved to be much more susceptible to FDV than expected, and its use in virtual monoculture had a crucial effect on the development and ultimate scale of the epidemic.

FDV cannot be inactivated in planting material by heat therapy. The vector has been controlled satisfactorily by insecticides in field trials (Bull, 1977), but widespread and repeated aerial sprayings are not feasible in an area containing many vegetable and fruit crops, as well as a large human population. In addition, the parasites and predators of leafhoppers could have been deleteriously affected. Disease control measures relied on inspection and roguing of diseased plants, ploughing out the more heavily diseased fields, and the use of "clean" plant sources and resistant cultivars (Leverington et al., 1977). As the epidemic developed, only the last two measures were at all effective. Unfortunately, resistant cultivars of satisfactory agronomic and economic worth were not available in the early 1970s. The first suitable, resistant cultivar (Q87), was not released until 1976, while the next, CP44-101, was issued in 1978. Subsequently the resistant cultivars Q108, Q109, Q110 and Q111 were planted widely for the first time in 1980. By 1981, NCo310 had been eliminated from over half the cane-growing localities and constituted less than 5% of the total crop except in the far west of the district where Fiji disease incidence was still small.

THE VECTOR

The known vectors are three species of *Perkinsiella*; in Australia the vector is the sugarcane leafhopper, *P. saccharicida*. The vector can acquire the virus only during its earliest nymphal stages, but remains infective for life (Hughes & Robinson, 1961). Transovarial transmission has been reported (Chang, 1977), but needs confirmation.

Sugarcane is the only important host of *P. saccharicida* in Australia, although probably it can exist on some other grasses. Populations in Bundaberg increase rapidly from early summer (November/December) as the cane starts to grow rapidly and usually reach maximum numbers in February or early March. They reach plague proportions in the worst infested fields, with numbers of adults plus nymphs sometimes >300 per stalk, i.e. 2000-3000/plant and >30 million/ha. "Swarming" of adults may occur at dusk over a period of 2-4 months. Mass swarming occurs in windless conditions on a few evenings each year, when vast numbers of adults may take wing almost simultaneously from hundreds or even thousands of hectares of cane. On such occasions during the late 1960s to the mid 1970s, it was possible to drive for 10 or even 20 km, through leafhopper swarms. Many were attracted by the lights of Bundaberg city. Some swarms appeared to move at random; swarms have been encountered up to 30 km off shore, while others have been found the morning after flight resting in native grassland and open forest

more than 10 km from the nearest commercial cane (Bull, 1981). These
enormous populations arose partly because of the suitability of NCo310
for the leafhoppers, and partly because parasites and predators were
unable to control them until just after maximum numbers were reached.
Populations then declined very rapidly to quite low levels in winter
and spring (Bull, 1972, 1981).

THE MONITORING SCHEME

The current epidemic at Bundaberg must have commenced in the mid-1960s,
although the first diseased stools of cane were not discovered until
1969. Information on the Bundaberg epidemic up to 1976 has been
provided elsewhere (Toohey & Nielsen, 1972; Egan & Fraser, 1977). Fiji
disease also built up in other districts of southern Queensland (Egan,
1976), mainly due to the large areas of NCo310. A disease eradication
campaign operated unsuccessfully during 1969-73 (Egan & Fraser, 1977),
but will not be considered further in this chapter.

A comprehensive annual survey of Fiji disease in sugarcane started
in 1976 throughout the Bundaberg district to help predict future
disease patterns, and to devise sound control recommendations. Some
information was also available from the 1973-75 crops. The Bundaberg
district was divided into 32 localities, delineated by similar topo-
graphy and levels of infection. Approximately 10% of the farms (102 out
of 1 030) were selected for monitoring as being representative of their
localities.

Annual inspections were made on these farms during January-June,
i.e. when the crop is half to three-quarters grown (Ledger & Egan, 1979).
The inspections were made in ratoon cane only and all results apply to
such crops. All planting material was virtually Fiji disease-free
(Egan & Toohey, 1977), although spread to plantings occurred during
December-April. However, symptoms in these newly infected stools were
seldom expressed clearly until very late in the life of the plant crop,
by which time inspections were very difficult or impracticable. These
diseased stools were of relatively minor importance epidemiologically
since few vectors were present by the time symptoms became obvious.

Infection on each farm was expressed as diseased stools/ha or as the
percentage of diseased stools in a field, based on the usual 15 000 cane
stools/ha. Farm, crop class, locality and district averages for
diseased stools were calculated using other crop statistics available
within the sugar industry. The field inspection work was undertaken by
staff of the Bingera, Fairymead and Millaquin-Qunaba Cane Pest and
Disease Control Boards in the Bundaberg district. Supervision, data
collation and interpretation were done by staff of the Bureau of Sugar
Experiment Stations (BSES), the Queensland sugar industry's agricultural
research and extension arm.

RESULTS

The large amount of data collected during the monitoring scheme has been examined in several ways, but only a few results have yet been published.

Disease incidence

The estimated numbers of diseased ratoon stools in Bundaberg district, and in each major sugar mill area, are shown in Fig. 1. This illustrates the very rapid increase in Fiji disease, which occurred despite extensive annual "ploughouts" of the older or more diseased ratoon crops, amounting to at least 20% of the area under cane. Infection increased 3-5 fold annually during 1974-77 despite control measures. This increase was due partly to the larger area of cane affected each year, and partly to increased intensity of infection.

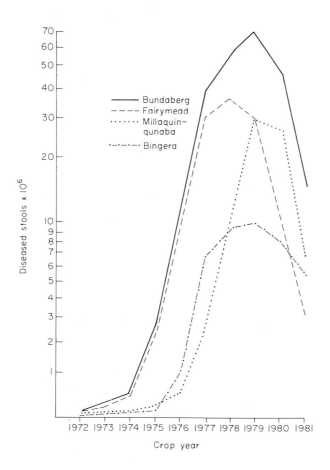

Figure 1. Estimated total numbers of Fiji-diseased stools in each mill area and in the Bundaberg district. (Note the use of a logarithmic scale to present the wide range of values encountered.)

The incidence of Fiji disease in successive NCo310 first ratoon crops was examined in 1974-80. The data for five different monitoring localities are shown in Fig. 2. The locality F3 contained the original epicentre of the epidemic, F4 was adjacent and B1 only slightly distant from the centre (<10 km), while the F6 and MQ7 localities were 12 km and 15 km away, respectively. The sigmoid curves for F3, F4 and B1 are steep and almost identical, while they are a little flatter for F6 and MQ7. The planting of NCo310 was prohibited in various localities at different times, hence the curves cease in different years.

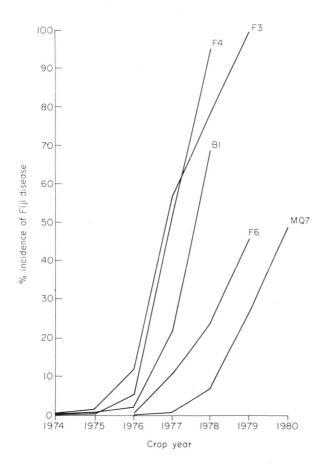

Figure 2. Average incidence of Fiji disease in NCo310 first ratoon crops in five monitoring localities.   (Arithmetic scale.)

Disease gradients

Fiji disease data from several transects through the Bundaberg district were examined. The transects originated at, or passed through, the centre of monitoring locality F3, the one most heavily affected by Fiji disease in the early-mid 1970s. Fig. 3 shows the infection percentages in NCo310 ratoons for 15 monitoring farms along a north-west/south-east transect of approximately 51 km, during 1974-80. The approximate location of the transect is given in Fig. 4. Farm No. 1 was taken as the epicentre of the epidemic, since it and a few adjacent farms showed the highest infection rates in 1973-76. It was also the first farm to show significant yield loss in 1975. The dramatic increase of Fiji disease with time is well illustrated.

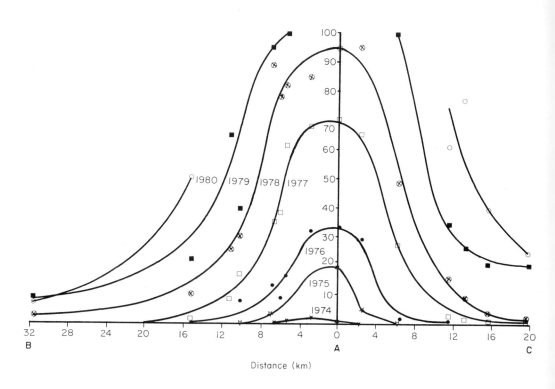

Figure 3. Gradients in the incidence of Fiji disease for the transect B-A-C 1974-80. The vertical axis shows the percentage of NCo310 ratoon stools with Fiji disease.

From data such as those presented in Fig. 3, it was possible to illustrate the spread of Fiji disease on maps of the Bundaberg district. Fig. 4 shows contour lines of the outer limits of 1% average Fiji disease infection in all NCo310 ratoons in 1974-80. The outer limits of the cane-growing area which was monitored are also shown in Fig. 4. The contour lines have been smoothed and are only approximate. Maps have also been prepared for each of the 1974-80 years, showing contour lines for Fiji disease infection percentages of 0.3->95% for that particular year's crop; this series of maps was published in the 1981 BSES Annual Report (Anon., 1981). Fig. 5 is one of the series and shows Fiji disease incidence near the peak of the epidemic in the 1978 crop.

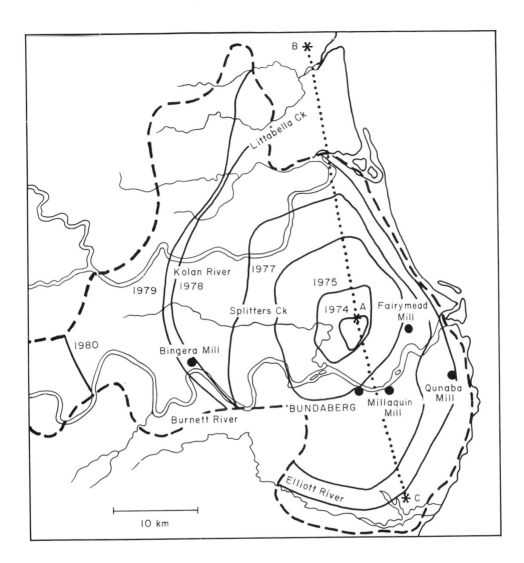

Figure 4. Contour lines of the approximate outer limits of 1% average incidence of Fiji disease in all NCo310 ratoons 1974-80. The outer limit of the main cane-growing area is shown as a broken line. The position of the transect illustrated in Fig. 3 is shown by the line B-A-C. (NB Although Site B is outside the main area, it was one of several such sites monitored.)

DISCUSSION

This study was made in an area where circumstances were very favourable for disease spread. For most of the 1970s, the sugarcane crop in the Bundaberg district consisted largely of the cultivar NCo310, which was

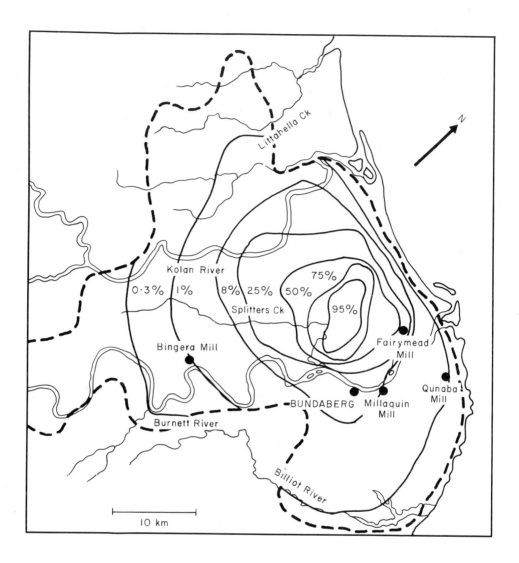

Figure 5. Contour lines for the 1978 NCo310 ratoon crop showing the approximate outer limits of average Fiji disease infection percentages varying from 0.3 to 95%. The outer limit of the main cane-growing area is shown as a broken line.

potentially more productive than others available, but which was very favourable for vector development and susceptible to Fiji disease. Significantly, NCo310 was also quite tolerant of the disease in the early stages of the epidemic, showing little loss of yield before 1976, even in fields which had up to 20% infected stools. However, once large numbers of Fiji diseased stools were present in a locality, young plant and first ratoon crops were subjected to large amounts of inoculum, many

infected plants developed conspicuous symptoms and yield losses became
very serious. Despite advice to the contrary, many farmers were
persuaded by the initial behaviour of NCo310 to continue planting it for
far too long, so that yield losses were inevitable.

During 1969-73, Fiji disease was present on many farms in the
district, but those to the north-west of Bundaberg city began to show
greater concentrations of diseased stools. This locality included the
area where disease was found originally in 1969. It had a history of
high populations of leafhoppers, and contained several farms (including
Farm No. 1 in Fig. 3) with consistently larger numbers of diseased
stools than any others, despite control measures. The locality also
made greater use of spring planting (August-September), which proved to
be more conducive than autumn planting to Fiji disease build-up (Egan &
Fraser, 1977). Factors responsible for this probably include the
smaller size of plants during the summer, peak transmission period, and
the 2-3 month longer period of growth before harvest, thus allowing
greater FDV build-up in the underground stool pieces. All these factors
probably resulted in the disease "hot-spot" from which the district-wide
Fiji disease epidemic developed.

In view of the enormous number of leafhoppers present each year and
their swarming behaviour, it is surprising that the epidemic did not
develop even faster. However, only the early nymphal stages of the
leafhoppers are able to acquire FDV and the percentage infective appears
to vary widely in populations bred on severely diseased cane under
glasshouse conditions (BSES unpublished data). In transmission tests
with individual *P. saccharicida* nymphs and/or adults caged on test
plants of a highly susceptible cultivar, the percentage of transmissions
varied from 5 to 49%. In many other experiments in the glasshouse,
using up to 20 insects/plant on at least 20 plants of mixed cultivars/
cage, quite low transmission rates were sometimes obtained, mainly in
trials carried out in autumn-spring (April-November). The incubation
period of the disease in the plant can also vary from 2 weeks to 8
months, depending on inoculum dose, size of plant, growing conditions
and time of year.

Fiji disease has two main types of disease gradient. Steep gradients
occur where spread is by nymphs from foci within fields or by non-
swarming adults within a field or a small group of fields, hence the
sharp increases in disease incidence shown in Fig. 2. There are also
shallow gradients due to swarming adults, and these can extend considerable distances (see Fig. 3). These matters, as well as Fiji disease
intensity and yield losses, have been discussed elsewhere (Egan, 1982).

Information obtained from the epidemic in Bundaberg district is
being used to predict the course of actual or potential Fiji disease
epidemics in several other districts in southern and central Queensland.
All involve NCo310 as the main cultivar, but the *Perkinsiella* populations are much smaller. The predictions have so far been reasonably
accurate.

REFERENCES

Anon. (1981) Monitoring the spread of Fiji disease in NCo310 at
    Bundaberg. *81st Annual Report of the Bureau of Sugar Experiment
    Stations, Queensland*, pp. 24-5.
Bull R.M. (1972) A study of the sugar cane leafhopper *Perkinsiella
    saccharicida* Kirk. (Hom : Delphacidae) in the Bundaberg district of
    south-eastern Queensland. *Proceedings of the 39th Conference of the
    Queensland Society of Sugar Cane Technologists*, pp. 173-83.
Bull R.M. (1977) Chemical control of the Fiji disease vector
    Perkinsiella saccharicida Kirk. in a Bundaberg cane plant production
    plot. *Proceedings of the 44th Conference of the Queensland Society
    of Sugar Cane Technologists* pp. 61-5.
Bull R.M. (1981) Population studies on the sugar cane leafhopper
    (*Perkinsiella saccharicida* Kirk.) in the Bundaberg district.
    *Proceedings of the 1981 Conference of the Australian Society of
    Sugar Cane Technologists*, pp. 293-303.
Chang V.C.S. (1977) Transovarial transmission of the Fiji disease
    virus in *Perkinsiella saccharicida* Kirk. *Sugarcane Pathologists'
    Newsletter* 18, 22-3.
Egan B.T. (1976) The fall and rise of Fiji disease in south Queensland.
    *Proceedings of the 43rd Conference of the Queensland Society of
    Sugar Cane Technologists*, pp. 73-7.
Egan B.T. (1982) The theoretical basis for the development of the
    Bundaberg Fiji disease epidemic. *Proceedings of the 1982 Conference
    of the Australian Society of Sugar Cane Technologists*, pp. 97-101.
Egan B.T. & Fraser T.K. (1977) The development of the Bundaberg Fiji
    disease epidemic. *Proceedings of the 44th Conference of the
    Queensland Society of Sugar Cane Technologists*, pp. 43-8.
Egan B.T. & Toohey C.L. (1977) The Bundaberg approved plant sources
    scheme. *Proceedings of the 44th Conference of the Queensland
    Society of Sugar Cane Technologists*, pp. 55-9.
Hughes C.G. & Robinson P.E. (1961) Fiji disease. *Sugar-cane Diseases
    of the World*. Vol. 1 (Ed. by J.P. Martin, E.V. Abbott & C.G. Hughes),
    pp. 389-405. Elsevier Publishing Company, Amsterdam.
Ledger P.E. & Egan B.T. (1979) Monitoring the spread of Fiji disease.
    Sugarcane Pathologists' Newsletter 22, 29-31.
Leverington K.C., Egan B.T. & Hogarth D.M. (1977) Breeding for
    resistance and other possible methods of control of Fiji disease at
    Bundaberg. *Proceedings of the 44th Conference of the Queensland
    Society of Sugar Cane Technologists*, pp. 93-5.
Toohey C.L. & Nielsen P.J. (1972) Fiji disease at Bundaberg. *Proceed-
    ings of the 39th Conference of the Queensland Society of Sugar Cane
    Technologists*, pp. 191-6.

Epidemiology and control of curly top diseases of sugarbeet and other crops

JAMES E. DUFFUS
Agricultural Research Service, US Department of Agriculture,
Salinas, California, USA

INTRODUCTION

Diseases due to beet curly top virus have been very important in the production of a number of crop plants in the western United States of America since the latter part of the last century. The diseases have caused severe and widespread losses in sugarbeet, tomato, *Phaseolus* bean, cucurbits and many other plants.

Curly top virtually destroyed the sugarbeet industry in the western USA in the years following World War I and continued to be the principal factor limiting sugarbeet production until World War II. In recent years, damage has been held to less than catastrophic proportions by a complex control programme including the use of:-

1.   cultivars resistant to or tolerant of the virus,
2.   cultural practices to delay infection,
3.   control of the leafhopper vector in crops,
4.   control of the vector in breeding areas,
5.   reduction of vector breeding areas,
6.   reduction in virus sources.

Despite the effective control of curly top in recent years, the virus is still present and changing as the ecology of the California agricultural areas changes. These developments mean that control procedures must be modified to cope with the new circumstances.

BEET CURLY TOP VIRUS

Beet curly top virus (BCTV) is a small (18-20 nm) nucleoprotein with particles that occur predominantly in pairs. It is considered to be a member of the geminivirus group (Goodman, 1981) and in North America is transmitted only by the beet leafhopper (*Circulifer tenellus*). The virus exists as a complex of strains that vary in virulence, host range, and other properties. Strains exhibit little evidence of interference or cross-protection in hosts or the vector (Bennett, 1971).

Plumb R.T. & Thresh J.M. (1983) *Plant Virus Epidemiology.*
Blackwell Scientific Publications, Oxford.

BCTV is widespread throughout the western USA and also occurs in south-western Canada and Mexico. It is found to a limited extent on the eastern slope of the Rocky Mountains and occasionally east of the Mississippi River. The virus is endemic in the Mediterranean Basin (Bennett, 1971) and curly top-like viruses, transmitted by various leaf-hopper species, have been reported from South America (Costa, 1952; Bennett, 1971) and Australia (Thomas & Bowyer, 1980).

BCTV causes dwarfing, yellowing in some species, rolling, curling or twisting of leaves, and swelling and distortion of veins. Phloem tissue often becomes necrotic, cracks develop and phloem exudates occur on stem and leaf tissue.

EPIDEMIOLOGY

The beet leafhopper is a very efficient vector of BCTV. It has an extensive host range and a high reproductive capacity, and can move hundreds of kilometres from breeding grounds to cultivated areas. It can acquire virus in feeding periods of minutes and can transmit within hours of virus acquisition. The virus is circulative in the insect's body and can be retained for more than 90 days.

The natural vegetation of the western USA does not favour the pro-duction of large populations of beet leafhoppers. However, the insect breeds readily on mustards (*Brassica* spp.), Russian thistle (*Salsola iberica*) and other weeds, and also on some crop plants, producing several generations during the summer. Harvesting and drying of crop and weed hosts in the autumn induce the leafhoppers to congregate on any remaining green summer annuals or perennials. As these plants dry, the leafhoppers move to breeding areas in the foothills and accumulate on perennials. During the late autumn there is a marked decrease in the incidence and virulence of BCTV in surviving leafhoppers.

Winter rainfall induces germination of various annuals, including filaree (*Erodium cicutarium*) and peppergrass (*Lepidium nitidum*) which grow in dense masses on the foothills. The leafhoppers move to these annuals and congregate and lay eggs on the plants on warm sunny slopes. Egg-laying and the emergence of nymphs continue into February.

The drying of plants in the breeding areas in the spring and the maturation and drying of the annual hosts forces the migration of the new generation to the agricultural lands of the valley. Relatively few, mostly avirulent, curly top isolates are found in the migrating spring generation.

The severity of the curly top attack depends on the climatic factors affecting the weed hosts of the virus, the prevalence and severity of the virus and the reproductive capacity and migration of the leafhopper. These factors are little understood and urgently need research atten-tion so that control measures can be developed that are consistent with modern methods of pest management.

CONTROL

Curly top virus has been effectively controlled in recent years by a complex series of measures. Although some of the control procedures are not consistent with modern principles of integrated pest management, they have apparently been effective. The gradual changes in the flora and agriculture of the San Joaquin Valley and adjacent rangelands (Magyarosy & Duffus, 1976), the increased virulence of BCTV (Magyarosy & Duffus, 1977) and perhaps the increased incidence of rhizoctonia root rot (Ruppel & Hecker, 1981) may have been partially the result of present control procedures.

Cultural practices

Resistance of sugarbeets to infection and to injury by BCTV increases with plant age. Young seedlings, even those of cultivars rated as resistant* are highly susceptible to infection and may be severely damaged by virulent strains. The avoidance of infection during the early part of the growing season has long been recognized as important in preventing serious losses (Carsner & Stahl, 1924). With the development of resistant cultivars the prevention of early infection was assumed to be less critical. During the last 20 years, however, curly top virus isolates have increased in severity to such an extent that isolates considered severe in the 1950s are now considered mild (Magyarosy & Duffus, 1977). With a high incidence of the more virulent strains, it is advantageous to plant sugarbeets in curly top areas early enough to minimize infection (Duffus & Skoyen, 1977).

   Cultural practices that promote rapid growth of the plants may reduce yield losses for several reasons. Diseased plants yield more under good cultural conditions and rapid growth of the tops may aid in decreasing virus spread. This is because the beet leafhopper is a sun-loving insect and it is so adversely affected by shade and high humidity that leafhopper populations decline to very low numbers under these conditions. Moreover, there is much yield compensation by the healthy neighbours of diseased plants and early infection of as much as 25% of the crop may cause little or no yield loss in well-maintained fields (Skoyen & Duffus, unpublished data).

Leafhopper control in crops

Early attempts to control curly top by controlling the insect vector with insecticides were largely unsuccessful. Although there may have been high mortality following treatments, they were ineffective because of the large numbers of leafhoppers invading the crop over an extended period and the inability of the insecticides to kill the leafhoppers before virus was transmitted.

---

* Editors' footnote: There is no general agreement on the use of this term and many British workers would refer to these cultivars as tolerant.

Good curly top control in sugarbeet fields has been achieved in recent years with phorate granules applied at sowing about 12.7 cm directly below the seed at the rate of 1.12 kg a.i./ha (Burtch, 1968). In experimental trials, the disease incidence is reduced dramatically in the rows treated with insecticide, but not in adjacent untreated areas. Since the insecticide does not kill the leafhoppers quickly enough to prevent infection, there is apparently a repellent effect of treated areas to the leafhopper. Also, the persistent systemic action of the insecticide effectively reduces curly top incidence in the critical early period of growth.

Leafhopper control outside crops

The concept of curly top control through the use of insecticides on weed stands outside the cultivated areas was tested as early as 1931 (Cook, 1933). The California State Department of Agriculture adopted and expanded this programme in 1943 and has continued it to the present at a cost of up to $850 000/year. The objective of the programme is to prevent leafhopper populations from reaching epidemic proportions. Control is attempted at four critical periods in the disease cycle (Fehlman, 1968). In the autumn, large areas (up to 81 000 ha) of beet leafhopper-infested stands of Russian thistle are sprayed from the air to reduce the overwintering populations. If winter annuals have not emerged, the remaining leafhoppers congregate on perennial brush in the foothills where their numbers may be further reduced by sprays. In January, the leafhoppers which have concentrated in the foothill areas are treated before eggs are laid. Finally, in the spring, at a critical time when most of the nymphs have emerged, but before migration to the agricultural areas, a further spray is applied.

It has become apparent in recent years that the vegetation in the leafhopper breeding areas is changing (Magyarosy & Duffus, 1976). Among such changes, the dominant species of Russian thistle (*S. iberica*) is being replaced by *S. paulsenii*. The ecological basis for this and other vegetation changes are not known but it is possible that the pesticide applications over the years have had effects on plant populations and thus on epidemiology. Recent evidence (Ruppel & Hecker, 1981) indicates that insecticides may increase the damage induced by root-rotting fungi. It is difficult to assess the effectiveness of the vector control programme since other measures are also in continual use and there are no equivalent untreated areas.

Reduction of leafhopper breeding areas

The undisturbed natural vegetation of the arid west seldom produces enough beet leafhoppers to threaten crops. The main breeding grounds are weed areas that have resulted from intermittent farming, land abandonment and over-grazing (Piemeisel, 1932). Replacing these weed areas with natural vegetation (annual and perennial grasses and shrubs) would greatly reduce the leafhopper breeding areas. In Idaho, attempts were made to reseed extensive leafhopper breeding areas with crested wheatgrass (*Agropyron cristatum*) on which beet leafhoppers do not reproduce (Knipling, 1979). More than 100 000 ha (77% of the total wild host area) were reseeded between 1959 and 1972. A thorough evaluation of the programme appears not to have been made but the general assessment was that leafhopper populations were greatly reduced and curly top became a less severe problem to vegetable and sugarbeet growers. Expenditure of just over $750 000 apparently reduced losses to sugarbeet and beans by over $1 million annually and increased meat production from livestock on the reseeded land by a similar amount.

The replacement of weed hosts with non-hosts, however, requires time and knowledge and the co-operation of the whole agricultural community, since the leafhopper breeding areas and crop areas are isolated geographically and economically and they are under separate ownership. In Idaho, the reseeding programme was not maintained properly following an initial decline in the severity of curly top attacks and these methods have not been used extensively elsewhere.

Reduction of virus sources

The effect of eliminating virus sources in decreasing the incidence of virus diseases has been dramatically shown with the yellowing complex of sugarbeet (Duffus, 1978; Duffus, this volume). The same principle has been suggested for the control of beet curly top (Hills *et al.*, 1968). It involves eliminating overwintering beet fields in and near the breeding areas and destroying ground-keepers and weed beets before the migration of leafhoppers from the breeding areas. This decreases the otherwise rapid increase in incidence of curly top in migrating leafhoppers.

The role of different populations of weed hosts in the breeding grounds, and/or agricultural lands, or for that matter the role of beets or other crop plants in the rapid increase in the proportion of viruliferous leafhoppers in the migrating population, is not known. Knowing the principal virus and leafhopper sources, coupled with more information on the effects of host species on the virulence of the prevalent strains, seems to be the key to future control.

Resistance

Sugarbeets resistant to BCTV were first reported in 1908, but the first real effort to develop this resistance was the work of Carsner in 1918 (Carsner, 1926). Selections of plants from commercial fields of European cultivars in the curly top areas of the western USA were made by government and sugarbeet company breeders. The most resistant of these selections were inter-crossed in 1929 and resulted in the first resistant cultivar US1 (Carsner, 1933). This cultivar was open-pollinated and had poor resistance to curly top virus. Yet its introduction in 1933 helped stabilize the western sugarbeet industry.

The need for seed of the new curly top-resistant cultivars led to a new seed production industry in the United States, as all sugarbeet seed had previously been imported from Europe.

The cultivar US1 was soon superseded by a series of open-pollinated cultivars more uniformly resistant to curly top and also resistant to bolting. Murphy (1942), in field tests in Idaho, demonstrated the dramatic improvement in curly top resistance within a few years in comparisons of the original susceptible cultivar (R. & G. Old Type) with a series of resistant selections. Yields in this test ranged from 0.0 for the susceptible cultivars to 37.3 t/ha for Improved US22, the most resistant cultivar at that time.

Curly top resistance has been incorporated into male-sterile inbred lines used in the production of hybrid seed and in monogerm lines. It has also been combined with non-bolting characteristics and resistance to *Cercospora* leaf spot, downy mildew and yellows (McFarlane, 1969).

The inheritance of curly top resistance in sugarbeet is poorly understood. Some evidence indicates this resistance may be dominant under moderate infection pressure (Abegg & Owen, 1936) but apparently breaks down under severe attack (Savitsky & Murphy, 1954). Disease resistance is associated with smaller virus concentrations in infected resistant plants compared with susceptible plants. Resistance is also associated with significantly longer incubation periods of the virus in infected plants rather than with differences in the ability of resistant cultivars to recover from the effects of virus infection (Duffus & Skoyen, 1977). Disease resistance has been important in several areas of curly top control.

A large range of resistance to BCTV occurs in the species of common *Phaseolus* bean and in small lima beans (Bennett, 1971). Resistance in these hosts is apparently associated with resistance to infection (Silbernagel, 1965). Interestingly, many of these resistant cultivars are also resistant to bean summer death which is an Australian disease induced by a geminivirus serologically related to BCTV (Thomas & Bowyer, 1980).

Good resistance to BCTV is available in wild species of tomato. Interspecific crosses between tomato and wild *Lycopersicon* spp. have resulted in the development of tomato cultivars with some resistance to the virus. This resistance is manifested as an ability of plants to escape infection (Thomas & Martin, 1971).

THE FUTURE

Beet curly top disease has been extensively studied since it was first reported in 1888. The control and epidemiological aspects of the disease have received attention, rivalling that given to any other virus disease, but ecological conditions are constantly changing in the areas of the western USA where curly top is prevalent. These changes include new irrigation systems, new land development and new agricultural chemicals that necessitate a fresh look at control measures (Magyarosy & Duffus, 1977).

Isolates of curly top virus are becoming increasingly virulent, but the reasons for this are not understood; cultivars with even greater resistance are required to combat these severe strains. In addition, much additional information is needed on the life history, host plants, flight habits, population dynamics and natural enemies of the vector. New knowledge is also needed on the host plants of the vector and of the virus, and of the interaction of these plant species on populations of the vector and incidence of severe forms of the curly top virus.

REFERENCES

Abegg F.A. & Owen F.V. (1936) A genetic factor for curly-top disease resistance in beets (*Beta vulgaris* L.) and linkage relationships. *American Naturalist* 70, 36.

Bennett C.W. (1971) The curly top disease of sugarbeet and other plants. *Phytopathological Monograph* No. 7, 81 pp. American Phytopathological Society, St. Paul, Minnesota.

Burtch L.M. (1968) The use of insecticides for curly top control. *Spreckels Sugar Beet Bulletin* 32, 15-7, 24.

Carsner E. (1926) Resistance in sugar beets to curly-top. *United States Department of Agriculture Circular* 388, 7 pp.

Carsner E. (1933) Curly-top resistance in sugar beets and tests of the resistant variety U.S. No. 1. *United States Department of Agriculture Technical Bulletin* 360, 68 pp.

Carsner E. & Stahl C.F. (1924) Studies on curly top disease of the sugar beet. *Journal of Agricultural Research* 28, 297-320.

Cook W.C. (1933) Spraying for control of the beet leafhopper in central California in 1931. *California Department of Agriculture Monthly Bulletin* 22, 138-41.

Costa A.S. (1952) Further studies on tomato curly top in Brazil. *Phytopathology* 42, 396-403.

Duffus J.E. (1978) The impact of yellows control on California sugarbeets. *Journal of the American Society of Sugar Beet Technologists* 20, 1-5.

Duffus J.E. & Skoyen I.O. (1977) Relationship of age of plants and resistance to a severe isolate of the beet curly top virus. *Phytopathology* 67, 151-4.

Fehlman D.R. (1968) Curly top virus control in California. *Spreckels Sugar Beet Bulletin* 32, 6-8.

Goodman R.M. (1981) Geminiviruses. *Handbook of Plant Virus Infections Comparative Diagnosis* (Ed. by E. Kurstak), pp. 879-910. Elsevier/North Holland Biomedical Press, Amsterdam.

Hills F.J., Ritenour G.L. & Burtch L.M. (1968) Curly top in the San Joaquin Valley. *California Sugar Beet* 68, 26-7.

Knipling E.F. (1979) The basic principles of insect population suppression and management. *United States Department of Agriculture Handbook* 512.

Magyarosy A.C. & Duffus J.E. (1976) Feeding preference and reproduction of the beet leafhopper on two Russian thistle plant species. *Journal of the American Society of Sugar Beet Technologists* 19, 16-8.

Magyarosy A.C. & Duffus J.E. (1977) The occurrence of highly virulent
strains of the beet curly top virus in California. *Plant Disease
Reporter* 61, 248-51.

McFarlane J.S. (1969) Breeding for resistance to curly top. *Journal of
the International Institute of Sugar Beet Research (IIRB)* 4, 73-83.

Murphy A.M. (1942) Production of heavy curly-top exposures in sugarbeet breeding fields. *Proceedings American Society of Sugar Beet
Technologists* 3, 459-62.

Piemeisel R.L. (1932) Weedy abandoned lands and the weed hosts of the
beet leafhopper. *United States Department of Agriculture Circular*
229, 24 pp.

Ruppel E.G. & Hecker R.J. (1981) Effect of three systemic insecticides
on severity of rhizoctonia root rot in sugarbeet. *Phytopathology*
71, 902.

Savitsky V.F. & Murphy A.M. (1954) Study of inheritance for curly top
resistance in hybrids between mono- and multigerm beets. *Proceedings
American Society of Sugar Beet Technologists* 8, 34-44.

Silbernagel M.J. (1965) Differential tolerance to curly top in some
snap bean varieties. *Plant Disease Reporter* 49, 475-7.

Thomas J.E. & Bowyer J.W. (1980) Properties of tobacco yellow dwarf
and bean summer death viruses. *Phytopathology* 70, 214-7.

Thomas P.E. & Martin M.W. (1971) Apparent resistance to establishment
of infection by curly top virus in tomato breeding lines. *Phytopathology* 61, 550-1.

Biology and identity of whitefly vectors of plant pathogens

LAURENCE A. MOUND

British Museum (Natural History), Cromwell Road,
London SW7 5BD, UK

INTRODUCTION

Whitefly are the only known vectors of several pathogens of crops in the tropics and also elsewhere. However, these vector species are usually not host specific; they disperse actively both for short distances and also over long distances on the wind, they reproduce much more rapidly than most of their natural enemies, and, as minute, immature scales on plant leaves, they are readily transported to new areas by man. Whitefly are thus ideal vectors of plant pathogens and, together with the debilitating effect of large populations (Mound, 1965b), they constitute a serious threat to some tropical and subtropical crops. Despite this, current efforts are minimal throughout the world in investigating their identity and biology, including host plant relationships, dispersive activity and natural population control. This chapter seeks to draw attention to aspects of whitefly biology which are of importance to plant pathologists and which require further investigation. Excellent reviews of the literature are given by Bird & Maramorosch (1978) and Muniyappa (1980). Mound (1973) summarized available information on whitefly and viruses, and a full taxonomic bibliography with host plant and natural enemy records is given by Mound & Halsey (1978).

LIFE CYCLE

An adult female whitefly may lay up to 200 eggs, but the mechanism controlling the sex ratio is not clear. An unfertilized female lays eggs which usually give rise only to males, although sometimes females may also be produced. The so-called "English Race" of greenhouse whitefly (*Trialeurodes vaporariorum*) is reputed to lack males completely. Control of sex ratio could be of considerable importance in understanding field biology and the ability of whitefly to transmit disease agents because of its relationship to the rate of population increase and the dispersive activity of adults. However, sex ratios cannot be determined simply from field samples of adults, because females usually live much longer than males.

Plumb R.T. & Thresh J.M. (1983) *Plant Virus Epidemiology.*
Blackwell Scientific Publications, Oxford.

The complete life cycle of *Bemisia tabaci*, from egg to adult under optimum conditions in the tropics, is often less than 3 weeks. Eggs of whitefly species of the subfamily Aleyrodinae usually associated with crops are often laid in small circles because females rotate about their feeding stylets whilst laying eggs. By contrast, females of the Aleurodicinae, which are found mainly on trees in South America, often wander over a leaf in a spiral as they lay eggs and deposit a twin trail of wax behind them, like a large fingerprint. Eggs are almost always laid on the undersurface of leaves, and each hatches to produce a minute, actively crawling larva. This first instar soon settles to feed, and the subsequent second, third and fourth instars remain in the same position. In many species, cast skins of earlier instars lie on top of the fourth instar, the so-called pupal case. An adult whitefly develops within each pupal case, from which it emerges through a dorsal, T-shaped split. The emerging adult has crumpled wings, and these take at least an hour to expand fully and be covered with powdery wax from glands on the body. Similarly, the dark markings found on the wings of some species may take several hours to develop.

In many species, eggs are laid on young apical leaves, so mature pupal cases are only found on older leaves. In this way, actively feeding adults and young larvae obtain their food from the youngest leaves, that is, leaves with the highest concentration of soluble nitrogenous compounds. Such feeding sites may also be the regions of greatest virus concentration. The apical leaves of tobacco bear numerous sticky, glandular hairs and adult whitefly usually cover these with a fine deposit of wax before ovipositing.

POPULATION CONTROL

Several species of whitefly appear to have only one generation/year. This univoltine life cycle, correlated with the growth cycle of the host plant, is particularly common on deciduous, woody plants and is probably the basic lifestyle of whitefly. More advanced species feeding on herbaceous hosts usually retain a relationship between their life cycle and the host plant condition, but often breed continuously by using a range of different plant species. Although whitefly have the potential for rapid population increase, such an increase frequently does not occur. Heavy rain may reduce the number of adults on a crop. Adult whitefly in large numbers do not always result in large numbers of eggs being laid. Heavy egg-laying may not lead to a large population of larvae because many first instars die, although later instars are not usually subject to such high mortality. Parasites and predators of pupae may become abundant but their population increase usually lags behind that of the whitefly until late in the season. In some conditions, entomophagous fungi may greatly reduce populations. Thus, the factors controlling populations appear to be largely associated with climate and the nutritional state of the host. A dry climate with irrigation, and nitrogenous fertilizers promoting vigorous plant growth facilitate whitefly breeding, but population size (Mound, 1965a) and disease spread (Shivanathan, 1981) can be affected by host cultivar.

Estimating population size by counting adults/leaf is difficult due to the activity of adults. There are also technical problems with estimating larval populations, i.e. number/leaf or number/unit area of leaf. Locally, comparative estimates of populations may be obtained by using sticky yellow traps (see below).

ADULT ACTIVITY

Adults appear to emerge from their pupal cases soon after sunrise, but in B. tabaci this is not simply a matter of light sensitivity. If plants are kept in the dark overnight, and not removed until 2-3 h after sunrise, many adults will by then have partially emerged but will not usually have expanded their wings. There is therefore some underlying diurnal rhythm, but whether this is internal to the insects or induced by host plant activity is not known.

Adults are able to fly within a few hours of emergence but details of their flight activity are almost unknown. For example, do females fly before or after mating, or both? Is there a single pre-oviposition dispersive flight, followed by a series of short intra-crop flights, or are adults capable of widespread dispersal throughout their life? Does the flight pattern of males differ from that of females?

The only published information on these questions relates to the European cabbage whitefly, Aleyrodes proletella (= brassicae), and this may not be typical of the group. El Khidir (unpublished thesis quoted by Johnson, 1969) found that adults of this species emerging in the spring and summer in southern England flew only short distances, but adults emerging in autumn flew out of the crop in large numbers and for considerable distances. Moreover, adults of this autumn generation could also be found flying at night. Comparable observations on the flight behaviour of B. tabaci under different conditions in tropical countries are now required, particularly as Costa (1975) has reported massed flights of this species drifting through small towns in southern Brazil.

Flight activity may be investigated by trapping, although the results are sometimes difficult to interpret. Suction traps sample a known volume of air under specified conditions and can thus produce an estimate of the number of whitefly flying at a particular height at a given time. By contrast, attractant traps, such as sticky yellow traps, depend on adult behaviour as much as population density, and they probably catch only those individuals which are ready to alight on plants. When both forms of trap were used among cassava plants in Nigeria, suction traps tended to catch most individuals in the morning, while yellow traps had two maxima of catch, an extended one in the morning and a short one in the late afternoon. This might reflect some difference in behaviour of adult B. tabaci, or it could be that the yellow colour is more attractive in afternoon sunlight. Cohen (1981) produced evidence that whitefly can only perceive yellow traps from a short distance. I have successfully used small black boards 25 cm square with a central, sticky yellow disc 10 cm in diameter. The black surround acts as a standard background and the disc should be a deep orange-yellow. Such traps are most effective when placed horizontally at a height of about 50 cm. During experiments in Nigeria it was noticed that yellow traps collected most whitefly on Mondays. This anomalous weekly periodicity was eventually traced to the weekly repainting of the traps; subsequent bleaching in sunlight reduced their attractiveness. Traps thus need careful standardization if they are to produce reliable estimates of whitefly population size and flight activity.

EVOLUTIONARY RELATIONSHIPS

Whitefly are sucking bugs of the order Hemiptera and, together with such plant-feeding insects as cicadas and leafhoppers, they are grouped in the suborder Homoptera. They are most closely related to aphids, coccids and psyllids, and these four together are given the group name Sternorrhyncha. In structure, whitefly most closely resemble psyllids, which often have sedentary scale-like immature stages and adults with the anus on the dorsal surface. This displacement of the anus seems, in whitefly, to be an adaptation to living on the lower surface of leaves, but the similar displacement in psyllid adults (it is rare in immature forms) seems to be due to the enlarged genitalia. Despite this structural similarity to psyllids, whitefly are more similar in their biology to coccids and aphids. Whereas only one psyllid species is thought to be parthenogenetic, and each species apparently breeds on a single species of plant, members of the other three groups are frequently both parthenogenetic and polyphagous. In a few aphids these biological characteristics are cyclic, involving host alternation and sexual/asexual phases. Whitefly and coccids do not exhibit such complex cycles, although some psyllids overwinter on plants which are not the hosts on which they breed. The capacity to breed on a wide range of host plants, and without males, is fundamental to the pest status often achieved by species of whitefly, aphids and coccids.

HOST PLANT ASSOCIATIONS

Even the most polyphagous of whitefly species, *T. vaporariorum*, is not known to feed on more than a small proportion of the available flowering plants (Mound & Halsey, 1978). Despite this, very few species appear to be strictly monophagous and recorded instances of host specificity are probably due in part to inadequate collecting, or poor taxonomic knowledge, or are based on introduced species, or species near the edges of their geographical range.

The host range of a species introduced to a new area may be limited by the absence of suitable hosts or because only a small fraction of the total genetic variability was available in the few individuals introduced. In contrast to the greenhouse whitefly, exotic species in Britain such as the rhododendron (*Dialeurodes chittendeni*), azalea (*Pealius azaleae*) or viburnum (*Aleurotrachelus jelinekii*) whiteflies

certainly have a very limited host range, but in their country of origin
they might be found on a wider range of plants. Most whitefly species
have a particular host species on which they are most abundant, their
remaining hosts usually supporting rather small populations. For
example, the jasmine whitefly (*Dialeurodes kirkaldyi*) occurs in large
numbers on the leaves of *Jasminum sambac* in many parts of the subtropics,
but it also occurs in smaller numbers on a range of other garden shrubs
such as *Allamanda, Plumeria, Lagerstroemia* and *Gardenia*. This phenomenon can also be found at higher taxonomic levels; certain genera of
whitefly may exhibit a preference for plants belonging to a particular
family of flowering plants, but will also occur in smaller numbers on a
few unrelated plants. A *full* list of all the known host plants of a
species may thus give a biased impression of its *typical* biology.

Moderate specificity (oligophagy) is probably the primitive host
relationship of Sternorrhyncha. From this has evolved the strict
monophagy found in many psyllids, and the moderate to extreme polyphagy
of several whitefly. Tree and shrub-feeding species appear to have
retained the more primitive relationship, possibly because such hosts
are available for longer periods both within and between years. Species
attacking herbs may have had to develop polyphagy because such hosts are
not available throughout the year. In such species the total population
of whitefly may fluctuate, but a reservoir population will always remain
available on some other plant. This type of biology also necessitates
an increase in dispersive behaviour.

The physiological and behavioural mechanisms underlying host recognition and host acceptance by whitefly are very little known. At least
three different types of stimuli are probably involved, although their
importance varies between species. Visual signals, such as colour and
shape, may be involved before landing. Olfactory, or possibly tactile,
stimuli may be important after landing but before feeding. Finally
gustatory stimuli will be important after the whitefly has landed and
started to probe. Strictly monophagous species, such as *Siphoninus
immaculatus* on *Hedera helix*, probably require all of these mechanisms
to prevent them feeding on the wrong host plant. *A. proletella*, the
cabbage whitefly, which also feeds on *Chelidonium majus* and several
other smooth-leaved plants with latex vessels, must have a rather less
strict searching behaviour and may recognize its host by olfactory means
after landing. The mechanism of host recognition is particularly
difficult to understand in *A. jelinekii* which normally feeds on the
small shrub *Viburnum tinus* but also occurs in very low numbers on the
tree *Arbutus unedo*. The searching behaviour which allows this whitefly
to find both a small shrub and a tree must be quite flexible, but the
acceptance behaviour which prevents the insect from feeding on anything
other than these two unrelated plants is presumably strictly controlled.

Some species of whitefly have a very wide host range. *B. hancocki* is
recorded from 47 species of plants in 19 families, *B. tabaci* from 300
species in 63 families, and *T. vaporariorum* from over 400 species in
82 families of plants. This last species is the only whitefly to have
been recorded from a non-Angiosperm host, the cycad *Dioon*. However,

just because a species is known to have the ability to accept a wide range of hosts, every population cannot be expected to have an equal ability to attack all these plants. Firstly, plant-feeding insects of widely differing groups frequently show a strong tendency to lay eggs on the same species of plant as that on which they themselves developed. In this way local populations exhibit host-plant preferences, although these may not be determined genetically. Secondly, individual populations differ genetically from each other, and these genetic differences will probably include differing abilities to breed on particular host plants.

B. tabaci shows both of these patterns. In Nigeria, experimental transfer of this species from one host plant to another was frequently difficult and could sometimes only be achieved by placing both plants in the dark or allowing them to wilt partially. Moreover, in southern Nigeria the predominant host is cassava, whereas in northern Nigeria and the Sudan it is cotton. In the Sudan Gezira, cassava plants grown amongst cotton supported very few whitefly, and conversely in southern Nigeria, cotton plants grown amongst whitefly-infested cassava were scarcely attacked. In Brazil, the situation appears to be even more complex in that the local biotype of B. tabaci apparently will not attack cassava at all, although a few eggs may be laid on this plant under experimental conditions (Costa & Russell, 1976). Similarly in Puerto Rico, the strain of B. tabaci on Jatropha will not breed on any other plants except Croton lobatus, and the most abundant strain of B. tabaci, which feeds on a wide range of plants, will not breed on any members of the Euphorbiaceae or Cucurbitaceae (Bird, 1981). To distinguish between non-genetically determined host preferences, genetically determined biotypes with limited host ranges, and host-specific species, requires careful experimentation, but such distinctions are essential for an understanding of whitefly biology.

VARIATION IN STRUCTURE

Whitefly species are recognized from the structure of their pupal cases; most species cannot yet be identified from adults. Unfortunately the structure of pupal cases of many species is variable, particularly in polyphagous species. The most confusing patterns of variation are induced by the form of the host-plant leaf, particularly the presence or absence of leaf hairs, although pupal cases of some species also vary in structure seasonally or because an internal parasite is present. Parasitized pupal cases usually develop a characteristic, internal submarginal line, and those of T. vaporariorum turn black. Similarly, pupal cases of B. tabaci sometimes develop more obvious cuticular glands in dry seasons, and the overwintering pupal cases of the European maple whitefly, Aleurochiton aceris, differ in structure from those of the summer generation.

When a population of B. tabaci, raised originally from a single virgin female, was developed on cassava in Nigeria and subsequently transferred to several other host plants (Mound, 1963), pupal cases on hairy leaves developed long dorsal setae, whereas pupal cases on glabrous leaves had short setae. This phenomenon is apparently directly induced by the presence of leaf hairs and cuticular rugosity, but it is not known how whitefly perceive the hairs. Eastop (1969) showed that these dorsal setae develop *after* the final moult from third to fourth instar, unlike other insects in which setae develop *before* moulting. The relationship between leaf surface and pupal-case structure also occurs in *B. hancocki*, but this species tends to have longer setae generally and also tends to differ in structure between localities.

T. vaporariorum bears a series of long, wax filaments on the dorsal surface of each pupal case. On hairy leaves these filaments project vertically, but on very smooth leaves they project horizontally and this difference in orientation is associated with considerable differences in size and disposition of the basal tubercles from which the filaments arise.

GEOGRAPHICAL ORIGIN OF WHITEFLY AND THE VIRUSES THEY TRANSMIT

B. tabaci, the most important whitefly vector species, is widespread in the tropics, particularly of the Old World, and also extends into North America, southern Europe and the USSR (Mound & Halsey, 1978). The country of origin was probably in the Orient where related species occur, and it seems likely that *B. tabaci* has been carried from India to America by man. Mound (1965c) suggested that dispersal by man might be reflected in the recorded spread of cassava mosaic across Nigeria from east to west between 1930 and 1940, and also in the apparent spread of cotton leaf curl from West Africa to Sudan between 1920 and 1930.

Similarly, *B. tuberculata* on cassava in South America is probably the same species as the African and Mediterranean whitefly, *B. hancocki*, because no structural differences between them have been found. Both *B. tabaci* and *B. hancocki* may have been transported across the Atlantic by man during the 17th or 18th century (Johnson & Bowden, 1973). In contrast, *T. vaporariorum* is considered to be of American origin (Russell, 1948), although it is now a cosmopolitan glasshouse pest. Within the last 10-15 years this species has been introduced to China and Japan, and in the latter country it has been reported as a virus vector on cucurbits and as a pest of field-grown tobacco (Nakazawa, 1981). Attempts at control with organophosphorus insecticides in Japan led to the conclusion that the populations were already resistant, suggesting that the species had been very recently introduced from a country where such insecticides are used frequently (Hosoda *et al.*, 1976).

The geographical origin of the viruses associated with whitefly is even more conjectural. However, it is interesting that neither cassava nor the whitefly vector of cassava mosaic is native to Africa. Whether the pathogen was in the crop originally in South America, in the absence of an insect vector, or whether it has been introduced into cassava from some native African plant or from an Oriental plant, is unknown. However, it may be relevant that the number of crop pathogens associated with *B. tabaci* in the Orient is particularly large (Narasimhan & Arjunan, 1977; Maramorosch & Muniyappa, 1981; Shivanathan, this volume). Similarly, leaf curl disease of cotton in Africa is a disease of *Gossypium barbadense* cultivars, which are derived from West Indian plants. This disease is probably different from the leaf crumple disease of *G. hirsutum* cultivars in California, the vector of which is the introduced whitefly, *B. tabaci*.

There seems to be a tacit assumption that whitefly/pathogen/host relationships are ancient. An alternative possibility might be that at least some viruses have originated as plant malfunctions and have only acquired an insect vector subsequently, perhaps recently. Given that the insect vectors disperse readily and widely, if the insect/pathogen relationship itself were ancient, such plant pathogens should be widespread naturally. If, however, a disease such as cassava mosaic developed in the absence of an insect vector, that is, the pathogen could disperse only through vegetative propagules, then a series of related strains might be expected to have evolved, and be detectable, in the country of origin of the crop.

REFERENCES

Bird J. (1981) Relationships between whiteflies and whitefly-borne diseases. *Abstracts of the 1981 International Workshop on Pathogens Transmitted by Whiteflies, Oxford, England.* Association of Applied Biologists, Wellesbourne.

Bird J. & Maramorosch K. (1978) Viruses and virus diseases associated with whiteflies. *Advances in Virus Research* 22, 55-110.

Cohen S. (1981) Control of whitefly vectors of viruses by non-conventional means. *Abstracts of the 1981 International Workshop on Pathogens Transmitted by Whiteflies, Oxford, England.* Association of Applied Biologists, Wellesbourne.

Costa A.S. (1975) Increase in the populational density of *Bemisia tabaci*, a threat of widespread virus infection of legume crops in Brazil. *Tropical Diseases of Legumes* (Ed. by J. Bird & K. Maramorosch), pp. 27-49. Academic Press, New York.

Costa A.S. & Russell L.M. (1975) Failure of *Bemisia tabaci* to breed on cassava plants in Brazil (Homoptera - Aleyrodidae). *Ciência e Cultura* 27, 388-90.

Eastop V.F. (1969) Post-ecdysal setal development in the Aleyrodidae (Hemiptera). *Proceedings of the Royal Entomological Society of London* (A) 44, 48.

Hosoda A., Naba K., Nakazawa K. & Hayashi H. (1976) *Bulletin Hiroshima Agricultural Experiment Station* 37, 63-8. (In Japanese)

Johnson C.G. (1969) *Migration and Dispersal of Insects by Flight.* Methuen, London.

Johnson C.G. & Bowden J. (1973) Problems related to the transoceanic transport of insects, especially between the Amazon and Congo areas. *Tropical Forest Ecosystems in Africa and South America.* Smithsonian Institution, Washington DC.

Maramorosch K. & Muniyappa V. (1981) Whitefly transmitted plant disease agents in Karnataka, India. *Abstracts of the 1981 International Workshop on Pathogens Transmitted by Whiteflies, Oxford, England*. Association of Applied Biologists, Wellesbourne.

Mound L.A. (1963) Host correlated variation in *Bemisia tabaci* (Gennadius) (Homoptera : Aleyrodidae). *Proceedings of the Royal Entomological Society of London (A)* 38, 171-80.

Mound L.A. (1965a) Effect of leaf hair on cotton whitefly populations in the Sudan Gezira. *Empire Cotton Growing Review* 42, 33-40.

Mound L.A. (1965b) Effect of whitefly (*Bemisia tabaci*) on cotton in the Sudan Gezira. *Empire Cotton Growing Review* 42, 290-4.

Mound L.A. (1965c) An introduction to the Aleyrodidae of Western Africa (Homoptera). *Bulletin of the British Museum (Natural History)* 17, 113-60.

Mound L.A. (1973) Thrips and whitefly. *Viruses and Invertebrates*. (Ed. by A.J. Gibbs), pp. 229-42. Van Nostrand Reinhold, Amsterdam.

Mound L.A. & Halsey S.H. (1978) *Whitefly of the World*. British Museum (Natural History), London.

Muniyappa V. (1980) Whiteflies. *Vectors of Plant Pathogens* (Ed. by K.F. Harris & K. Maramorosch), pp. 39-85. Academic Press, New York.

Nakazawa K. (1981) Control of greenhouse whitefly, *Trialeurodes vaporariorum* (Westwood), in Japan. *Japan Pesticide Information* No. 39, 8-11.

Narasimhan V. & Arjunan G. (1977) Whitefly population in relation to the incidence of mosaic in cassava varieties at Salem (Tamil Nadu). *Indian Journal of Mycology & Plant Pathology* 6, 187-8.

Russell L.M. (1948) The North American species of whiteflies of the genus *Trialeurodes*. *Miscellaneous Publications of the US Department of Agriculture* 635, 1-85.

Shivanathan P. (1981) Transmission efficiency of whiteflies in the epidemiology of three diseases. *Abstracts of the 1981 International Workshop on Pathogens Transmitted by Whiteflies, Oxford, England*. Association of Applied Biologists, Wellesbourne.

Epidemiology and control of
tomato yellow leaf curl virus

K.M. MAKKOUK* & H. LATERROT†

*National Council for Scientific Research and Faculty of Agricultural and
Food Sciences, American University of Beirut, Lebanon
†INRA, Amélioration des Plantes Maraîchères, Centre de Recherches,
d'Avignon, Montfavet, France

INTRODUCTION

Tomato yellow leaf curl virus (TYLCV), which affects tomato production
in many Middle Eastern countries, is transmitted by the whitefly
Bemisia tabaci. It has been reported from Israel (Cohen & Nitzany, 1966;
Nitzany, 1975), Jordan (Makkouk, 1978), Lebanon (Makkouk et al., 1979),
Iraq (personal communications) and Saudi Arabia (Mazyad et al., 1979).
Tomato leaf curl virus (TLCV) reported from the Sudan (Yassin & Nour,
1965b) could be the same, or a closely related, virus. The only
reported difference between TYLCV and TLCV is that TYLCV cannot infect
Nicotiana tabacum cv. White Burley or *Hibiscus esculentum* (okra) (Cohen
& Nitzany, 1966; Nitzany, 1975), whereas the latter can (Yassin & Nour,
1965b). However, such variations often exist among different strains
of the same pathogen and Makkouk (1978) reported two isolates of TYLCV
from the Jordan Valley that differed in their effects on tomato. Losses
due to TYLCV or TLCV reach 50-75% in many regions (Yassin & Nour, 1965a;
Makkouk et al., 1979), making tomato production during the autumn
unprofitable.

Other whitefly-transmitted pathogens have been reported to cause leaf
curl in tomato in other regions of the world (Varma, 1963; Verma et al.,
1975; De Uzcátegui & Lastra, 1978). However, available information on
the host range of these pathogens indicates that they differ from TYLCV.
This paper deals mainly with TYLCV reported in Israel, Jordan and
Lebanon, and to a lesser extent with TLCV reported in the Sudan.

No virus has yet been isolated from TYLCV or TLCV-infected plants,
but symptoms were not suppressed when infected plants were sprayed with
a tetracycline antibiotic (Makkouk, 1978) which suggests that mycoplasmas are not involved. A whitefly-transmitted disease of tomato in
Brazil (tomato golden mosaic) is caused by a geminivirus (Matyis et al.,
1975), and Russo et al. (1980) found a geminivirus in ultrathin
sections of petioles, main veins and secondary veins of TYLCV-affected,
but not of healthy, tomato plants.

Plumb R.T. & Thresh J.M. (1983) *Plant Virus Epidemiology.*
Blackwell Scientific Publications, Oxford.

EPIDEMIOLOGY

Sources of inoculum

Malva nicaensis and *Datura stramonium* are hosts of TYLCV (Nitzany, 1975)
and *D. stramonium* is also a host of TLCV (Yassin & Nour, 1965b). Both
plants are common weeds where TYLCV occurs and serve as "oversummering"
hosts of the virus during the hot months of July and August. Whiteflies
acquire the virus from these plants during late summer or early autumn.
The importance of each species as a reservoir of TYLCV or TLCV varies
between localities and it is possible that other plant species are also
reservoirs. At many locations almost all tomato plants are infected by
TYLCV soon after transplanting, which suggests that there are numerous
sources of inoculum around tomato fields.

   Crop plants can also act as virus sources. Okra is a host for TLCV
(Yassin & Nour, 1965b), but not TYLCV (Nitzany, 1975). *Nicotiana
glutinosa, N. tabacum* cv. Samsun, *Lycopersicon esculentum, L. hirsutum,
L. peruvianum, L. pimpinellifolium* (Nitzany, 1975) and *Solanum penellii*
(Makkouk, 1978) are hosts of TYLCV but, on current evidence, tobacco
and tomato are the only crops that could be sources of TYLCV.

Transmission

Wherever TYLCV or TLCV occur, the only known vector is the tobacco
whitefly *Bemisia tabaci* (Yassin & Nour, 1965b; Cohen & Nitzany, 1966;
Makkouk, 1978; Makkouk *et al.*, 1979). The whitefly *Trialeurodes
vaporariorum* failed to transmit TYLCV (Makkouk, 1978).

   Spread of TYLCV or TLCV between tomato plants by contact seems
unlikely as the disease agent(s) have not been mechanically transmitted
from tomato to tomato (Yassin & Nour, 1965b; Cohen & Nitzany, 1966;
Makkouk, 1978). More recently, Makkouk *et al.* (1979) reported a few
successful mechanical transmissions to tomato from *D. stramonium, N.
glutinosa* and *S. penellii*. However, this is unlikely to be significant
in the spread of TYLCV in tomato fields. Transmission between tomato
plants by natural root grafting has not been reported.

   Both TYLCV and TLCV were not transmitted through seed (Yassin & Abu
Salih, 1972; Nitzany, 1975; Makkouk, 1978) and tomato plants were not
infected when grown in soil collected from around TYLCV-infected plants
(Makkouk, 1978). Consequently the spread of TYLCV or TLCV in tomato
fields is due entirely to the whitefly vector.

   Comparative data are not available on the efficiency with which
TYLCV or TLCV are acquired by whiteflies from different wild hosts as
compared with tomato. It has been found for other whitefly-transmitted
disease agents (Pruthi & Samuel, 1939; Costa, 1954) that whiteflies
were more efficient vectors when they acquired the causal agent from
weeds rather than from crop plants. Yassin & Abu Salih (1976) did not
observe spread round primary foci of TLCV infection in tomato crops in
which spread subsequently occurred, possibly because virus was intro-

duced from plants outside the crop which are better sources of virus
than tomato.

Influence of weather on disease spread

Vector populations are large in late summer when sowings are made for
autumn tomato crops. These are the only ones in which there is much
infection by TYLCV in Lebanon (Makkouk, 1976; Makkouk et al., 1979),
Israel (Nitzany, 1975), Jordan (Makkouk, 1978) and Saudi Arabia (Mazyad
et al., 1979) and by TLCV in the Sudan (Yassin & Abu Salih, 1972).
Vectors are few when further sowings are made in February-April in
Lebanon, Jordan and Israel; the incidence of TYLCV is negligible or
non-existent in such crops.

No correlation was found between TLCV incidence and wind direction
when observations were made in the Sudan for five growing seasons on
tomato cultivars of the bushy type, which may protect vectors from the
wind (Yassin, 1975).

Nitzany (1975) suggested that in Israel TYLCV outbreaks always
follow months with a mean relative humidity <60% and mean maximum
temperatures of *c*. 30°C. However, in Lebanon TYLCV outbreaks were only
reported in the coastal region after months with a mean relative
humidity >60%. More work is needed to relate TYLCV or TLCV incidence
to weather at different locations.

CONTROL

Chemical control

Attempts to reduce the incidence of TYLCV or TLCV by chemical control of
the vector have been made in different locations. Nitzany (1975)
obtained the best results with endrin, methidathion and cutnion. In
trials in Lebanon a slight delay in TYLCV infection was observed, and a
slight increase in yield obtained, when the insecticides azinphos methyl,
methidathion and methomyl were used twice weekly for 8-10 weeks after
transplanting. However, the small yield increase does not justify,
either economically or ecologically, the use of 16-20 pesticide applications. Yassin (1975) reported significant decreases in TLCV
incidence and significant increases in yield after weekly sprays of
omethoate, malathion or mevinphos when used either in the seedbed or
after transplanting.

Sharaf & Allawy (1981) reported two to threefold increases in tomato
yield and a reduction in TYLCV incidence when the insecticides
permethrin, methidathion and pirimphos-methyl were applied with oils
(Hi Par or Sunoco). The advantages of mixing oils with insecticides to
control other whitefly-transmitted diseases have also been reported
(Singh et al., 1975, 1979).

Whiteflies very rapidly develop resistance to insecticides, so the benefits of these chemicals may decrease greatly in successive years (Cohen et al., 1974). Moreover, experience suggests that chemical measures can at best provide only partial control of TYLCV or TLCV and other means are needed to contain outbreaks effectively.

Husbandry

Date of planting   In Lebanon and the Jordan Valley a larger proportion of tomato crops sown in August and September are infected by TYLCV than later plantings. However, the advantages of late planting are counteracted by the deleterious effects on growth of low temperatures in January and February.

Crop mulching   Covering the soil with fresh wheat straw decreased TYLCV incidence (Cohen et al., 1974). The mulch was most effective in seedbeds and at early stages of growth, and gave a 2-week delay in TYLCV spread. If the straw mulch was replaced by yellow polyethylene sheets, spread was delayed for 4 weeks (Cohen, 1981).

Mixed cropping   Crops known to be good hosts for the vector but not the virus might be expected to attract the vector and hence reduce the incidence of the disease in tomato when grown in mixed stands. In Jordan, Al-Musa (1981) showed that TYLCV infection was delayed when cucumber was inter-planted with tomato. TLCV incidence has also been reduced by inter-planting lubia bean (*Cajanus cajan*) with tomato (Yassin & Abu Salih, 1976).

Eliminating virus sources   A significant decrease in disease incidence is to be expected if sources of infection are eliminated from within and near crops. However, in practice it is not easy, and is sometimes impossible, to eliminate common weed hosts such as *D. stramonium* and *M. nicaensis*. More information is required on wild and cultivated hosts of TYLCV and on their relative importance as sources of TYLCV for *Bemisia tabaci*. Only then will it be possible to evaluate the effect of removing sources of TYLCV infection in different areas.

Varietal tolerance and breeding for disease resistance

All the many tomato cultivars tested for resistance to infection with TYLCV or TLCV were susceptible (Yassin & Abu Salih, 1972; Pilowsky & Cohen, 1974; Abu-Gharbieh et al., 1978). Thus there is a great need for tomato cultivars with increased tolerance and since 1975 we have been developing TYLCV-tolerant cultivars using the two wild species *L. pimpinellifolium* and *L. peruvianum*.

Tolerance derived from L. pimpinellifolium   Line LA121 of *L. pimpinellifolium* is tolerant of TYLCV and this character is controlled by a single, incompletely dominant gene (Pilowsky & Cohen, 1974). Our tests with this line showed that upon infection with TYLCV plant growth is not seriously affected and symptoms appear 8-10 weeks after inoculation compared with only 3-4 weeks in the sensitive *L. escul-entum* cultivars. This long latent period could contribute to better plant growth and yield. Tolerance is now being introduced into commercial cultivars and tolerant lines from the F_4 generation were selected in autumn 1980. Plants obtained from two selfings following the first backcross and others obtained from one selfing following the second backcross are being evaluated. The tolerance in the selected lines does not appear to be sufficient to permit high yield, but fruit set and plant growth of such lines when planted in the autumn is better than in the sensitive cultivar Marmande RAF.

Tolerance derived from L. peruvianum   The TYLCV tolerance derived from a population obtained by crossing two sources of *L. peruvianum* seems greater than that of line LA121 of *L. pimpinellifolium*. Inoculated plants did not produce any symptoms in the 4 months after inoculation, although the disease agent could be detected by back-inoculation to the sensitive cultivar Marmande RAF. Thus our line of *L. peruvianum* is not immune but a symptomless carrier of TYLCV. The disadvantage of working with *L. peruvianum* is the difficulty of obtaining inter-specific hybrids with *L. esculentum* because of incompatibility. This problem was overcome by culturing immature seeds *in vitro* and by the "pollen mixture" technique. This technique consists of pollinating the parent *L. escul-entum* with a mixture of pollen from the same cultivar of *L. esculentum* and from the parent *L. peruvianum*. By using as the mother plant a cultivar possessing a recessive gene to act as a marker at the seedling stage it would be easy to sort the F_1 hybrid seedlings. Many F_2 seeds from this inter-specific crossing were obtained by pollinating the F_1 hybrids with the pollen mixture of all of these F_1 hybrids (total of eight hybrids). Selection of TYLCV-resistant F_2 plants was carried out in Lebanon and now is being done on the F_4 obtained from inter-crossing F_3 families obtained from the F_2 resistant plants. Back-crosses will be made later with *L. esculentum*.

THE FUTURE

Each of the measures investigated so far provides only partial control but combining chemicals and husbandry can be expected to lead to improvements (Nitzany, 1975). In future, the use of cultivars having tolerance derived from *L. pimpinellifolium* together with a few insecticidal sprays might improve control of TYLCV. Ultimately, the incorporation of genes for tolerance from both *L. pimpinellifolium* and *L. peruvianum* may prove even more beneficial.

More work is needed on the epidemiology of the diseases, their distribution, natural hosts, transmission by vectors and the relationship of epidemics to weather so that better control strategies can be developed.

More work is also required on the aetiology of the disease. Once the causal agent is isolated and purified, antisera can be prepared and comparative studies become feasible with the agents transmitted

by whiteflies and causing leaf curl symptoms in tomato in different
regions. Such information should lead to an increased exchange of
materials and ideas, and so facilitate progress.

The seriousness of TYLCV or TLCV in many countries of the Middle
East requires collaboration to exploit all possible control measures to
produce acceptable tomato yields in areas where epidemics due to TYLCV
or TLCV are common.

REFERENCES

Abu-Gharbieh W.I., Makkouk K.M. & Saghir A.R. (1978) Response of
    different tomato cultivars to the root-knot nematode, tomato yellow
    leaf curl virus, and orobanche in Jordan. *Plant Disease Reporter*
    62, 263-6.
Al-Musa A. (1981) Incidence, economic importance and prevention of
    tomato yellow leaf curl virus in Jordan. *Abstracts of the 1981
    International Workshop on Pathogens Transmitted by Whiteflies,
    Oxford, England*, p. 47. Association of Applied Biologists,
    Wellesbourne.
Cohen S. (1981) Control of whitefly vectors of viruses by non-
    conventional means. *Abstracts of the 1981 International Workshop
    on Pathogens Transmitted by Whiteflies, Oxford, England*, p. 51.
    Association of Applied Biologists, Wellesbourne.
Cohen S., Melamed-Madjar V. & Hameiri J. (1974) Prevention of the
    spread of tomato yellow leaf curl virus transmitted by *Bemisia
    tabaci* in Israel. *Bulletin of Entomological Research* 64, 193-7.
Cohen S. & Nitzany F.E. (1966) Transmission and host range of the
    tomato yellow leaf curl virus. *Phytopathology* 56, 1127-31.
Costa A.S. (1954) Identidade entre o mosaico comum do algodoeiro e a
    clorose infecciosa das malvaceas. *Bragantia* 13, XXVII.
De Uzcátegui R.C. & Lastra R. (1978) Transmission and physical
    properties of the causal agent of mosaico amarillo del tomate
    (tomato yellow mosaic). *Phytopathology* 68, 985-8.
Makkouk K.M. (1976) Reaction of tomato cultivars to tobacco mosaic
    and tomato yellow leaf curl viruses in Lebanon. *Poljopriveredna
    Znanstvena Smotra.* 39, 121-6.
Makkouk K.M. (1978) A study on tomato viruses in the Jordan Valley
    with special emphasis on tomato yellow leaf curl. *Plant Disease
    Reporter* 62, 259-62.
Makkouk K.M., Shehab S. & Majdalani S.E. (1979) Tomato yellow leaf
    curl: Incidence, yield losses and transmission in Lebanon.
    Phytopathologische Zeitschrift 96, 263-7.
Matyis J.C., Silva D.M., Oliveira A.R. & Costa A.S. (1975)
    Purificação e morfologia do virus do mosaico dourado do tomateiro.
    Summa Phytopathologica 1, 267-74.
Mazyad H., Omar F., Al-Taher K. & Salha M. (1979) Observations on the
    epidemiology of tomato yellow leaf curl disease on tomato plants.
    Plant Disease Reporter 63, 695-8.
Nitzany F.E. (1975) Tomato yellow leaf curl virus. *Phytopathologia
    Mediterranea* 14, 127-9.

Pilowsky M. & Cohen S. (1974) Inheritance of resistance to tomato
   yellow leaf curl virus in tomatoes. *Phytopathology* 64, 632-5.
Pruthi H.S. & Samuel C.K. (1939) Entomological investigations on the
   leaf curl disease of tobacco in northern India. III. The transmission of leaf curl by white-fly, *Bemisia gossypiperda*, to tobacco,
   sunn-hemp, and a new alternate host of the leaf-curl virus.
   Indian Journal of Agricultural Sciences 9, 223-75.
Russo M., Cohen S. & Martelli G.P. (1980) Virus-like particles in
   tomato plants affected by the yellow leaf curl disease. *Journal of
   General Virology* 49, 209-13.
Sharaf N.S. & Allawy T.F. (1981) Control of *Bemisia tabaci* Genn. a
   vector of tomato yellow leaf curl virus disease in Jordan.
   Zeitschrift für Pflanzenkrankheiten und Pflanzenschutz 87, 123-31.
Singh S.J., Sastry K.S.M. & Sastry K.S. (1975) Effect of alternate
   spraying of insecticides and oil on the incidence of tomato leaf curl
   virus. *Pesticides, India* 9, 45-6.
Singh S.J., Sastry K.S. & Sastry K.S.M. (1979) Efficacy of different
   insecticides and oil in the control of leaf curl virus disease of
   chillies. *Zeitschrift für Pflanzenkrankheiten und Pflanzenschutz*
   86, 253-6.
Varma P.M. (1963) Transmission of plant viruses by whiteflies.
   Bulletin of the National Institute of Sciences, India 24, 11-33.
Verma H.N., Srivastava K.M. & Mathur A.K. (1975) A whitefly-transmitted
   yellow mosaic virus disease of tomato from India. *Plant Disease
   Reporter* 59, 494-8.
Yassin A.M. (1975) Epidemics and chemical control of leaf curl virus
   disease of tomato in the Sudan. *Experimental Agriculture* 11, 161-5.
Yassin A.M. & Abu Salih H.S. (1972) Leaf curl of tomato. *Technical
   Bulletin, Agriculture Research Corporation, Sudan* No. 3, 31 pp.
Yassin A.M. & Abu Salih H.S. (1976) *Annual Report 1975/76, 1976/77,
   Gezira Research Station*. Agriculture Research Corporation, Sudan.
Yassin A.M. & Nour M.A. (1965a) Tomato leaf curl disease, its effect
   on yield and varietal susceptibility. *Sudan Agricultural Journal* 1,
   3-7.
Yassin A.M. & Nour M.A. (1965b) Tomato leaf curl diseases in the Sudan
   and their relation to tobacco leaf curl. *Annals of Applied Biology*
   56, 207-17.

The epidemiology of three diseases caused by whitefly-borne pathogens

P. SHIVANATHAN
Central Agricultural Research Institute, Gannoruwa,
Peradeniya, Sri Lanka

INTRODUCTION

The whitefly (*Bemisia tabaci*) is an important vector of pathogens that
seriously damage a range of crops in Sri Lanka (Shivanathan, 1976).
Whitefly-borne pathogens cause leaf curl diseases of chilli pepper
(*Capsicum frutescens, C. annuum*), tomato (*Lycopersicon esculentum*) and
tobacco (*Nicotiana tabacum*), and mosaic diseases of mungbean (*Vigna
radiata*), bean (*Phaseolus vulgaris*), pigeon pea (*Cajanus cajan*), soyabean
(*Glycine max*), okra (*Hibiscus esculentus*) and pumpkin (*Cucurbita
maxima*). There are large differences in the spread of leaf curl and
mosaic diseases in different seasons and climatic zones of Sri Lanka.
This paper reports epidemiological studies on chilli leaf curl disease
(CLCD), mungbean yellow mosaic disease (MYMD) and okra yellow mosaic
disease (OYMD). The aetiology of these diseases is uncertain but they
are assumed to be caused by viruses.

EXPERIMENTAL SITES

Studies were made near Maha Illuppallama, which is approximately
$8^\circ 10'$ N : $80^\circ 30'$ E., at an altitude of 150 m. The mean annual rainfall
of 1480 mm is distributed bimodally with maxima in October-November
(north-east monsoon) and April-May (south-west monsoon). February,
July, August and September are dry with little or no rain. The annual
mean temperature is 27°C (range $22-32^\circ$C). Agriculture is of the
subsistence type with mixed cropping in small (1-2 ha) farms. Cost and
lack of equipment limit the use of pesticides.

DISEASE SPREAD

The three diseases studied occur following an influx of infective
whiteflies from weeds and other alternative hosts of both the disease
agents and the vector. The movement and multiplication of vectors
within the crop lead to epidemics of the "compound interest" type
(Vanderplank, 1963). The epidemics that occurred in 1976 typify the

Plumb R.T. & Thresh J.M. (1983) *Plant Virus Epidemiology.*
Blackwell Scientific Publications, Oxford.

pattern of disease spread in non-drought years. Data on the spread of
CLCD were obtained from 18 plots each of 0.25 ha (20 000 plants/ha).
The plants were grown on mounds ("hills") at 0.3 m x 0.6 m spacing
with 2-3 plants per mound. For MYMD and OYMD, data were obtained from
single 0.5 ha plots of mungbean (120 000 plants/ha) and okra (32 000
plants/ha). All plants were scored for disease and vectors.

Wet season 1976

The chilli and okra crops were scored every 3 weeks and the mungbean
crop as necessary. There was no epidemic of OYMD, and MYMD spread much
faster than CLCD (Fig. 1a).

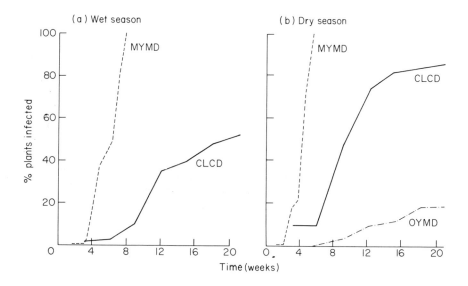

Figure 1. The progress of leaf curl disease of chilli pepper (CLCD)
and yellow mosaic diseases of mungbean (MYMD) and okra (OYMD) during
the wet (a) and dry (b) seasons.

Dry season 1976

MYMD and CLCD spread more quickly than in the wet season and although
OYMD developed it was late and spread only slowly (Fig. 1b). Thus
epidemics of the three diseases, although spread by the same vector and
in the same locality, showed differences in the rate and period of
spread, and in response to seasonal changes.

THE PROGRESS OF EPIDEMICS

Vector populations

The dominant feature of the monsoon climate at Maha Illuppallama is
rainfall, and populations of many insect species are correlated with the
rains (Fellows & Amarasena, 1977). However, systematic surveys showed
that whitefly populations in crops varied more with cultivar, age of
crop, vigour of growth, irrigation frequency and other factors than with
season, and increased in dry periods after the monsoon rains and in the
dry season. Vectors were found on crop plants throughout the year,
except late in the dry season when populations diminish as crops senesce
and sudden changes occur in temperature, humidity and wind (Shivanathan,
1979).

Onset of disease

Initial foci of infection were difficult to determine because whiteflies
can move directly into crops from their original breeding hosts or they
can first feed on other plants during the dispersal flight. Nevertheless, the introduced fodder plant *Phaseolus lathyroides*, which is now a
common weed, has been found to be the main source of MYMD in both the
wet and dry seasons (Table 1). The disease is absent or unimportant in
parts of the island where the plant has not become established. Two
other common weeds, *Acanthospermum hispidum* (Compositae) and *Abelmoschus
esculentus* (Malvaceae), are not hosts of MYMD. The successful transmissions obtained by transfer from these species show that the whiteflies had previously fed on *P. lathyroides*.

Table 1. Pathogens detected in whiteflies collected from weeds.

Source	Season	No. of samples*	Percentage of samples infective[†]		
			CLCD	MYMD	OYMD
Phaseolus	wet	12	8	100	0
lathyroides	dry	9	11	100	0
Acanthospermum	wet	14	14	7	0
hispidum	dry	10	40	20	0
Abelmoschus	wet	16	0	0	0
esculentus	dry	17	6	12	24

* Each sample consisted of 20 whiteflies.
† Determined by transfers to indicator hosts: chilli cv. Skantha for
  CLCD, mungbean cv. H101 for MYMD, okra cv. H10 for OYMD.

Ac. hispidum is the primary source of CLCD. Other weeds become infected but they are seldom colonized by whiteflies and they are not good sources of the pathogen. This suggests that the successful transmissions from *P. lathyroides* were by whiteflies that had previously fed on *Ac. hispidum*.

OYMD was not transmitted from any of the weeds tested during the wet season and the disease was not seen in crops at this time. Successful transmissions were made during the dry season but only from *Ab. esculentus*, which is assumed to be the main weed host (Table 1).

Host acceptability

In experiments to determine the acceptability of different hosts to whiteflies, batches of 200 whiteflies from laboratory-reared colonies, maintained on the cotton cultivar HC101, were released into cages containing different hosts. The number of whiteflies on each plant was counted daily for 14 days (Table 2). Whiteflies tended to avoid chilli, preferring mungbean and *P. lathyroides*. However, although more adults were found on *P. lathyroides*, more nymphs developed on mungbean.

One reason for the larger population on mungbean than on chilli may be that mungbean is heliotropic and the underside of leaves is always shaded. Thus they provide a favourable habitat for whitefly oviposition even if the host is affected by MYMD, as this disease does not cause an upward leaf curling. Chilli is not heliotropic and if infected by CLCD the undersides of the leaves are exposed throughout the day.

Table 2. Average numbers of whitefly adults (and nymphs) per plant at different times after infestation.

Host	Days after infestation				
	1	3	4	11	14
Mungbean cv. MI1	35	39	24	13	(296)
Chilli cv. MI1	7	0	0	1	(18)
Okra cv. H10	14	19	9	4	(63)
Phaseolus lathyroides	63	77	68	71	(172)

Disease spread

Infective potential of vectors   The infective potential of whitefly vectors was assessed by determining the total number of host plants single adults could infect in successive 30 or 60 min infection periods after a maximum acquisition period of 24 h. The values obtained varied according to pathogen, season, host and the sex of the vector (Table 3). The pathogen causing MYMD was the one most readily transmitted. Females were consistently more successful than males, mainly because they lived longer.

Table 3. The number of plants infected* in successive inoculation feeding periods of 30 or 60 min.

Disease	Whitefly sex (no. tested)	Infective potential† 30 min		60 min	
		range	mean	range	mean
CLCD	male (16)	6 - 32	21	8 - 34	26
	female (18)	27 - 54	38	40 - 63	45
MYMD	male (24)	28 - 66	54	33 - 57	43
	female (34)	94 - 158	142	72 - 95	81
OYMD	male (12)	12 - 39	28	17 - 36	21
	female (8)	54 - 114	88	63 - 102	74

The host cultivars used as sources of each pathogen (and in tests for infective potential) were as in Table 1.

* Total number of plants infected by transferring individual white-flies from infected source plants to fresh test plants at 30 min or 60 min intervals throughout each working day (0600-1600 h) for as long as they lived (3-6 days for males and 9-16 days for females). The whiteflies were left overnight on a healthy test plant.
† Comparable values for the infective potential obtained using the resistant cultivars referred to at the end of the chapter were 6 for okra and 18 for mungbean (mean of figures obtained by testing males and females separately).

Whitefly movements   The frequency and duration of visits made by whiteflies to plants is important in disease spread and detailed observations were made daily (0600-1600 h) on the movement of adult whiteflies in a 0.25 ha chilli planting in two contrasting seasons. In the overcast conditions of the wet season, whiteflies moved 4-16 times (mean 6) in each 10 h period at 22-26°C. They spent 14-196 min (mean 94 min) on each plant visited and there were 0.2 hill-to-hill movements per insect. During the hotter (26-32°C), sunny periods of the dry season there were 27-85 (mean 52) movements in each 10 h period, with 3.8 hill-to-hill movements per whitefly. However, the probability of disease spread was reduced by the short duration of each visit (range 2-64 min, mean 11 min). Wind often initiated whitefly movement, and bright, warm and breezy days therefore facilitate disease spread.

Spread within crops   Whiteflies both spread disease and multiply within crops. In the early stages of the epidemics, counts were made of the nymphs produced by whiteflies entering the crops. On mungbean, whitefly emergence from nymphs was counted on a 10% sample of the plants selected at random 14 and 21 days after seedling emergence. In chilli and okra the counts were made 3 and 5 weeks from planting. Five times as many nymphs were found on mungbean as on okra and twice as many as on chilli.

The amount of spread that could have occurred was calculated from estimates of the infective potential of vectors (Table 3) and on the assumption that the whiteflies infecting chilli, mungbean and okra came from *Ac. hispidum*, *P. lathyroides* and *Ab. esculentus*, respectively. These calculations (Table 4) showed that disease spread in okra is restricted by the low infectivity of the whiteflies reaching the crop and by their limited reproduction within plantings. By contrast, each of the many whiteflies produced on mungbean is likely to be capable of transmitting MYMD to several plants and the rapid progress of epidemics (Fig. 1) is readily explicable.

Table 4. Generation of secondary inoculum in the three epidemics.

Disease	Season	Nymphs on crop ($\times 10^3$)	Infective potential*	Potential number of infections† ($\times 10^3$)	Adjusted number of infections¶ ($\times 10^3$)
CLCD	wet	4.66	35	163.10	22.84
	dry	1.28	35	44.80	15.68
MYMD	wet	8.74	98	856.52	856.52
	dry	3.40	98	333.20	333.20
OYMD	wet	1.86	58	107.88	0.00
	dry	0.64	58	37.72	9.05

* Mean values assuming mixed populations of males and females, and optimum inoculation feeding period (Table 3).
† Total number of whiteflies multiplied by infective potential.
¶ Adjusted according to the infectivity of the colonizing whiteflies, assuming that those infesting chilli, mungbean and okra come from *Ac. hispidum*, *P. lathyroides* and *Ab. esculentus*, respectively (Table 1).

The calculations also suggest that the potential for spread in mungbean and chilli is greater in the wet season, when whiteflies are
numerous, than in the dry. However, in practice, spread is least in the
wet season when the diminished activity of vectors limits spread. The
importance of behavioural factors is also indicated by the observation
that epidemics in mungbean and chilli spread much less rapidly than
suggested by the calculations.

DISCUSSION

The studies reported here help to explain some of the observed features
of epidemics caused by whitefly-borne pathogens of chilli, okra and
mungbean in Sri Lanka. They have also indicated features that should be
sought in any programme of breeding for some form of resistance to
pathogen and/or vector.

In work on chilli, all the commercial cultivars grown in Sri Lanka
were found to be susceptible to whiteflies and to CLCD. Accordingly,
collections were made of wild, medicinal and "bird" chillies and of the
semi-wild types that are grown in shifting systems of cultivation.
Four semi-wild and thirteen other accessions have shown promise and
they fall into four categories:-

1. whitefly-resistant types that cannot be infected by grafts or
    by whiteflies,
2. whitefly-resistant types that can be infected by grafts but
    not by whiteflies,
3. whitefly-resistant types that can be infected by grafts and
    by whiteflies,
4. types with better resistance to whiteflies and to CLCD than
    standard commercial cultivars.

All such plants that are not immune to CLCD are difficult to infect,
slow to develop symptoms and poor sources of CLCD. Attempts are being
made to incorporate adequate levels of such resistance into acceptable
commercial cultivars. Work is also in progress on the development of
resistant varieties of mungbean and okra. The need is urgent because
whitefly-borne pathogens are causing serious losses in Sri Lanka and
in the worst-affected areas crops cannot be grown successfully at times
when conditions are otherwise particularly favourable for growth.

ACKNOWLEDGEMENTS

I am grateful to Dr. S.N. De S. Seneviratne, Plant Pathologist, Depart-
ment of Agriculture, for useful suggestions in preparing this paper.
The financial and other assistance by the Department of Agriculture,
Sri Lanka, and the FAO Project on Strengthening of Plant Protection -
Phase II is gratefully acknowledged.

330   P. SHIVANATHAN

REFERENCES

Fellows R.W. & Amarasena J. (1977) Entomology. *Half Yearly Reports, Maha 1976/1977.* Agricultural Research Station, Maha Illupallama.

Rapilly F., Fournet J. & Shajanikoft M. (1970) Etudes sur l'épidemiologie et la biologie de la rouille jaune du blé *Puccinia striiformis* Westend. *Annales de Phytopathologie* 2, 5-31.

Shivanathan P. (1976) Virus diseases of crops in Sri Lanka. *Tropical Agriculture Research Series* 10, 65-8.

Shivanathan P. (1979) Chilli virus disease in the dry zone of Sri Lanka. *Proceedings Sri Lanka Association Advancement of Science* 35, 22.

Vanderplank J.E. (1963) *Plant Diseases: Epidemics and Control.* Academic Press, New York.

Epidemiology of yellow mosaic disease of horsegram (*Macrotyloma uniflorum*) in southern India

V. MUNIYAPPA

Department of Plant Pathology, University of Agricultural Sciences,
Hebbal, Bangalore 560024, India

INTRODUCTION

Yellow mosaic disease is a serious constraint on the cultivation of horsegram (*Macrotyloma uniflorum*) in India (Williams *et al.*, 1968; Muniyappa *et al.*, 1975, 1978). The causal agent is transmitted by the whitefly *Bemisia tabaci* and is assumed to be a virus. The recent introduction of horsegram cultivars that mature in only 12 weeks (Shivashankar *et al.*, 1976) has made it possible to grow additional crops a year with sowings in late summer (April) and early kharif (May-June). This has led to the disease becoming more serious than hitherto.

MATERIALS AND METHODS

In 1978 surveys were carried out to record the incidence of horsegram yellow mosaic (HYM) in the southern districts of Karnataka State. Four or five fields were examined in each district and the proportion of infected plants determined in four or five 3 x 4 m quadrats per field.

To determine the effect of sowing date on disease incidence three horsegram cultivars, the early-maturing PLKU32 and two late-maturing cultivars Palladam and BGM, were sown fortnightly from June 1979 to May 1980. Each cultivar was sown in three 3 x 4 m plots. The incidence of symptoms of HYM was recorded weekly until just before harvest. Whitefly were collected fortnightly from June 1979 to May 1980 using a suction sampler moved over the whole experimental area for 2 min. Weather data were recorded at the Main Research Station, University of Agricultural Sciences, Hebbal, Bangalore, where all the experiments were done.

For glasshouse transmission tests the whitefly *B. tabaci* was maintained on cotton (*Gossypium hirsutum*). Both acquisition and inoculation feeds were of 24 h.

Plumb R.T. & Thresh J.M. (1983) *Plant Virus Epidemiology.*
Blackwell Scientific Publications, Oxford.

RESULTS

Surveys

Disease incidence varied with season and cultivar (Table 1). Almost all plants of the early cultivars PLKU32 and EC7460, but only 45-50% of late local cultivars, sown in May-June, became infected. Crops sown in October were less extensively infected, with 4-16% and 1-7% infection of early and late cultivars, respectively.

Table 1. Incidence of HYM in early and late horsegram cultivars in southern districts of Karnataka State during 1978.

District	Average disease incidence (%)			
	May-June sowings		October-November sowings	
	Early cultivars	Late cultivars	Early cultivars	Late cultivars
Bangalore	95.0	50.5	12.5	7.3
Chitradurga	-	-	7.5	3.5
Hassan	-	-	-	0.5
Kolar	100.0	45.4	16.5	6.0
Mandya	-	-	5.5	1.5
Mysore	-	-	4.0	3.0
Tumkur	-	-	10.5	3.2

Sowing date

In the sequential sowings, all plants were infected in plots sown between the end of January and the end of May (Fig. 1). Infection in plots sown later was progressively less, reaching a minimum of 5% in plots sown from August to October. There was little difference in disease incidence in the three cultivars tested but symptoms appeared 4-7 days sooner in the early cultivar than in the late cultivars.

Whitefly populations

The only whitefly species caught was *B. tabaci*. Adults were caught from late January to early June and few were caught later in the year (Fig. 1). There was a positive correlation between whitefly catches and the final incidence of HYM in plantings.

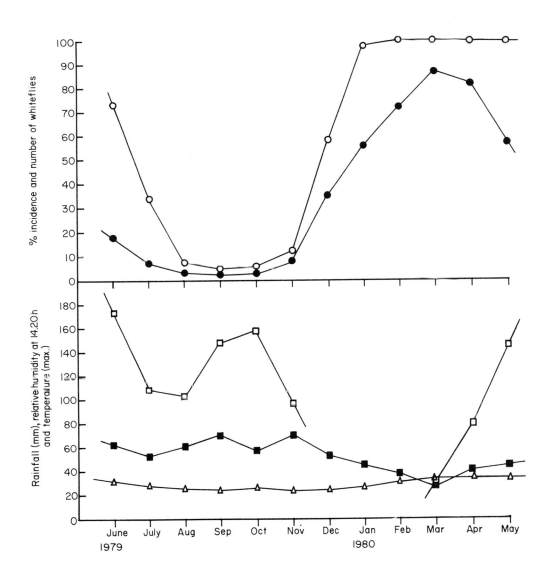

Figure 1. Incidence of HYM in PLKU32 in relation to time of sowing, whitefly population and weather conditions.
O % incidence   ● whitefly population
□ rainfall (mm)   ■ relative humidity at 14.20 h
Δ temperature (max.)

Disease spread

The spread of HYM and populations of *B. tabaci* were monitored in three crops of PLKU32 and BGM sown either in July or October 1979 or February 1980.

In the February-sown crop, HYM symptoms first appeared 15 days after sowing in the early cultivar PLKU32 (Fig. 2a) and after 20 days in the later-maturing cultivar BGM. Initially disease spread slowly but from 5 weeks after sowing incidence rapidly increased to reach 100% by the ninth week for PLKU32 (Fig. 2a) and 2 weeks later in BGM. *B. tabaci* was first caught 1 week after the crop germinated and populations increased rapidly to a maximum 7 weeks later (Fig. 2a).

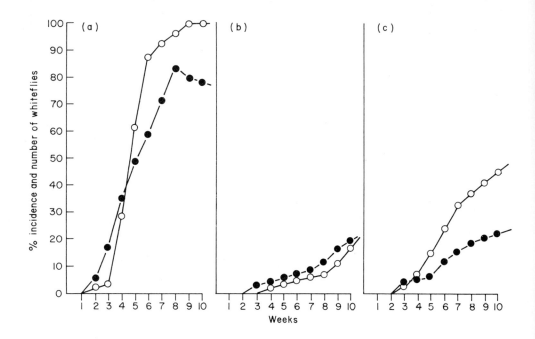

Figure 2. Spread of HYM in PLKU32 in relation to whitefly populations in different seasons during 1979/80.
O % incidence   ● whitefly population
(a) February-April crop;   (b) July-September crop;
(c) October-December crop

In the July-sown crop symptoms first appeared 15-20 days after sowing and incidence slowly increased to harvest, when 20% of PLKU32 was infected (Fig. 2b).

Disease incidence in the October-sown crop increased slowly for 6 weeks but then increased rapidly reaching 45% in PLKU32 at harvest (Fig. 2c).

Host range of HYM

Twenty-five of 40 leguminous species showed yellow mosaic symptoms when *B. tabaci* previously fed on HYM-diseased horsegram plants were allowed to feed on them for 24 h. Most important was the discovery that *M. axillare*, a wild species, is immune to HYM.

Yellow mosaic diseases were also seen occurring naturally on various legumes. *B. tabaci*, when fed on 12 of these species (*Cajanus cajan*, *Centrosema* sp., *Glycine max*, *Indigofera hirsuta*, *Phaseolus aconitifolius*, *P. lathyroides*, *P. lunatus*, *P. vulgaris*, *Teramnus uncinatus*, *Vigna mungo*, *V. radiata* and *M. uniflorum*), caused a yellow mosaic, identical to HYM, when transferred to horsegram.

Host range of B. tabaci

B. tabaci has a very wide host range (Mound, this volume) and it was recorded on 41 species from 10 plant families growing near Bangalore.

CONCLUSIONS

HYM is evidently widespread in southern Karnataka and there are numerous sources of both the causal agent and its vector. Populations of *B. tabaci* are most affected by rainfall, humidity and temperature and the worst effects of HYM can be avoided by sowing horsegram from July to October when rainfall and humidity are high and temperatures relatively low. In the future, HYM-resistant cultivars may contribute to disease control if the immunity of *M. axillare* can be transferred to the early-maturing cultivars. This will permit the successful cultivation of horsegram in seasons when whiteflies are numerous.

REFERENCES

Muniyappa V., Reddy H.R. & Mustak Ali J.M. (1978) Studies on the yellow mosaic disease of horsegram (*Dolichos biflorus*). IV. Epidemiology of the disease. *Mysore Journal of Agricultural Sciences* 12, 277-9.
Muniyappa V., Reddy H.R. & Shivashankar G. (1975) Yellow mosaic disease of *Dolichos biflorus* Linn. (Horsegram). *Current Research* 4, 176.
Shivashankar G., Chikkadevaiah, Sreekantaradhya R. & Shanta R. Hiremath (1976) Hebbal Hurali 1 and Hebbal Hurali 2, two new promising varieties of horsegram (*Macrotyloma uniflorum* L.). *Current Research* 5, 203-4.
Williams F.J., Grewal J.S. & Amin K.S. (1968) Serious and new diseases of pulse crops in India in 1966. *Plant Disease Reporter* 52, 300-4.

Epidemiology of cassava mosaic disease in Kenya

K.R. BOCK
Overseas Development Administration/Kenya Agriculture Research Institute
Crop Virology Research Project, P.O. Box 30148, Nairobi, Kenya

INTRODUCTION

Bock & Guthrie (1977) reported experiments on the rate of spread of cassava mosaic disease (CMD) into and within mosaic-free plots. The results of these and other similar experiments, the information they give on the epidemiology, and their implications for the control of CMD in Kenya are described and discussed below.

Until the recent investigations, only one experiment on CMD epidemiology had been done. Storey & Nichols (1938) studied seasonal variation in rate of spread and found that in the eastern foothills (altitude 180 m) of the Usambara Mountains of north-east Tanzania, spread was fastest from February to May, and then slowed rapidly and remained at a low level until December. Jennings (1960) suggested that this sequence corresponded with the increase and decrease of populations of the whitefly vector (*Bemisia tabaci*).

Storey (1936) observed that the incidence of CMD at Amani, Tanzania, (1000 m) was significantly less than in coastal districts and commented "it is perfectly feasible, in so far as mosaic is concerned, to establish healthy plots and to maintain them, by inspection and roguing, practically disease free". Although Storey clearly recognized the possibility of controlling CMD in this way, he did not pursue it further. Instead, effort was directed entirely towards breeding for resistance (Nichols, 1947; Jennings, 1957; Doughty, 1958).

Storey and his colleagues were well aware of the relationship between altitude, vector populations and incidence of CMD in East Africa. For example Doughty (1958) refers to the "exacting conditions at the coast" and to "up-country areas, where mosaic is less virulent".

SELECTION AND MAINTENANCE OF MOSAIC-FREE CULTIVARS

Generally the incidence of CMD is high in Kenya (> 80% in some districts) and may approach 100% in individual smallholdings, yet it seems that in

Plumb R.T. & Thresh J.M. (1983) *Plant Virus Epidemiology.*
Blackwell Scientific Publications, Oxford.

all cultivars some plants are healthy. Mosaic-free plots were readily
established by selecting and propagating from apparently healthy plants,
and subsequently roguing promptly any diseased individuals. Such plots
were on agricultural research stations and were usually several hundred
metres from mosaic-affected cassava. Initial selection of healthy
material and all subsequent propagation of mosaic-free plants was done
at the beginning of the rains, during April and May.

GENETIC BACKGROUND OF CULTIVARS

Local and imported true cassavas (Manihot esculenta)

In western Kenya and in coastal districts collections were made of
several local cultivars, of which Tamisi and Kibandameno are probably
the most widely grown in the two areas, respectively. Their resistance
to infection with CMD was unknown at the time of selection.

Two exotic cultivars, F279 (Java) and Aipin Valenca (Zaire),
imported early in the East African cassava breeding programme, were
included in many experiments in coastal districts. Both have been
described as "possessing some resistance" to CMD, and Aipin Valenca
has been described as "possessing a degree of resistance superior to
local cassavas" (Jennings, 1957).

East African hybrids

Two hybrid cultivars which had been selected for CMD resistance were
used extensively. Hybrid 46106/27 was derived from an inter-specific
cross between *M. esculenta* and *M. glaziovii*, backcrossed three times
to cassava. Jennings (1957) indicated that while the yields of 46106/
27 and similarly derived hybrids were in general "much higher than for
diseased local cultivars", resistance was not considered good enough
"for the very severe conditions occasionally encountered". Resistance
of 46106/27 was apparently very variable; in at least one trial, over
50% of plants developed mosaic (Jennings, 1960) and the cultivar was
classified as only "moderately resistant".

The other hybrid, 5543/156, was derived from an inter-cross between
M. glaziovii and *M. melanobasis* hybrids. Such inter-crosses were
considered the most promising material to emerge from the East African
breeding programme (Jennings, 1957).

Hybrid 5318/34 is also of interest because of the apparent effect
of CMD on its yield (see below), and was derived from an inter-cross
between two *M. glaziovii* breeding lines. It was described by Jennings
(1957) as "very highly resistant" and remained unaffected in a comparative trial where almost all plants of other hybrid lines became
diseased.

East African cassava breeders initially based selection for
resistance on the total number of months that plants of a cultivar

remained healthy, and measured both time taken for symptoms to appear and the total number of plants affected. Among other things, this would presumably be a measure of resistance to inoculation. Apparently hybrid 46106/27 was selected in this way. Later in the breeding programme, resistance was assessed on tolerance or absence of symptoms. Although hybrid 5543/156 was presumably selected in this way, it reacts severely to infection.

EPIDEMIOLOGY OF CMD

With one exception, all epidemiology experiments were planted at the beginning of the rains, in April or early May, and were stopped after 12 months. Plant spacing was always 1 x 1 m, and, in accordance with local practice, no fertilizer was applied. Incidence of CMD was recorded weekly.

Spread of CMD into plots of healthy cassava

Initially, 100 mosaic-free plants of each of two cultivars (46106/27 and F279) were planted in 10 rows of 20 plants with the cultivars alternating between rows; plants that became diseased were rogued. The results of these earlier studies (1974-76) have already been reported (Bock & Guthrie, 1977).

The scope of these simple experiments was subsequently broadened considerably to include local cultivars as they became available. The experiments thus took the form either of randomized block trials and compared both CMD susceptibility and yield of local and hybrid types, or they were designed as time-of-harvesting and rotation trials to demonstrate the practicability of detailed agronomic research in a disease-free crop. Experiments were sited on research stations where the nearest known source of infection was 100-300 m away.

The incidence of CMD in experiments between 1977 and 1981 was 0.0-2.0% (mean *c.* 0.5%) (Table 1).

In April 1978 a small plot of Brazilian cassava cultivars was established in the field at Mtwapa, after their release from quarantine. This plot was closely surrounded by multiplication plots of mosaic-free East African cassava cultivars. CMD incidence in the Brazilian cultivars was *c.* 50% (64/126) but < 1% in the adjacent plots. Healthy survivors of the Brazilian cultivars were planted in April 1979 and CMD incidence in these was again *c.* 50% (93/180) while adjacent plots of cultivars 46106/27 and 5543/156 were unaffected (0/360).

Spread of CMD within healthy cassava

In each of the trials from 1973 to 1976, seven mosaic-affected cuttings were surrounded by a total of 156 mosaic-free plants arranged in concentric hexagons; plants which became diseased were recorded but not rogued.

Table 1. Spread of CMD into mosaic-free cassava plots during
12-month crop cycles, 1974-1981 (Bock & Guthrie, in preparation).

Year	Site	Cultivars*	No. plants exposed	% Incidence
	W. Kenya			
1977/78	Alupe	1,2,3,4	1650	0.6
1977/78	Alupe	4	2770	0.3
	Coast			
1977/78	Mtwapa	14,15	180	1.1
1977/78	Mtwapa	14,16	200	0.0
1977/78	Mtwapa	13,16,17	1023	2.0
1977/78	Mtwapa	17	2592	0.3
1978/79	Mtwapa	14,15	320	0.6
1978/79	Magarini	16,17	1531	0.4
1978/79	Mtwapa	17	2592	0.0
1978/79	Mtwapa	13,16,17	768	1.2
1979/80	Mtwapa	5,6,13,15,16,17	896	0.0
1979/80	Mtwapa	17	2592	0.0
1979/80	Msabaha	6,13,15,16,17	761	0.7
1979/80	Msabaha	17	1728	0.3
1980/81	Msabaha	17	1820	0.4
1980/81	Mtwapa	5,7,8,13,14,16,17	412	0.0
1980/81	Mtwapa	5,6,7,8,11,12,13,14,16,17	2048	0.4
1980/81	Msabaha	5,6,7,8,9,10,11,12,13,14,16,17	2172	0.3

$$\bar{x} = 0.48$$

* Local cultivars: Sifwembe (1), Dodo (2), Serere (3), Tamisi (4),
Kibandameno (5), Mwakazanza (6), Kasimbiji (7),
Chokorokote (8), Nusu Rupia (9), Shelisheli (10),
Gari Moshi (11), Mpira (12)

Imported cultivars: Aipin Valenca (13), F279 (14), C756B (15)

Hybrids: 46106/27 (16), 5543/156 (17)

The 1980/81 experiment was designed to study directional spread
within square, 100-plant plots from scattered and point sources of
infection amounting to 1, 2, 4 and 8% of the population; when diseased,
plants were or were not rogued.

Table 2. Incidence of CMD after one year in plots planted in Kenya with different proportions of diseased plants placed centrally or scattered throughout the plot.

Year	Region	Mosaic-free cultivar	Planted level of infection (%)	No. plants exposed	% Incidence	Rainfall (mm)
1973/74*	Coast	46106/27	4	156	53.8	1261
1974/75	Coast	46106/27	4	156	1.3	683
1975/76	Coast	46106/27	4	135	10.4	1129
1976/77	Coast	46106/27	4	156	7.7	937
1976/77	West	Tamisi	4	156	0.0	1606
1976/77	West	Sifwembe	4	139	4.3	
1980/81	Coast	46106/27	8	173	1.2	1105
1980/81	Coast	46106/27	4	355	0.8	
1980/81	Coast	46106/27	2	198	0.0	
1980/81	Coast	46106/27	1	191	0.5	
1980/81	Coast	46106/27	0	190	0.0	
1980/81	Coast	5543/156†	0	1536	1.2	

$\bar{x}$ = 2.49¶

*   1973/77 results from Bock & Guthrie, 1977
†   Guard lines between plots of the 1980 experiment; mean incidence in both varieties was 25/2633 (0.9%)
¶   All experiments were planted in April except the 1973/74 experiment which was planted in November.

The results of these experiments are given in Table 2. With the exception of the plot planted in 1973, incidence of disease was low to very low, and infection appeared to be entirely at random with no evidence for directional spread, or for spread between neighbouring plants. The 1973 experiment, unlike all others, was planted in November and monitored for 15 months. Here, 20 of the 84 plants that became infected showed symptoms after 12 months and 53 of the 84 were in the northern half of the experiment, suggesting directional spread. There was no evidence of a gradient away from the source of infection. The results of this experiment are difficult to explain, particularly as the greatest number of new CMD infections (30 plants out of 84) was recorded during August-October, which was previously regarded as the period of least spread (Storey & Nichols, 1938).

Spread of CMD into large blocks of mosaic-free cassava

Before starting a mosaic-free cassava multiplication scheme, the incidence of CMD in comparatively large blocks (0.2-1.0 ha) of mosaic-free cassava was monitored during the 1977-1980 seasons. The mean incidence was at the very low level of 0.15% (range 0.0-0.6%) (Table 3).

Table 3. Incidence of CMD in initially mosaic-free blocks after 12 months, Coast Province, 1977-1980.

Site	Year	Cultivar	Area (ha)	% Incidence
Mtwapa	1977/78	5543/156	0.50	0.00
	1978/79	5543/156	0.50	0.00
	1978/79	46106/27	0.20	0.00
Msabaha	1977/78	46106/27	0.25	0.60
	1978/79	46106/27	0.50	0.40
	1978/79	5543/156	0.35	0.20
	1979/80	46106/27	0.94	0.02
	1979/80	5543/156	0.25	0.00

$\bar{x} = 0.15$

Spread of CMD into plots sited in farmers' fields

It is obviously important to study the rate of re-infection of mosaic-free cassava in farmers' fields as it enables an informed estimate to be made of the area of mosaic-free stock necessary to satisfy the requirements of the region. Six farms were selected and 100 mosaic-free cuttings of each of the two hybrid cultivars 46106/27 and 5543/156 were planted by the farmer at a site selected and maintained by him. At Gede and Takaungu the elite, mosaic-free cassava was planted adjacent to mosaic-affected local cultivars, while at Tezo there were scattered CMD-affected plants within 20 m of the plot. Plots were inspected every 1-2 months and mosaic-affected plants rogued. Because of erratic rainfall at the start of these experiments, establishment was variable. Table 4 summarizes the incidence of mosaic in the 6 plots during the 12-month growing cycle in 1980/81. It is intended that the farmers will propagate from existing plants for the following year's plot and CMD will be monitored in successive years to estimate rates of re-infection. Meanwhile it is obvious that there was no significant infection in the first year of exposure, even where abundant sources of infection were nearby.

COMPARATIVE YIELDS OF LOCAL AND HYBRID CASSAVAS

When the experimental yields of local, imported and hybrid cassavas

Table 4. Incidence of CMD after one year in farmers' fields,
Coast Province, April 1980 - March 1981.

Site	No. plants exposed	% Incidence
Gede	200	0.0
Chumani	146	0.7
Tezo	159	0.6
Mbuyuni	193	0.0
Mavueni	175	0.0
Takaungu	157	0.6
Total	1030	$\bar{x} = 0.3$

obtained from five experiments at two sites in the 1979/80 and 1980/81
seasons were compared, the best seven cultivars and their yields (t/ha)
meaned over sites and years were: Kibandameno, 32; Aipin Valenca, 29;
Mwakazanga, 27; Kasimbiji, 26; 5543/156, 24; F279, 23; and 46106/27, 23.
It seems, therefore, that there is little incentive for the farmer to
use improved cultivars, particularly because, when affected, yield
losses in the two types (hybrids and true cassavas) are similar after
infection.

EFFECT OF CMD ON YIELD

Estimates of the effect of CMD on yield were made by comparing the
weight of tubers from healthy plants harvested after the standard 12-
month growing period with that of mosaic-affected plants derived from
mosaic-infected cuttings. The experimental design was either single-
line plots or conventional randomized blocks; all data were analysed
statistically and all comparisons were highly significant (Table 5). Of
particular interest is the reduction in yield caused by the disease in
the two hybrid cultivars and the greater loss shown by the cultivar
Serere for 1977/78 compared with 1976/77.

DISCUSSION

Clearly, in Kenya, spread of CMD into mosaic-free plots of cassava is
usually very slow. Spread from affected to healthy plants within plots
is also slow. The data indicate that the rate of re-infection of mosaic-
free plots is < 2%, and frequently < 1%, in a 12-month growing period,
irrespective of plot size (0.02-1.0 ha), location (on agricultural
research stations or in farmers' fields), and annual or regional climatic
differences (coast and western Kenya, 1974-80). The results are consistent with the suggestion that man is the principal vector of CMD in

Table 5. Effect of CMD on yield of local, imported and hybrid cultivars.

Region	Cultivar	Tuber yield (kg/plant) Healthy	CMD-affected	% Reduction
W. Kenya	Sifwembe* (1976/77)	5.10	1.68	67
	Sifwembe* (1977/78)	3.07	1.15	63
	Tamisi* (1976/77)	3.24	0.94	71
	Tamisi* (1977/78)	3.58	1.41	60
	Serere* (1976/77)	2.50	1.40	44
	Serere* (1977/78)	3.42	1.27	63
	Dodo* (1977/78)	3.46	1.09	68
Coast	Kibandameno*	3.95	0.81	79
	F279†	3.67	0.52	86
	46106/27¶	3.86	1.19	69
	5318/34¶	2.68	0.59	78

$$\bar{x} = 68$$

* local cultivars † imported true cassava clone also
   totally infected with cassava brown streak virus ¶ hybrids

Kenya through the inadvertent use of infected cuttings as planting material. Farmers do not discriminate between affected and healthy plants, which is hardly surprising where 80% or more plants are affected.

No obvious pattern of spread was discerned in any of the experiments done after 1973, and there was no evidence of spread from affected to adjacent healthy plants. Affected plants appeared to occur at random and the evidence suggests that, for crops planted in April, the probabilities of infection being derived from external sources (mean incidence 0.5%, Table 1) and from internal sources (mean incidence 2.5%, Table 2) do not differ greatly.

The extreme susceptibility to inoculation shown by the Brazilian cassavas was predictable as CMD does not occur in South or Central America and there is no "selection pressure" for resistance. Incidence in these cultivars indicates the infection pressure at a more or less isolated site in two successive seasons. Local and hybrid cassavas, presumably subjected to similar infection pressures, either remain healthy, or only a few plants become infected (mean incidence 0.5%, Table 1). All the cultivars used in the experiments must therefore be considered resistant to infection and able to withstand the infection pressures found locally in Kenya.

The two hybrids used in the experiments were selected for resistance to CMD, and Jennings considered the two exotic cassavas to have at least "some resistance". However, the cultivar 46106/27 was described as only "moderately resistant", and apparently the use of the cultivars Aipin Valenca and F279 on their own was not considered because of their susceptibility. Although there was evidence for variation in susceptibility among local and imported cultivars, it seems they were all classed as "susceptible" (Nichols, 1947; Jennings, 1957).

The locally adapted cultivars appear to be no more susceptible to inoculation under natural conditions than the hybrids which were selected for resistance. Neither group is tolerant of infection, as both groups react more or less similarly and severely to CMD infection when inoculated by grafting or by whiteflies. Bock & Guthrie (1977) suggested that three factors might contribute to the slow rate of spread of CMD to healthy plants in Kenya: comparatively inefficient vector transmission, seasonally low population densities of the vector, and cassava growth patterns. Resistance to inoculation seems to be an additional factor.

Whatever the relative importance of these factors, it is certain that CMD can be largely avoided in Kenya by the use of mosaic-free propagation material, planted in April. Moreover, it would seem advantageous that cultivars used in mosaic-free multiplication schemes should be sensitive to CMD and react rapidly to produce clear and unequivocal symptoms when affected as this would facilitate effective roguing.

This approach seems to be diametrically opposed to that of plant breeders, who tend to equate resistance with tolerance or absence of symptoms and select accordingly. Thus, Jennings (1960) states that "varieties are described as resistant if they do not readily show disease symptoms when exposed to infection". In the East African cassava breeding programme an elaborate scoring system was devised to estimate mosaic "resistance" and this involved assessing symptom intensity (Jennings, 1957). Selection for resistance was thus deliberately based on lack of symptoms, it being assumed that, with resistance, "symptoms when present, are invariably of such low intensity as to have no influence on yield" (Jennings, 1957). Curiously, no comparative data on yields of infected and healthy plants of the same "resistant" (= tolerant) cultivar were reported during the 21 years of the East African breeding programme. It seems that only recently has attention been paid to yield in studies at the International Institute of Tropical Agriculture on resistance to CMD (Terry & Hahn, 1980).

The hybrid 5318/34 was classed as "very highly resistant" but it seems that while it may be resistant to infection and may not develop conspicuous symptoms, it is not tolerant if this is measured by yield (Table 5). The yield loss of 78% in plants derived from infected cuttings is of the same order as that of the cultivar 46106/27 (69%) and of several local cultivars (67-79%). Unlike the improved cultivar studied by Terry & Hahn (1980), there is no evidence of significant tolerance, as measured by yield, in any of the cultivars tested in

Kenya and they must be considered sensitive. The data suggest therefore
that symptom expression is possibly an unreliable measure of the effect
of CMD on yield. Moreover the variation between years in yield loss of
the cultivar Serere (Table 5) suggests studies on tolerance of CMD
should extend over at least two and preferably three years.

Experience in both coastal and upland cassava areas of Kenya therefore indicates that control of CMD is readily achieved with mosaic-free
planting material of cultivars which possess some degree of resistance
to inoculation. While the degree of this resistance is difficult to
define, it is quite clear that it need only be "moderate".

As well as possessing some resistance to inoculation, recommended
cultivars should also react unequivocally when infected to facilitate
roguing, not only in official virus-free multiplication plots but also
by farmers. Cultivars bred and selected for suppression or absence of
leaf symptoms (= tolerance of infection) would not be appropriate in
virus-free propagation schemes. Terry & Hahn (1980) do not indicate
whether the tolerant cultivar TMS 30395 reacts severely when affected,
or whether leaf symptoms are inconspicuous. Whatever the situation
with this cultivar, the yield reduction of 32% in affected plants
suggests that the use of mosaic-free planting material would in any
event prove highly advantageous.

It seems that breeding for a high degree of tolerance of infection
by CMD is a strategy which is neither relevant nor appropriate to
conditions in Kenya, although it may be appropriate elsewhere. The
results from epidemiological studies of this disease have also shown
how desirable such studies are, especially before embarking on lengthy
and often costly breeding programmes.

REFERENCES

Bock K.R. & Guthrie E.J. (1977) African mosaic disease in Kenya.
    Proceedings Cassava Protection Workshop, CIAT, Cali, Colombia,
    7-12 November 1977 (Ed. by Brekelbaum, Bellotti & Lozano), pp. 41-4.
    Series CE-14, May 1978.
Doughty L.R. (1958) Cassava breeding for resistance to mosaic and
    brown streak viruses. *Annual Report, East African Agriculture and*
    Forestry Research Organisation, pp. 48-51.
Jennings D.L. (1957) Further studies in breeding cassava for virus
    resistance. *East African Agricultural Journal* 22, 213-9.
Jennings D.L. (1960) Observations on virus diseases of cassava in
    resistant and susceptible varieties. 1. Mosaic disease. *Empire*
    Journal of Experimental Agriculture 28, 23-34.
Nichols R.F.W. (1947) Breeding cassava for virus resistance. *East*
    African Agricultural Journal 12, 184-94.
Storey H.H. (1936) Virus diseases of East African plants. VI. A
    progress report on studies of the diseases of cassava. *East*
    African Agricultural Journal 2, 34-9.
Storey H.H. & Nichols R.F.W. (1938) Virus diseases of East African

plants. VII. A field experiment in the transmission of cassava
mosaic. *East African Agricultural Journal* 3, 446-9.
Terry E.R. & Hahn S.K. (1980) The effect of cassava mosaic disease on
growth and yield of a local and an improved variety of cassava.
Tropical Pest Management 26, 34-7.

Plant virus epidemiology and control: current trends and future prospects

J.M. THRESH

East Malling Research Station, Maidstone, Kent, UK

INTRODUCTION

The contributors to this volume provide a representative assessment of current attitudes and approaches to plant virus epidemiology. They also illustrate the continuing influence of the eminent pioneers who established the subject and showed that there is seldom any adequate substitute for simple, meticulous, long-term field observation and recording to determine the main features of epidemics. Nevertheless, as Harrison (this volume) points out it is disappointing that there have been few innovations or developments of the type that have so transformed aetiology and work on the physico-chemical properties of viruses.

"Molecular" and "taxonomic" aspects of virology now tend to predominate and short-term laboratory studies have obvious advantages for post-graduate students and their supervisors compared with the vagaries of field work which is notoriously unpredictable and demanding in time, land and experienced support staff. Moreover, for field projects to be effective the close collaboration of vector entomologists and others in multidisciplinary teams is needed and such teams are difficult and expensive to maintain.

These considerations largely explain why plant virus epidemiology is currently neglected and why only limited and slow progress has been made towards effective virus control. A change of attitudes is required if virologists are to respond satisfactorily and provide adequate solutions to the increasing problems of virus disease in many crops in many regions of the world. Mycologists have already made this change and work on the epidemiology of fungal diseases has entered a period of intense activity. There is close collaboration between mycologists, plant breeders, population geneticists, fungicide chemists, physicists, meteorologists, statisticians and modellers and progress has been rapid. Plant virus epidemiology *could* proceed equally rapidly *if* sufficient personnel and resources were made available. Some of the main topics that merit study are discussed in the following sections.

Plumb R.T. & Thresh J.M. (1983) *Plant Virus Epidemiology.*
Blackwell Scientific Publications, Oxford.

NEW METHODS OF VIRUS DETECTION

The development of new methods of virus detection and their influence
on epidemiology is impossible to predict but is likely to be considerable. This is evident from recent experience with enzyme-linked immunosorbent assay (ELISA) which was not used for plant virus detection until
1976 (Bar-Joseph & Garnsey, 1981). ELISA has since been used widely:

1. to follow the pattern and sequence of spread of viruses that
    do not cause diagnostic symptoms,
2. to improve the effectiveness of eradication measures against
    citrus tristeza, plum pox and other diseases,
3. to survey weed and wild hosts of viruses,
4. to facilitate the production of virus-free stocks of seed and
    vegetative planting material,
5. to detect viruses in vectors, although the sensitivity
    required is even greater than that needed to detect viruses
    in plants.

These uses are likely to continue and be extended as further refine-
ments are introduced that increase the sensitivity and versatility of
ELISA for detecting viruses in plants and vectors (Clark, 1981).
Another possibility is that "fingerprinting" oligonucleotides in
ribonuclease digests of virus preparations will provide a very precise
method of typing virus strains. This method has been used already to
provide important information on the origin of recent outbreaks of foot
and mouth disease of livestock in western Europe and the technique has
far wider applicability (King *et al.*, 1981). Many other innovations
are possible provided that there is adequate collaboration between
laboratory and field workers in exploiting new developments.

DISEASE ASSESSMENT AND REMOTE SENSING

Accurate information is required on the losses due to viruses to determine the need for control and to analyse the cost/benefit ratio for the
available options for control. It is easier to assess the effects of
viruses on the growth and yield of plants that are inoculated or
infected naturally in field trials than to determine the incidence of
viruses in a representative range of commercial plantings. One of the
main problems is that it is difficult and expensive to do comprehensive
surveys as they require numerous trained personnel and adequate transport for long periods.

In Britain, field officers employed by the British Sugar Corporation
have for *c.* 40 years made routine assessments, throughout the growing
season, of the incidence of beet yellowing viruses (Watson *et al.*,
1975). Regular surveys are also made of the incidence of sugarcane
Fiji disease in Queensland (Egan & Hall, this volume) and of virus and
virus-like diseases of rice in Japan (Kiritani, this volume). Few
other crops have received such detailed attention and there is an
almost total lack of information on the prevalence of viruses in many
important crops. There is obviously a need for quicker and more

efficient methods of disease assessment that will also provide information on the pattern and sequence of spread and lead to more effective
control.

One possibility is to exploit the great advances being made in the
collection and interpretation of data obtained using remote sensing
techniques.   In recent years piloted and remotely controlled aircraft,
helicopters, satellites, kites, balloons and rockets have all been used
in a wide range of studies for military and civilian purposes. Some of
the information obtained is secret, but it seems clear that little
attention has been given to plant diseases compared with the work on
land utilisation, pollution and mineral resources. Pathologists have
seldom had access to the special equipment required and it is expensive to mount a large programme of aerial surveillance linked with
the necessary ground observations.

Despite these difficulties some progress has been made and in
Britain the work of the Ministry of Agriculture Fisheries and Food
Aerial Photography Unit has contributed to the increased awareness of
the losses caused by barley yellow dwarf virus in autumn-sown cereals
(Hooper, 1978). Aerial surveys also facilitated rapid assessments of
the distribution of barley yellow mosaic virus after it was discovered
in Britain for the first time in 1980. Elsewhere, aerial photography
has been used to locate citrus orchards and diseased trees in the
Israeli tristeza eradication programme (Bar-Joseph et al., this volume).
Citrus and various other crops have also been photographed from aircraft in the United States following the pioneer work of Colwell (1956)
on barley yellow dwarf and other diseases of cereals. These studies
have included detailed analyses of the spectral reflectance patterns
from healthy and diseased foliage in attempts to develop automatic
methods of data collection and processing for disease assessment (Toler
et al., 1981). This work has shown the value of remote sensing
techniques in plant virus epidemiology and the need for further studies
on methods of realizing their full potential.

MATHEMATICAL TECHNIQUES AND MODELLING

The publication of Plant Diseases: Epidemics and Control (Vanderplank,
1963) was an event of seminal importance in the development of epidemiology as a quantitative science. Data on the spread and control of
fungal diseases have since been subject to increasingly sophisticated
mathematical analyses and the development of models and computer
simulations has given new insights into the dynamics of disease progress. The influence of Vanderplank has been less evident in publications on virus diseases. With a few notable exceptions virologists are
not very numerate or mathematically minded and only some field data
have been subject to the rigorous scrutiny and analysis adopted in
early studies on tomato spotted wilt virus and other vector-borne
viruses in Australia (Bald, 1937; Bald & Norris, 1943). Subsequent
work on sugarbeet yellowing viruses in Britain and rice dwarf virus in
Japan are further examples of what can be achieved and the progress in

disease forecasting that computer analysis makes possible (Watson et al., 1975; Kiritani & Sasaba, 1978).

There is increasing interest in the potential of mathematical techniques and in the use of modern methods of data collection and processing. For example, mathematical approaches to three completely different epidemiological problems are discussed in this volume (Allen; Gibbs; Madden et al.). Elsewhere, Frazer (1977) has presented a computer simulation of the spread of alfalfa mosaic virus with time and Gutierrez et al. (1974) modelled the spread of subterranean clover stunt virus and its aphid vector (Aphis craccivora) in Australia. Mathematical techniques have also been used to assess dispersal gradients of insects (Taylor, 1978) and in a detailed cost/benefit analysis of the effectiveness and consequences of the eradication measures used to control citrus tristeza disease in Israel (S. Fishman et al., personal communication).

These examples illustrate the wide range of problems that can be tackled with the mathematical and computing techniques now available or being developed. There is obviously the potential for great advances in current methods of forecasting and assessing spatial and temporal patterns of virus spread. This is likely to bring greater understanding of the dynamics of disease development and to improve the effectiveness of control measures.

EPIDEMIOLOGICAL CONCEPTS

Southwood (1981) writing on the history of ecology commented that "there comes a time in the development of most phenomenological sciences when the apparent uniqueness of everything is seen as but a component of pattern". There are signs that this critical stage is approaching or has been reached in plant virus epidemiology. The steady rate of accumulation of experience and data since the early years of the century has recently increased. There is now more effort on more crops and in more geographic areas than ever before and this has made it possible to develop unifying concepts of general validity.

Some of the terms used are derived from ecology; hence cucumber mosaic, sugarbeet curly top, tomato spotted wilt, tobacco rattle and other viruses with a very wide host range can be regarded as "generalist" pathogens compared with "specialists" such as strawberry crinkle and wheat soil-borne mosaic viruses that naturally infect a restricted range of closely related species. A further distinction has been made between "CULPAD" viruses that seem particularly well adapted for survival in crops and "WILPAD" viruses that are better adapted for survival in wild plants and seem to infect crops incidentally (Harrison, 1981).

The behaviour of viruses or the diseases they cause has also been considered recently in relation to the distinction made by ecologists between "opportunist" or "r" -selected species and "equilibrium" or

"*K*" -selected species (Thresh, 1980; Harrison, 1981; Gibbs, 1982). Opportunist species multiply rapidly and readily invade and exploit new sites. They have great resilience and recover quickly from drastic decreases in their numbers of the type encountered frequently in characteristic "boom and bust" cycles of population growth and decline in transient habitats. By contrast, equilibrium species tend to be long-lived, slow to reproduce, large, and of limited mobility. They frequent durable environments and usually maintain extremely stable populations that are close to the carrying capacity of the environment. There are obvious analogies in the behaviour of viruses, some of which spread rapidly to cause serious epidemics of short-lived annual crops, whereas others are much less mobile and spread slowly in long-lived perennials.

These concepts give new insights into the behaviour of viruses and it is now clear that their survival strategies are as diverse as their size, shape and physico-chemical characteristics. The different methods of perennation and spread can be considered, in ecological terms, as means of invading new habitats and of exploiting those already colonized. At one extreme are the nematode or fungal-borne viruses that do not readily invade new sites, but persist for long periods and reappear in successive plantings of susceptible crops. Such behaviour contrasts with that of viruses with active, arthropod vectors that repeatedly search out and colonize new plantings.

Studies of the pattern and sequence of virus spread have shown considerable differences between the different groups of arthropod-borne viruses. These differences can be related to their mode of transmission and to the behaviour of their vectors. However, in evaluating the available results it is important to appreciate that a disproportionate amount of research has been on aphid vectors and particularly on a few well known species such as *Myzus persicae* (Eastop, this volume). Other aphid species are now receiving more attention and some that are inefficient vectors of non-persistently transmitted viruses have been shown to be of considerable importance because they fly in very large numbers (Madden *et al.*; Raccah, this volume). Increased work is also being done on other types of vector, especially whiteflies that seem to have become increasingly important in recent years in many regions and especially in the tropics and sub-tropics (Mound, this volume). This increased effort should lead to a better appreciation of the main features of spread by the different types of vector and so provide a rational basis for devising the most appropriate control measures.

CROP HUSBANDRY AND VIRUS SPREAD

There has been a growing realization that advances in crop protection have not always kept pace with progress in other aspects of agriculture. Pathologists have responded in various ways and an important development has been the attempt to explain the underlying causes of recent major epidemics. Considerable attention has been given to crop husbandry practices and ways in which they affect the spread of pests and pathogens, including viruses and their vectors (Thresh, 1981a, 1982;

Bos, this volume). The frequency with which damaging epidemics occur in crops compared with natural vegetation has been stressed by various workers (e.g. Browning, 1974, 1981). This has led to the view that epidemics are largely man-made phenomena, caused by the disruption of the equilibria that develop in nature when pathogens coevolve with their hosts (Harlan, 1976).

Mycologists in Israel and elsewhere have become aware of the benefits to be gained from studying the coevolution of pathogens with crop species and their close relatives at centres of diversity (Browning, 1974, 1981). These studies have located new sources of resistance and shown the complexity of the various resistance mechanisms operating in natural stands compared with crops. Virologists have begun to exploit this approach and Jones (1981) has compared the wide range of viruses and virus strains that infect potato and related species in the Andean centre of diversity with the far simpler situation in the main potato-growing areas of North America and Europe. New sources of resistance and the potential threat posed by viruses or virus strains not yet established outside the Andean region have been revealed. Similar studies are in progress at the Mexican centre of origin on leafhopper-borne viruses and other pathogens of maize and their vectors (Nault, this volume). There are likely to be substantial benefits from an extension of such studies to other crop species and other regions such as *Citrus* spp. in the Orient and *Phaseolus* spp. in Central America (Bar-Joseph *et al.*; Gámez & Moreno, this volume).

It should also be profitable to extend the work initiated by mycolo-gists on the differences between crop stands and natural vegetation and the implications of planting in dense arrays. The latter often involves planting a single genotype over large areas, with little or no interval between successive crops. Such practices greatly aid the spread of fungal pathogens, but the effects are complex and not always predict-able. A similar situation is emerging in comparable work on the spread of viruses and their vectors. For example, some aphid vectors when they invade crops tend to alight and accumulate on the outer rows of plantings and in the lee of hedges or other barriers. With other species the effect is less pronounced (Dean & Luuring, 1970; Plumb, this volume) and there is an equally varied response to crop spacing. *Aphis craccivora* and *M. persicae* alight preferentially on widely spaced plants that stand out clearly against a background of bare soil but not all species behave in this way (Halbert & Irwin, 1981). These findings indicate the complex implications of changes in crop spacing, weed control and thinning practices and in field size, shape, disposition and boundary (Thresh, 1982). For some crops, sowing dates or cultural practices can be changed to make plantings less vulnerable to incoming vectors and so decrease virus spread as discussed by Plumb (this volume) in relation to barley yellow dwarf. Similarly growers of sugarbeet and groundnut are advised to achieve a continuous canopy of foliage as early as possible to decrease the losses due to viruses with leafhopper, aphid or thrips vectors (Heathcote, 1972; Amin & Reddy, this volume). Elsewhere growers of early-sown vegetables who use plastic mulches to warm the soil and conserve moisture can use a colour that will repel

incoming whitefly or aphid vectors (Lecoq & Pitrat, this volume). There
is undoubtedly much more that can be achieved by further work on these
lines, but the development of a fully integrated approach to pest and
disease management will require the close collaboration of virologists,
entomologists, mycologists and agronomists.

A multidisciplinary approach is also required for work on mixed
cropping systems, using two or more species planted together or in overlapping sequence. Such systems are traditional in many regions including large areas of Africa, South and Central America (Gámez & Moreno,
this volume). However, they have, until quite recently, been little
studied by agricultural scientists who have been mainly concerned with
promoting monocropping. It is now realized that mixed cropping can be
an effective means of land use and gives stable yields that do not
necessarily depend on expensive artificial fertilizers and pesticides.

Part of the advantage of mixed cropping systems over monocropping is
their relative freedom from pests and diseases and this is now being
investigated in studies of the type described by Gámez & Moreno (this
volume). The advantages of monocropping in modern agriculture are so
great that they cannot readily be discarded. Nevertheless, it may be
possible to introduce at least some diversity that will decrease the
vulnerability of monocultures and enable them to acquire some of the
advantageous features of mixed stands and natural vegetation.

RESISTANCE BREEDING

Breeding for resistance to viruses or their vectors has already been
used successfully with several important crops and the approach is
likely to be used increasingly in the future (Russell, 1978). However,
any further work will require greater involvement from virologists,
and better collaboration than hitherto with breeders, if resistant
cultivars are to be produced and deployed most effectively. The inte-
gration of resistant cultivars with other measures, as for the control
of beet curly top virus in California (Duffus, this volume) and
cucumber mosaic virus in southern France (Lecoq & Pitrat, this volume)
is likely to be advantageous. Thus it should be possible to decrease
the "infection pressure" on plantings and the likelihood of the emergence of resistance-breaking strains of virus or vector.

Only limited attention has been given to the epidemiological consequences of using resistant cultivars and to methods of utilizing them
most advantageously. Much of the work has been done by breeders who
have adopted the pragmatic approach of selecting plants that yield well
after exposure to infection by virus strains that may be mixed or of
uncertain provenance. It is not always known whether the plants
selected are very tolerant of infection or difficult to infect. This
distinction is of great epidemiological importance yet is seldom made.
There is ambiguity in many publications; cultivars are often referred
to simply as "resistant" with no indication whether they support virus
multiplication and become infected systemically.

The situation is further complicated by the continuing inconsistencies and lack of agreement on resistance terminology. There are large, and possibly insuperable, difficulties in devising a uniform system for use by all plant pathologists and it will be even more difficult to obtain the agreement of plant breeders and those working on pathogens of animals. Nevertheless, some progress is possible and a considerable advance would be the general adoption of the terms *insensitive* or *tolerant* for plants that continue to grow and yield satisfactorily when they are infected systemically. Many virologists and breeders seem reluctant to use these terms, yet tolerance is the main feature of many "resistant" cultivars (Posnette, 1969; Buddenhagen, this volume).

The ability of tolerant cultivars to support at least some virus multiplication may, in one sense, be advantageous because their use seems less likely to lead to the appearance of resistance-breaking strains than the introduction of cultivars that are immune or not readily infected. This may explain why resistance-breaking strains of virus have been encountered less frequently than in breeding for resistance to fungal pathogens for which far greater use is made of genes conferring hypersensitivity (Russell, 1978). Another possibility is that viral genomes are so small that the range of biologically effective variants is limited and a change towards increased virulence could be associated with deleterious effects on their epidemiological competence (Harrison, 1981).

Far more evidence is required on the durability of the different types of resistance, on possible gene-for-gene relationships, on the relative merits of major and minor genes and on the possible advantages of using mixtures of different genotypes. At least some of this information may soon be forthcoming because of the increased attention being given to breeding for some form of resistance at many establishments and especially at the international research institutes.

The international institutes are likely to become increasingly important in agricultural development because they have responsibility for research on the main food crops in some of the most densely populated regions of the world (Bos, this volume). One of their principal research objectives is the development of high-yielding cultivars for use over very large areas in many different countries. This involves establishing and distributing large germplasm collections, and the increasing traffic in seeds and vegetative propagules necessitates strict quarantine. Moreover, potential cultivars must be tested adequately, and in different areas, before they are generally released. Otherwise there is the risk that the new cultivars will soon succumb to pathogens that previously caused little damage to the traditional, locally adapted cultivars being displaced.

The serious epidemics of rice tungro disease that occurred in the Far East in the 1960s and 1970s, soon after the release of the first high-yielding cultivars, illustrate the problems that can occur (Thresh, 1981; Ling, this volume). Some of the new cultivars were more sensitive to tungro than those grown previously and they were also exposed to far

greater "infection pressure" because of the more intensive cropping
practices adopted. There are many other examples of serious epidemics
of virus that have followed the introduction of new cultivars (Thresh,
1981a, 1982) and some of these are discussed elsewhere in this volume
(Autrey & Ricaud; Bos; Egan & Hall; Kiritani).

Comparable experience with fungal diseases has led to the concept of
"genetic vulnerability" (Day, 1977) and breeders are now well aware of
the need for adequate testing and of the danger of relying on a very
restricted range of genetic material. The current aim is to produce
high-yielding cultivars with at least some of the resilience and
diversity of the traditional land-races. Thus it should be possible to
exploit the great potential of modern plant breeding techniques without
creating, or aggravating, pest and disease problems.

CHEMICAL CONTROL

The introduction of eradicant and systemic fungicides and new groups of
compounds has transformed control of fungal diseases. By contrast,
chemicals still have only a limited role in controlling viruses,
although they are used successfully against some insect, mite, nematode
and fungal vectors. Despite the prolonged and continuing search for
virus chemo-therapeutants and the emergence of several potentially useful compounds from glasshouse screening programmes none has been introduced commercially and there appears to be no immediate prospect of
their use. Nevertheless, it is likely that chemicals will eventually
be developed that restrict virus multiplication and/or symptom
expression. It is certainly not necessary to search exclusively for
materials toxic to vectors, as shown by the effectiveness of foliar
sprays of mineral oils in restricting the spread of non-persistently
transmitted and some other aphid-borne viruses (Simons, 1981).
Problems of phytotoxicity have been overcome and oils are now being
used extensively in Florida, Israel and elsewhere without the environmental hazards associated with the frequent use of pesticides. The use
of oils is likely to increase still further because non-persistent
viruses cause serious problems in many countries and are difficult to
control with insecticides.

This emphasises the importance of recent glasshouse results which
showed that the quick-acting, synthetic pyrethroid insecticide deltamethrin decreased the transmission of potato virus Y to and from
treated plants (Gibson et al., 1982a). Synthetic pyrethroids have also
been more effective than insecticides that were more toxic to aphids in
restricting the spread of beet yellows and other aphid-borne viruses
in field trials (Highwood, 1979; Barrett et al., 1981).

There are great opportunities for influencing the behaviour of
virus vectors by using repellents, synthetic pheromones or chemicals
that influence the attractiveness of crops and the alighting or feeding
response of vectors. Promising results have been obtained recently in
glasshouse tests on the acquisition of potato virus Y and other aphid-

borne viruses from plants treated with the non-toxic insect repellent/
anti-feedant polygodial (Gibson *et al.*, 1982b). Such materials merit
further study in field trials, although there will be formidable
problems of experimental design in avoiding interference between treatments and in determining their epidemiological effects.

Integrating the use of chemicals with other control measures may have
the advantage of decreasing the likelihood of breakdown due to the
emergence of insensitive strains of the vector. Accurate forecasts of
disease are also required so that chemicals are used only when necessary
and at the optimum time. Hence the importance of work on the infectivity of migrating vectors and the timing, magnitude, distance and
direction of dispersal (Ling; Plumb; Raccah; Rosenberg & Magor; Taylor,
this volume).

CONCLUSIONS

Projects of the type outlined above will require a large commitment of
manpower and resources for many years. The work can be justified on
economic, social, political or humanitarian grounds because of the
urgent need to increase food production. Current shortages and
inadequacies of diet and the anticipated threefold increase in human
population in the next 100 years indicate a need for a fourfold
increase in food and timber production (Bunting, 1981). The present
tendency is to stress the need for land reform and the social, political and economic problems that must be solved to increase yields and to
achieve a more equitable distribution of world resources. However,
this overlooks the limitations of the agricultural techniques currently
available and the major constraints imposed by pests and pathogens
including viruses.

Some virus diseases that have been known for a long time continue to
cause serious losses or they have become increasingly difficult and
expensive to control. Moreover, some hitherto minor diseases have
become important and new ones have been encountered. These developments emphasise the need for far more work on plant virus epidemiology,
but the immediate prospect is not encouraging because of current
attitudes to epidemiological studies and financial constraints. Nevertheless, the strong support for the recently formed Plant Virus
Epidemiology Committee of the International Society for Plant Pathology
demonstrates that an influential group of virologists and others
concerned with vectors is active and likely to make further substantial
contributions. This provides encouragement for the future and suggests
that it may eventually be possible to exploit the full benefits of the
continuing improvements and innovations being made in agricultural
technology by decreasing the serious losses now caused by viruses.

REFERENCES

Bald J.G. (1937) Investigation on 'spotted wilt' of tomatoes. III.
   Infection in field plots. *Bulletin 106, Council for Scientific and
   Industrial Research, Australia*, 32 pp.

Bald J.G. & Norris D.O. (1943) Transmission of potato virus diseases.
   I. Field experiment with leaf roll at Canberra, 1940-41. *Bulletin
   163, Council for Scientific and Industrial Research, Australia*, 19 pp.

Bar-Joseph M. & Garnsey S.M. (1981) Enzyme-linked immunosorbent assay
   (ELISA): Principles and applications for diagnosis of plant viruses.
   Plant Diseases and Vectors: Ecology and Epidemiology. (Ed. by K.
   Maramorosch & K.F. Harris), pp. 35-39. Academic Press, New York.

Barrett D.W.A., Northwood P.J. & Horrelou A. (1981) The influence of
   rate and timing of autumn applied pyrethroid and carbamate insecticide sprays on the control of barley yellow dwarf virus in English
   and French winter cereals. *Proceedings, 1981 British Crop Protection
   Conference - Pests and Diseases 2*, 405-12.

Browning J.A. (1974) Relevance of knowledge about natural ecosystems
   to development of pest management programmes for agro-ecosystems.
   American Phytopathological Society Proceedings 1, 191-9.

Browning J.A. (1981) The agro-ecosystem - natural ecosystem dichotomy
   and its impact on phytopathological concepts. *Pests, Pathogens and
   Vegetation* (Ed. by J.M. Thresh), pp. 159-72. Pitman, London.

Bunting A.H. (1981) Changing patterns of land use. *Pests, Pathogens
   and Vegetation* (Ed. by J.M. Thresh), pp. 23-37. Pitman, London.

Clark M.F. (1981) Immunosorbent assays in plant pathology. *Annual
   Review of Phytopathology 19*, 83-106.

Colwell R.N. (1956) Determining the prevalence of certain cereal crop
   diseases by means of aerial photography. *Hilgardia 26*, 223-86.

Day P.R. (1977) The genetic basis of epidemics in agriculture.
   Annals of New York Academy of Sciences 287, 1-400.

Dean G.J. & Luuring B.B. (1970) Distribution of aphids in cereal crops.
   Annals of Applied Biology 66, 485-96.

Frazer B.D. (1977) Plant virus epidemiology and computer simulation of
   aphid populations. *Aphids as Virus Vectors* (Ed. by K.F. Harris &
   K. Maramorosch), pp. 413-31. Academic Press, New York.

Gibbs A.J. (1982) Virus ecology - struggle of the genes. *Physiological
   Plant Ecology. III. Responses to the Chemical and Biological Environment.* (Ed. by O.L. Lange *et al.*) Springer Verlag, Berlin (in press).

Gibson R.W., Rice A.D. & Sawicki R.M. (1982a) Effects of the pyrethroid deltamethrin on the acquisition and inoculation of viruses
   by *Myzus persicae*. *Annals of Applied Biology 100*, 49-54.

Gibson R.W., Rice A.D., Pickett J.A., Smith M.C. & Sawicki R.M. (1982b)
   The effects of the repellents dodecanoic acid and polygodial on the
   acquisition of non-, semi- and persistent plant viruses by the aphid
   Myzus persicae. *Annals of Applied Biology 100*, 55-9.

Gutierrez A.P., Havenstein D.E., Nix H.A. & Moore P.A. (1974) The
   ecology of *Aphis craccivora* Koch and subterranean clover stunt virus
   in south-east Australia. II. A model of cowpea aphid population in
   temperate pasture. *Journal of Applied Ecology 11*, 1-20.

Halbert S.E. & Irwin M.E. (1981) Effect of soybean canopy closure on
   landing rates of aphids with implications for restricting spread of

soybean mosaic virus. *Annals of Applied Biology* 98, 15-9.

Harlan J.R. (1976) Diseases as a factor in plant evolution. *Annual Review of Phytopathology* 14, 31-51.

Harrison B.D. (1981) Plant virus ecology: ingredients, interactions and environmental influences. *Annals of Applied Biology* 99, 195-209.

Heathcote G.D. (1972) Influence of cultural factors on incidence of aphids and yellows in beet. *IIRB* 6, 6-14.

Highwood D.P. (1979) Some indirect benefits of the use of pyrethroid insecticides. *Proceedings, 1979 British Crop Protection Conference - Pests and Diseases* 2, 361-9.

Hooper A.J. (1978) Aerial photography. *Journal of the Royal Agricultural Society of England* 139, 115-23.

Jones R.A.C. (1981) The ecology of viruses infecting wild and cultivated potatoes in the Andean region of South America. *Pests, Pathogens and Vegetation* (Ed. by J.M. Thresh), pp. 89-107. Pitman, London.

King A.M.Q., Underwood B.O., McCahon D., Newman J.W.I. & Brown F. (1981) Biochemical identification of viruses causing the 1981 outbreaks of foot and mouth disease in the U.K. *Nature, London*, 293, 479-80.

Kiritani K. & Sasaba T. (1978) An experimental validation of the systems model for the prediction of rice dwarf virus infection. *Applied Entomology and Zoology* 13, 209-14.

Posnette A.F. (1969) Tolerance of virus infection in crop plants. *Review of Applied Mycology* 48, 113-8.

Russell G.E. (1978) *Plant Breeding for Pest and Disease Resistance.* Butterworth, London.

Simons J.N. (1981) Use of mineral oils for the control of plant virus diseases. *Proceedings, 1981 British Crop Protection Conference - Pests and Diseases* 2, 413-20.

Southwood T.R.E. (1981) The rise or fall of ecology. *New Scientist* 92, 512-4.

Taylor R.A.J. (1978) The relationship between density and distance of dispersing insects. *Ecological Entomology* 3, 63-70.

Thresh J.M. (1980) An ecological approach to the epidemiology of plant virus diseases. *Comparative Epidemiology* (Ed. by J. Palti & J. Kranz), pp. 57-70. Centre for Agricultural Publishing and Documentation, Wageningen.

Thresh J.M. (1981a) The origins and epidemiology of some important plant virus diseases. *Applied Biology* 5, 1-65.

Thresh J.M. (1981b) *Pests, Pathogens and Vegetation.* Pitman, London.

Thresh J.M. (1982) Cropping practices and virus spread. *Annual Review of Phytopathology* 20, 193-218.

Toler R.W., Smith B.D. & Harlan J.C. (1981) Use of aerial color infrared photography to evaluate crop disease. *Plant Disease* 65, 25-31.

Vanderplank J.E. (1963) *Plant Diseases: Epidemics and Control.* Academic Press, New York.

Watson M.A., Heathcote G.D., Lauckner F.B. & Sowray P.A. (1975) The use of weather data and counts of aphids in the field to predict the incidence of yellowing viruses of sugar-beet crops in England in relation to the use of insecticide. *Annals of Applied Biology* 81, 181-98.

Index

inoculation pressure 159-67
insects 27,225,273, see also
  individual names
insecticides 99,155,192,200,
  240-1,243,246,254,273,288
  combined control 319
  oils 317
  organophosphorus 311
  resistance 120-9,317-8
  synthetic pyrethroid 192,357
insensitive 356,358, see also
  resistant, tolerant
inspection, crop 12-14,51-7,79,
  138,288-9,295-6,337,342
inter-cropping 100-1,106-7,109-10,
  277, see also mixed cropping
interaction, host/virus/vector 2,
  7-8,30,211,217,303, see also relationship
International Bureau for Plant
  Genetic Resources (IBPGR) 18-19
International Center for Agricultural Research in Dry Areas
  (ICARDA) 18
International Center for Improvement of Maize and Wheat (CIMMYT)
  18,191,261
International Crops Research
  Institute for the Semi-Arid
  Tropics (ICRISAT) 18,93-100
International Institute for Tropi-
  cal Agriculture 18
International organizations 16-18
International Plant Protection Con-
  vention 16
International Rice Research Insti-
  tute (IRRI) 18,26
International Seed Testing Associ-
  ation 17
International Service for National
  Agricultural Research (ISNAR) 18
introduction, species 9,20,62,119,
  308, see also quarantine
Iran 121,127
Iraq 315
irrigation 4,93,118,189,225,302,306
isolation 4,14,54,58,81,200,223,262
Israel 13,51,57,62-3,69,149-55,315,
  317,351,352,357

Italy 25,26,142
Ivory Coast 17,73

Japan 26,121,124,128,229-30,232,
  239-46,350-1
Jasminum sambac 309
Jatropha 310
Java 62,338
Jordan 315,317
June yellows 222

Kenya 17,33,122,127,128,278,337-46
Korea 229-30

Lactuca 119
Lactuca sativa 212, see lettuce
Lactuca serriola 212
Lagasca mollis 95
Lagerstroemia 309
Laodelphax striatellus 240,244-5
latent period 53-8,80,212,319, see
  also incubation period
latent infection 80
latent virus 30
leaf curl 93,323-9
leaf spot 29,31
leaf yellowing 249
leafhoppers 1,26,54-5,225,239-46,
  249-54,267-73,277-84,288-96,297-
  303,308,354
Lebanon 315,317
leek 10
leek yellow stripe 10
legislation 11-13
legumes 9,93-101,103-111,116,
  180,199
  leguminous species 121,335
  trees or shrubs 120,180
lemon 61,63,68
Lepidium nitidum 298
lettuce 14,9-10,202-3,207,211-7,
  221-3,226
lettuce big-vein 10
lettuce mosaic virus (LMV) 14,208,
  222-3
lettuce necrotic yellows virus
  (LNYV) 124,202-3,211-7
lettuce speckles 222
lettuce speckles mottle virus 226